Quality Breeding in Field Crops

Asif M. Iqbal Qureshi • Zahoor Ahmad Dar
Shabir Hussain Wani
Editors

Quality Breeding in Field Crops

Springer

Editors
Asif M. Iqbal Qureshi
Genetics, Plant Breeding & Biotechnology
MRCFC, Khudwani
Sher-e-Kashmir University of Agricultural
Sciences & Technology of Kashmir
Srinagar, Jammu and Kashmir, India

Zahoor Ahmad Dar
Genetics, Plant Breeding & Biotechnology
DARS, Budgam
Sher-e-Kashmir University of Agricultural
Sciences & Technology of Kashmir
Srinagar, Jammu and Kashmir, India

Shabir Hussain Wani
Genetics, Plant Breeding & Biotechnology
MRCFC, Khudwani
Sher-e-Kashmir University of Agricultural
Sciences & Technology of Kashmir
Srinagar, Jammu and Kashmir, India

ISBN 978-3-030-04608-8 ISBN 978-3-030-04609-5 (eBook)
https://doi.org/10.1007/978-3-030-04609-5

Library of Congress Control Number: 2018968016

This Springer imprint is published by the registered company Springer Nature Switzerland AG
The registered company address is: Gewerbestrasse 11, 6330 Cham, Switzerland

Foreword

Improving crop plants for enhanced quality of produce through conventional and modern plant breeding approaches is apt at this moment. Most of the conventional breeding experiments indicated that yield and quality traits are negatively correlated. However, with the advent of genomic resources and next-generation sequencing technologies, research can be directed towards precise understanding of the target genes responsible for controlling important quality traits. Systematic research and deployment of modern technologies including molecular breeding, genetic engineering, and genome editing will lead to development of high-yielding crop varieties with quality improvement. This informative book provides state-of-the-art information on improving nutritional quality in field crops such as maize, rice, wheat, pearl millet, soybeans, legumes, potatoes, and oilseed crops. With contributions from leading authorities in the field, this book will bring you up to date on the uses of conventional plant breeding and modern biotechnologies for improving the quality of important food, feed, and fibre products.

This book is a timely reference material and will be of great importance for a large number of scientists, students, and policymakers, who will find a common reference to discuss ways that plant breeding can be beneficial to all. I appreciate the untiring efforts made by Dr Asif Mohammad Iqbal Qureshi and his co-editors for bringing out this outstanding book on *Quality Breeding in Field Crops* for the reputed Springer Publishers. The authors deserve commendation and congratulations for their efforts. I am sure that the contents covered in this book will serve to satisfy the needs of scientists and scholars engaged in upgrading the quality of agricultural produce.

Nazeer Ahmed
Vice Chancellor
Sher-e-Kashmir University of Agricultural
Sciences & Technology of Kashmir
Srinagar, Jammu and Kashmir, India

Preface

The long-term objective of plant breeding remains at increasing productivity to meet the food requirements of people; however, with today's world of nutraceuticals, an essential component of economic yield lies in its quality. The breeding for improved nutritional quality has played a pivotal role in solving the problem of malnutrition especially for the community, including animals. However, with the advent of modern plant biotechnological tools, high precision has been achieved with higher-quality standards particularly since last decade; numerous accomplishments have been made in developing crop varieties with improved nutritional quality, which are summarized in this book. Thirteen chapters are written by globally reputed researchers and academicians in the field of crop improvement research (specific crop).

Chapter 1 is an overview of general nutritional aspects and genomic interventions for biofortification of field crops. The concept of markers and various commonly used molecular markers in cooking and eating quality of rice is discussed in detail in Chap. 2, while genetic modification for improving nutritional value of potatoes is presented in Chap. 3. Chapter 4 addresses conventional and molecular perspectives, highlighting bio-fortification of pearl millet, while Chap. 5 is focussed on common bean nutritional quality using genetic approaches. The beta-carotene-rich maize hybrids pursued by breeders using MAS are detailed in Chap. 6. The historical perspectives, highlighting the contributions by researchers for improving the fatty acid profile of soybean oil, are presented in Chap. 7. Chapter 8 is devoted to issues pertaining to breeding for cooking and canning quality traits in dry beans. Chapter 9 provides insight on genomic approaches in wheat for improved iron and zinc content, and the discussion includes the dependence of plant breeding on heritable variation. Improved grain and nutritional qualities are discussed in Chap. 10, whereas Chap. 11 is devoted to discussing the development of high tryptophan maize, and end-use quality in wheat is presented in Chap. 12. Lastly, quality value of oilseed Brassicas using molecular tools is discussed and detailed in Chap. 13.

Through this multi-authored book, an effort has been made to assimilate the most topical results about quality improvement in crop plants, which will be prodigious as reference material for researchers, teachers, and graduate students involved in quality breeding in crop plants using conventional and modern biotechnological tools by unfolding principles of lately developed technologies and their application in improvement of crop plants. We express sincere thanks and gratefulness to our revered authors; without their untiring efforts this book would not have been possible. We are also thankful to Springer Nature for providing such opportunity to complete this book project.

Srinagar, Jammu and Kashmir, India

Asif M. Iqbal Qureshi
Zahoor Ahmad Dar
Shabir Hussain Wani

Contents

Contributors

Francis Osei Amoako-Andoh CSIR-Crops Research Institute, Kumasi, Ghana

Maxwell Darko Asante CSIR-Crops Research Institute, Kumasi, Ghana

Vanya Bawa Division of Plant Breeding and Genetics, Sher-e-Kashmir University of Agricultural Sciences and Technology, Jammu, Jammu and Kashmir, India

Matilda Ntowaa Bissah CSIR-Plant Genetic Resources Research Institute, Bunso, Ghana

Abhishek Bohra ICAR-Indian Institute of Pulses Research (IIPR), Kanpur, Uttar Pradesh, India

John E. Bradshaw Retired Plant Breeder and Geneticist, James Hutton Institute, Dundee, UK

Subhash Chandra ICAR-Indian Institute of Soybean Research, Indore, Madhya Pradesh, India

Joshua N. Cobb International Rice Research Institute, Los Baños, Philippines

Eleonora Cominelli Institute of Agricultural Biology and Biotechnology (IBBA), Italian National Research Council (CNR), Milan, Italy

Zahoor Ahmad Dar Genetics, Plant Breeding & Biotechnology, DARS, Budgam, Sher-e-Kashmir University of Agricultural Sciences & Technology of Kashmir, Srinagar, Jammu and Kashmir, India

Antonio M. De Ron Biology of Agrosystems, Misión Biológica de Galicia (MBG), Spanish National Research Council (CSIC), Pontevedra, Spain

M. Govindaraj International Crops Research Institute for the Semi-Arid Tropics (ICRISAT), Patancheru, Telanagana, India

Hari Shanker Gupta ICAR-Indian Agricultural Research Institute, New Delhi, India

S. K. Gupta International Crops Research Institute for the Semi-Arid Tropics (ICRISAT), Patancheru, Telanagana, India

Firoz Hossain ICAR-Indian Agricultural Research Institute, New Delhi, India

Asif M. Iqbal Qureshi Genetics, Plant Breeding & Biotechnology, MRCFC, Khudwani, Sher-e-Kashmir University of Agricultural Sciences & Technology of Kashmir, Srinagar, Jammu and Kashmir, India

Rintu Jha ICAR-Indian Institute of Pulses Research (IIPR), Kanpur, Uttar Pradesh, India

Uday Chand Jha ICAR-Indian Institute of Pulses Research (IIPR), Kanpur, Uttar Pradesh, India

Hanif Khan ICAR-Indian Institute of Wheat and Barley Research, Karnal, Haryana, India

Mohammad Abdul Latif Bangladesh Rice Research Institute, Gazipur, Bangladesh

Wendy Chui Phing Lau Independent researcher, Kuching, Sarawak, Malaysia

S. S. Mahdi MRCFC, Khudwani, Sher-e-Kashmir University of Agricultural Sciences & Technology, Kashmir, Jammu and Kashmir, India

Alok Kumar Maurya ICAR-Indian Institute of Pulses Research (IIPR), Kanpur, Uttar Pradesh, India

Vignesh Muthusamy ICAR-Indian Agricultural Research Institute, New Delhi, India

S. J. Satheesh Naik ICAR-Indian Institute of Pulses Research (IIPR), Kanpur, Uttar Pradesh, India

Prakash G. Patil ICAR-National Research Centre on Pomegranate, Solapur, Maharashtra, India

S. K. Rai Division of Plant Breeding and Genetics, Sher-e-Kashmir University of Agricultural Sciences and Technology, Jammu, Jammu and Kashmir, India

A. Paula Rodiño Biology of Agrosystems, Misión Biológica de Galicia (MBG), Spanish National Research Council (CSIC), Pontevedra, Spain

Rie Sadohara Department of Plant, Soil and Microbial Sciences, Michigan State University, East Lansing, MI, USA

Konsam Sarika ICAR Research Complex for Northeast Hill Region, Umiam, Meghalaya, India

M. Shivakumar ICAR-Indian Institute of Soybean Research, Indore, Madhya Pradesh, India

Ravi P. Singh Global Wheat Program, International Maize and Wheat Improvement Center, Texcoco, Estado de México, Mexico

N. R. Sofi MRCFC, Khudwani, Sher-e-Kashmir University of Agricultural Sciences & Technology, Kashmir, Jammu and Kashmir, India

Francesca Sparvoli Institute of Agricultural Biology and Biotechnology (IBBA), Italian National Research Council (CNR), Milan, Italy

R. K. Srivastava International Crops Research Institute for the Semi-Arid Tropics (ICRISAT), Patancheru, Telanagana, India

Akshay Talukdar Division of Genetics, ICAR-Indian Agricultural Research Institute, New Delhi, India

Valentin Stanislas Edgar Traore Institute of Environment and Agricultural Research (INERA), Ouagadougou, Burkina Faso

Govindan Velu Global Wheat Program, International Maize and Wheat Improvement Center, Texcoco, Estado de México, Mexico

O. P. Yadav Central Arid Zone Research Institute, Jodhpur, Rajasthan, India

Rajkumar Uttamrao Zunjare ICAR-Indian Agricultural Research Institute, New Delhi, India

About the Editors

Asif M. Iqbal Qureshi received his Ph.D. in Genetics and Plant Breeding from SKUAST Kashmir, India, in 2010. After his Ph.D., he worked as a senior research fellow in the NAIP sponsored project on saffron for two and half years. He was selected as an Assistant Professor in Genetics, Plant Breeding, and Biotechnology in Bihar Agricultural University, Sabour, Bihar, in 2012. Subsequently, Dr. Qureshi was selected as an Assistant Professor (GPB) and took over as principal investigator of the All India Coordinated Research Project Rapeseed-Mustard at SKUAST Kashmir in 2013. His work was recognised in the form of best centre award under the All India Coordinated Research Project Award (2015) by the Directorate of Rapeseed-Mustard, Bharatpur, Rajasthan (ICAR, New Delhi, India) in 2015. Dr. Qureshi was selected for the prestigious Raman Fellowship Programme for advanced post-doctoral research on "Marker Assisted Breeding for Quality Improvement in Dry Beans" at Michigan State University, USA for one year (2016–2017). He is also the recipient of INSA Summer Research Fellowship of Indian National Science Academy, Bengaluru and has handled five research projects from different funding agencies. He has published more than 60 research papers in journals of national and international repute. He is a member of many professional societies involved in crop improvement and was associated as a breeder in the development and release of two oilseed brassica varieties, Shalimar Sarson-2 and Shalimar Sarson-3 (*B. rapa var.* brown sarson), and two rice varieties, Shalimar Rice-4 (low altitude) and Shalimar Rice-5 (high altitude).

Zahoor Ahmad Dar is Professor of Plant Breeding and is working as Principal Investigator in the AICRP (Maize) Srinagar Centre at DARS, Budgam, SKUAST-K and and has been a gold medalist at his bachelor's and doctorate levels. He has published more than 200 Research papers in various international and national journals. He is involved in guiding various students at master's and doctorate levels. He has been involved in the development of 15 varieties in maize, oats, pulses, and brown sarson. He has handled nine projects from different funding agencies. He is the recipient of the INSA Summer Research and Visiting Scientist Fellowship. He has registered more than 75 germplasm accessions of maize with NBPGR and is a member of various societies at national and international levels involved in crop improvement. He has authored five books on plant breeding.

Shabir Hussain Wani is a senior assistant professor at Mountain Research Centre for Field Crops, Khudwani, Sher-e-Kashmir University of Agricultural Sciences and Technology of Kashmir, J&K, India. He received his B.Sc. in Agriculture from Bhim Rao Agricultural University Agra, India; M.Sc. in Genetics and Plant Breeding from Central Agricultural University, Manipur, India; and Ph.D. in Plant Breeding and Genetics on "transgenic rice for abiotic stress tolerance" from the Punjab Agricultural University Ludhiana, India. After obtaining his Ph.D., he worked as a research associate in the Biotechnology Laboratory, Central Institute of Temperate Horticulture (ICAR), Srinagar, India. He then joined the Krishi Vigyan Kendra (Farm Science Center) as programme coordinator at Senapati, Manipur, India. He teaches courses related to plant breeding, seed science and technology, and stress breeding and has published more than 100 papers/chapters in journals and books of international and national repute. He served as a guest editor and review editor for the journal *Frontier in Plant Science* (2015–2018). He has also edited several books on current topics in crop improvement for abiotic stress tolerance published by Springer Nature and CRC press USA. His Ph.D. research won first prize in the North Zone Competition, at national level, in India. He was awarded a Young Scientist Award from the Society for Promotion of Plant Sciences, Jaipur, India, in 2009. He is a fellow of the Society for Plant Research, India. Recently, he received the Young Scientist Award (Agriculture) 2015 from the Society for Plant Research, Meerut, India. He also served as visiting scientist in the Department of Plant Soil and Microbial Sciences, Michigan State University, USA, under the UGC-Raman Post-Doctoral Fellowship programme. He has attended several international and national conferences, presenting his research.

Chapter 1
Genomic Interventions for Biofortification of Food Crops

Abhishek Bohra, Uday Chand Jha, Rintu Jha, S. J. Satheesh Naik, Alok Kumar Maurya, and Prakash G. Patil

1.1 Introduction

Global food security covers not only the quantity of food but also the quality of food that is consumed. Nutrition is a cause of concern to confer good health to new world generation. Although a 27% decrease has been witnessed in the level of world hunger over the last seventeen years, millions of people still experience chronic hunger; this may be attributed to occurrence of famine due to changing climatic factors and geopolitical conflicts (http://www.globalhungerindex.org/). The situation where the food, deficient in vitamins and minerals, remains insufficient to meet the nutritional needs of the people is referred to as hidden-hunger or malnutrition. The prevalence of the malnutrition reflects from the fact that more than 30% of women of reproductive age worldwide suffer from anaemia, which also renders the children venerable from nutrition and health (WHO 2017). Equally importantly, world is still home for 154.8 million stunted and 52 million wasting children under five years of age group. Rural and semi-urban areas are more vulnerable to hidden-hunger. The intermittent crop failures and lack of other remunerative means to buy the increasing cost of healthy foods are the prime cause for increase in hunger and hidden-hunger. Deficiencies of iron (Fe), zinc (Zn) and/or other micronutrient are reported to plague more than two billion people worldwide (De Valença et al. 2017). Of the total deaths occurring among 6–60 months aged children in developing countries, a staggeringly high proportion (41%) is attributed to malnutrition (Schroeder and Brown 1994). The cereal-based food makes the dominant portion of the diets of the people suffering from micronutrient deficiency, especially in developing world.

A. Bohra (✉) · U. C. Jha · R. Jha · S. J. S. Naik · A. K. Maurya
ICAR-Indian Institute of Pulses Research (IIPR), Kanpur, Uttar Pradesh, India

P. G. Patil
ICAR-National Research Centre on Pomegranate, Solapur, Maharashtra, India

A. M. I. Qureshi et al. (eds.), *Quality Breeding in Field Crops*,
https://doi.org/10.1007/978-3-030-04609-5_1

The growing problem of nutrient deficiency worldwide calls for implementing timely interventions at the community level. In the context, potential ways suggested to alleviate nutrient deficiencies include (1) Direct or nutrition-specific interventions that involve altering the consumption behaviour (dietary diversification, micronutrient supplementation, etc.) and (2) Indirect or nutritional-sensitive interventions or biofortification. Biofortification remains the most sustainable means for increasing the nutrient density of crop plants during plant growth through genetic or agronomic practices (De Valença et al. 2017; Bouis and Saltzman 2017). Biofortification offers a way to reach larger target population whereas supplementation and conventional fortification activities might be difficult to implement and/or limited. The genetic interventions facilitating biofortification include plant breeding protocols and transgenic techniques or genetic engineering. In this chapter, we cover crop biofortification using breeding techniques, a process that involves assessment of genetic variation for mineral nutrients in crop's gene pool, understanding the genetic architecture of the nutrient trait, and eventually the introgression of the genes responsible for greater nutrient density to deliver nutrient-rich crop cultivars. The immense potential of the large-scale elemental profiling techniques or ionomics and modern breeding methods like genomic selection (GS) is also discussed in relation to biofortification breeding of food crops. Comprehensive reviews on nutritional enhancement of crops have been published in recent years (Dwivedi et al. 2012; Graham et al. 2001; Hirschi 2009; White and Broadley 2009). Therefore, we will be focusing on the latest findings on crop biofortification that have been reported over the last 5 years. More recently, Bouis and Saltzman (2017) reviewed the progress of biofortification witnessed during 2003–2016 with a focus on HarvestPlus programme.

1.2 Genetic Variation for Various Micronutrients in Crop Plants

Crop genetic resources particularly landraces and wild relatives serve as reservoir of natural variation for improving breeding traits including mineral concentration. Genetic variability for mineral content has been evident in various crop species from various studies. Examples include grain Zn showing variation up to 11.6-fold among various cereal crops (Bänziger and Long 2000; Gregorio et al. 2000) and 6.6-fold in grain legumes (Raboy et al. 1984). Adequate genetic variation for grain Zn was recorded in rice following survey of 1763 accessions under multiple water regimes (Pinson et al. 2015). The variation ranged from 15.72 to 65.01 mg/g under flooded condition, whereas the range under non-flooded regime was noted to be 19.34–63.13 mg/g. Additionally, concerning grain Fe content, sufficient genetic variability was noted in this large panel of rice accessions, with variation ranging from 1.55 to 16.58 mg/g (under flooded condition), and 0.09 to 25.89 mg/g (under non-flooded condition). Importantly, various rice genotypes such as Nagina 22,

Honduras, Jeerigesanna, Kalabath, Pusa Basmati (Mahender et al. 2016) were identified as potential sources for increasing grain Zn content. In case of basmati rice, Zn content was found to be varying between 25 and 165 μg/g, whereas Fe content varied from 32 to 218 μg/g (Renuka et al. 2016). Additionally, the authors also examined the aromatic rice lines for grain β-carotene, revealing significant variation in β-carotene ranging from 1.23 to 9.9 μg/g in brown rice, and 0.08–1.99 μg/g in milled rice. Trijatmiko et al. (2016) reported up to 16 μg/g grain Zn content in high yielding elite polished rice. Moreover, to mitigate the challenge of Zn deficiency in human population worldwide, "Harvest Plus" breeding programmes have set the target of increasing Zn up to 28 μg/g in biofortified rice (Trijatmiko et al. 2016).

In wheat, Zhao et al. (2009) reported up to 2.6-fold variation for grain Zn content, ranging from 13.5 to 34.5 mg/kg. Apart from cultivated hexaploid wheat, wild emmer, einkorn, and landraces (Cakmak et al. 2000; Ortiz-Monasterio et al. 2007) and *T. dicoccoides*, *Ae. tauschii*, *T. monococcum* (Cakmak et al. 2000) act as natural storehouse for improving grain Zn content in wheat. Similarly, substantial amount of genetic variability was reported in spring and winter wheat ranging from 20 to 39 mg/kg and spring wheat possesses higher grain Zn than the winter wheat (Morgounov et al. 2007). Additionally, the authors also examined the genetic variation for grain Fe content that ranged from 25 to 56 mg/kg. With high concentration of Zn (up to 70 kg/mg) and Fe (up to 70 kg/mg) in grain, spelt wheat was identified as potential source for improving mineral traits in the crop (Gomez-Becerra et al. 2011). Another study in wheat uncovered substantial genetic variability for grain Zn concentration (7.4–59.4 mg/kg) from a larger set of diverse genotypes (Pandey et al. 2016). More recently, Manickavelu et al. (2017) surveyed variation of grain Zn content among 269 Afghan wheat landraces, and the authors found grain Zn varying from 15.56 to 87.29 ppm.

In maize, significant amount of genetic variation for grain Zn concentration (5.41–30.85 mg/kg) was recorded in 50 genotypes grown across various agro-climatic zones in India (Mallikarjuna et al. 2015). Similarly, in sorghum, Jambunathan (1980) observed considerable range of grain Zn (1.10–5.02 mg/100 g) and grain Fe content (3–11.30 mg/100 g). Badigannavar et al. (2016) also observed wide variation in grain Zn content (1.12–7.58 mg/100 g) in sorghum. More recently, after testing 336 individuals derived from the cross 296B × PVK801 under multi-location trials, the range of genotypic variation for grain Zn content was found to be 10.2–58.7 mg/kg in sorghum (Phuke et al. 2017).

Millet crops remain instrumental in offering essential micronutrients to human foods basket. Pearl millet, an important member of millet crop family, demonstrates significant genetic variation ranging from 25 to 65 mg/kg (Velu et al. 2007). Additionally, A wide range for grain Zn was noted from 319 diverse genotypes ranging from 10 to 86 μg/g and the two genotypes GEC164 and GEC543 showed higher accumulation of grain Zn (Yamunarani et al. 2016).

Several potential donors viz., Annada, ASD 16, CH 45, HKR 126, Nagina 22 and wild relatives such as *Oryza nivara* and *O. rufipogon* (see Anuradha et al. 2012) have been reported in rice for improving grain Fe content. Significant genetic variation (from 0.25 to 34.8 μg/g) was also found for grain Fe content in rice landraces, with Swetonunia showing the highest Fe content of 34.8 μg/g (Roy and Sharma 2014). Zhao et al. (2009) recorded a 1.8-fold genetic variation for grain Fe content in wheat. Wild emmer wheat showed high variability for grain Fe content (Gomez-Becerra et al. 2010) followed by durum wheat (Velu et al. 2011). Importantly, spring wheat possesses higher concentration of grain Fe than winter wheat (Liu et al. 2014). Examination of a large set of Indian and Turkish genotypes revealed wide variation in grain Fe content from 9.2 to 49.7 mg/kg. Significant amount of genetic variability varying from 55.14 to 122.2 ppm has been registered in a large set of 269 Afghan wheat landraces (Kondou et al. 2016; Manickavelu et al. 2017).

Higher grain Fe content extending up to 9.54 mg/100 g has been suggested in sorghum (Badigannavar et al. 2016). In addition, grain protein content varying from 3.50% to 12.60% became evident from 112 landraces and varieties assessed from this study. In sorghum, Phuke et al. (2017) reported significant genetic variability for grain Fe content ranging from 10.8 to 76.4 mg/kg conducting trails in various locations and different years. Likewise, large variation for grain Fe content varying from 18.88 to 47.65 mg/kg reflected from analysis of 50 diverse maize genotypes grown across various locations in India (Mallikarjuna et al. 2015).

Selenium is an important micronutrient for combating various diseases in human (Adams et al. 2002). A daily uptake of 55–200 μg of Se is beneficial for human health (Monsen 2000). Wheat also serves an important source of Se for combating Se deficiency in human (Guerrero et al. 2014; Eiche et al. 2015). Poblaciones et al. (2014) suggested that the accumulation of Se in wheat grain is up to 5.53 mg/kg. High variability of grain Se content reaching up to 7.2-fold (33–238 mg/kg) has been reported in wheat (Zhao et al. 2009). Among grain legume crops, lentil is an important source of Se content ranging from 166 to 858 μg/kg apart from carrying high protein (Ates et al. 2016). Earlier, Rahman et al. (2013) also demonstrated wide variation in seed Se content of lentil, ranging from 74 to 965 μg/kg.

Grain protein content in rice remains low (8.5%) in comparison to wheat (12.3%), barley (12.8%) and millets (13.4%) (Mahender et al. 2016). On the basis of dry weight, Mohanty et al. (2011) reported 16.41% crude protein in ARC 10063 rice genotype and 15.27% crude protein in ARC 10075 rice genotype. In case of wheat existence of significant genetic variability for grain protein has been assessed (Peleg et al. 2008; Amiri et al. 2015). Grain protein ranging from 10.1% to 17.1% has been assessed in diverse wheat genotypes collected from India and Turkey (Pandey et al. 2016). Likewise, substantial amount of genetic variability for grain protein ranging from 10.9% to 13.6% was predominantly higher in bicolor-guinea race (Rhodes et al. 2017).

Significant genetic variation for grain Mn content was found under both flooded (14.35–46.51 mg/g) and unflooded (15.37–76.00 mg/g) conditions in rice (Pinson et al. 2015). In wheat also, ample genetic variability for grain Mn was recorded that ranged from 5.82 to 66.5 mg/kg (Khokhar et al. 2018).

β-carotene precursor of pro-vitamin A remains an important micronutrient for human health for ameliorating vitamin A deficiency in world human population (Giuliano 2017). Among cereals, rice, wheat and maize harbour a certain amount of genetic variability for grain carotene content (Zhai et al. 2016). In case of wheat, variability for β-carotenes at endosperm level has been investigated by different research groups (Qin et al. 2012; Qin et al. 2016). Similarly, preponderance of β-carotene in maize has been examined by various researchers (Harjes et al. 2008; Yan et al. 2010). Considerable zeaxanthin (1.2–13.2 μg/g dry weight), β-cryptoxanthin (1.3–8.8 μg/g DW) and β-carotene (1.3–8.0 μg/g DW) in maize has been suggested (Muzhingi et al. 2017). In chickpea, the total carotenoid content ranged from 22 μg/g (yellow cotyledon kabuli type) to 44 μg/g (green cotyledon desi type) at post-anthesis stage (Rezaei et al. 2016).

1.3 The Genetic Structure of Mineral/Nutrient Content in Crop Plants

1.3.1 Discovery of QTL Controlling Micronutrient Content in Crops

Understanding the genetic nature of grain micronutrient has been met with limited success, and this may be attributed to complex quantitative inheritance of this trait coupled with substantial influence of genotype × environment (G × E) interactions upon elemental concentration (Mahender et al. 2016). Low to moderate estimates of heritability underlying nutrient (in particular, minor elements) content in plants adds to poor understanding of its genetic nature (Manickavelu et al. 2017). Given this, finding genomic regions or QTL that explain substantial phenotypic variation is key to elucidate the genetic basis of various traits controlled by several gene(s) including grain micronutrient content. Several studies in different crops, particularly in cereals and grain legumes, have reported QTL for micronutrient content viz., for grain Zn (Zhou et al. 2010; Qin et al. 2012), Fe (Lung'aho et al. 2011), Mn (Zhou et al. 2010; Zhang et al. 2014), protein (Blanco et al. 2006), β-carotene (Kandianis et al. 2013; Jittham et al. 2017) contents and so forth.

In rice, several QTLs controlling grain Zn have been reported (Garcia-Oliveira et al. 2009; Zhang et al. 2014; Swamy et al. 2016). Readers are referred to recent literature for greater details (Bohra et al. 2015, 2016). Ishikawa et al. (2017) identified a total of 4 QTLs controlling grain Zn on LGs 2, 9 and 10 from a back-cross recombinant inbred population (*O. sativa* 'Nipponbare' × *O. meridionalis* W1627) (see Table 1.1). A more recent study on fine mapping of *qGZn9* QTL revealed two tightly linked loci (*qGZn9a* and qGZn9b) and the study led to pinpointing a candidate gene *Os09g0384900*. Similarly, in wheat, various genomic regions associated with grain Zn and Fe content were detected on various chromosomes through QTL

Table 1.1 List of QTLs controlling various micronutrients in crop plants

Crop	Nutrient element	Mapping population	QTL/loci/genomic region	Type of marker	LG/chromosome no.	PV%	Reference
Rice	Zn	*O. sativa* 'Nipponbare' × *O. meridionalis* W1627, BIL (151)	*qGZn2-1* and *qGZn2-2* *qGZn9*, *qGZn10*	*RM24085-RM566* *RM171-RM590* *RM573*, *RM6*	2, 9, 10	15–21.9	Ishikawa et al. (2017)
Rice	Mn	93-11 × PA64s RIL (132), CSSL	A major QTL *qGMN7.1*	SNP7-53 and SNP7-64 RM427 and RM11	7	15.6–22.8	Liu et al. (2017)
Maize	Zn and Fe		A total of 22 QTLs	SSR	1, 2, 3,4, 5,6,7, 8, 9 and 12	–	Jin et al. (2015)
Wheat	Zn Fe	*Triticum spelta* L. × synthetic hexaploid RIL (188) Triticum spelta L. × synthetic hexaploid RIL (188)	*QGZn.cimmyt-7B_1P2* *QGZn.cimmyt-7B_P1* *QGFe.cimmyt-4A_P2* *QFe.cimmyt-3A_P1*	DArT	7B 4A	10.3–32.7	Crespo-Herrera et al. (2017)
Wheat	Zn	Saricanak98 × MM5/4 (RIL) Adana99 × 70,711(RIL)	*2QTLs*	DArT	1B and 6B 6A and 6B		Velu et al. (2016a, b)
Wheat	Fe, Zn	'Berkut' × 'Krichauff', BC	2 QTLs for Zn and one QTL for Fe	gwm120–wpt2430 wmc036– cfa2129	1B and 2B	22.2–35.9	Tiwari et al. (2016)
Wheat	Fe, Zn	WH542 × synthetic derivative	*QGFe.iari-2A*, *QGFe.iari-5A* QGFe.iari-7A, QGFe. iari-7B *QGZn.iari-2A*, *QGZn.iari-4A*, *QGZn.iari-2A*, *QGZn.iari-4A* *QGZn.iari-7B*	SSR	2A, 5A, 7A, 7B	20 32	Krishnappa et al. (2017)

Wheat	Fe, Zn	Seri M82 × SHW CWI76364	*One major QTL*	–	4BS	19.6	Crespo-Herrera et al. (2016)
Wheat	Protein	Yumechikara × Kitahonami DH (94)	*QGpc.2B-yume*	SSR, Xgpw4382	2B	32.1	Terasawa et al. (2016)
Wheat	Se	Tainong18 × Linmai6, RIL	16 QTLs	D-3033829 and D-1668160	1B, 2B, 4B, 5A, 5B 5D, 6A, and 7D	7.71–20.22	Wang et al. (2017)
Maize	Fe, Zn, Mn	RIL	–	–	–	–	Zhang et al. (2017a)
Maize		Ye478 × Wu312, RIL (2012)	*qMnCC1–1*, *qMnCC1–2* *MnCC2–1*, *qMnCC2–1* *MnCC4–1*, *qMnCC4–2* *qZnCT4–1*, *qZnCT4–2* *qZnCC5–1*, *qZnCC5–1*	SSR	–	6.22–27.7	Gu et al. (2015)
Maize	Zn and Fe	–	mMQTL2.1, mMQTL3, mMQTL5 mMQTL9.2	SSR	2, 3, 5 and 9	–	Jin et al. (2015)
Maize	β-carotene	By804 × B73, RIL (178)	A total of 62 QTLs	SSR, SNP, InDel	1–10.	4.21–47.53	Jittham et al. (2017)
Sorghum	Protein	–	Alpha-amylase 3 gene	SNP	2	–	Rhodes et al. (2017)
Sorghum	Protein	(BTx642 × BTxARG-1) RIL (BTxARG-1/P850029) RIL	One QTL	SNP	1 and 2	–	Boyles et al. (2017)
Soybean	Protein	Nannong94–156 × Bogao RIL(152)	32 QTLs for water soluble protein content and grain protein content	SNP	3, 4, 5, 6, 8, 9, 10, 11, 12, 13, 14, 15, 16, 17, 18, 19 and 20	2–29.2	Zhang et al. (2017b)
Pearl millet	Zn and Fe	130 accessions	16 significant MTAs for Zn and Fe	*SSRs*, *Xpsmp* 2261 *Xipes* 0180 *Xipes* 0096	3, 4, 5, and 7	11.3–13.4	Anuradha et al. (2017)

(continued)

Table 1.1 (continued)

Crop	Nutrient element	Mapping population	QTL/loci/genomic region	Type of marker	LG/chromosome no.	PV%	Reference
Pearl millet	Fe and Zn	841-P3 × 863B-P2 RIL (106)	2 QTLs for Zn and Fe on Same genomic region	DArT	3, 5, 7	42	Kumar et al. (2016)
Brassica napus	Protein	KenC-8 × N53–2	68 QTLs	SNP	A02, A03, A04, A07, A09 C01, C03, C05, C06 C07, C08 and C09	27.5	Chao et al. (2017)
Chickpea	Fe and Zn	ICC4958 × ICC 8261, RIL (277)	QTL (*CaqFe1.1*, *CaqZn2.1*) *CaqFe3.1 CaqZn3.1* *CaqFZ4.1 CaqFe4.1 CaqFZ5.1* *CaqFZ7.1*	SNP	1, 2, 3, 4, 5 and 7	18.7–21.8 1.1–23.4	Upadhyaya et al. (2016a)
Chickpea	Protein	ICC 12299 × ICC 4958, RIL (F7)	Seven genomic region	SNP	1, 2, 4, 6 and 7	10–20.	Upadhyaya et al. (2016b)
Chickpea	Protein		19 significant MTAs/5QTLs	SSR	1, 2, 3, 4 and 5	8.64–16.85	Jadhav et al. (2015)
Common bean	Mn	DOR364 × G19833, RIL	3 QTLs		1, 5, 8		Blair et al. (2016)
Lentil	Fe and Zn	138 accessions	Two significant MTA for Fe One significant MTA for Zn	SNP	–	9–21.	Khazaei et al. (2017)
Lentil	Fe and Zn	96 accessions	4 significant MTA for Fe 4 significant MTA for Zn	SSR	–	9–11. 14–21.	Singh et al. (2017a)
Lentil	Se	PI 320937″ × "Eston," RIL (96)	*SeQTL2.1*, *SeQTL5.2* *SeQTL5.3* and *SeQTL5.1*	SSR, SNP	2 and 5	6.3–16.9	Ates et al. (2016)

mapping experiments (Krishnappa et al. 2017; Srinivasa et al. 2014; Velu et al. 2016a, b). Crespo-Herrera et al. (2016) discovered recently a QTL *QGZn.cimmyt-7B_1P2* governing 32.7% of the phenotypic variation (PV) for grain Zn on chromosome 7B in wheat. Another QTL (*QGFe.cimmyt-4A_P2*) in wheat on chromosome 4A explained more than 20% PV for grain Fe content (Crespo-Herrera et al. 2017). Additionally, two major QTLs associated with grain Zn content could be located on chromosomes 1B and 6B from two recombinant inbred populations (Velu et al. 2016a). Importantly, one QTL on chromosome 2B controlling grain Fe content coincided with the genomic region harbouring QTL for grain Zn content. In a recent study in wheat, four QTLs (*QGFe.iari-2A*, *QGFe.iari-5A*, *QGFe.iari-7A* and *QGFe.iari-7B*) for grain Fe content with PV up to 20% PV and five QTLs for grain Zn accounting for 32% PV were reported by Krishnappa et al. (2017).

Several QTLs have been mapped on rice chromosomes 3, 7 and 8 for grain Mn concentration. Liu et al. (2017) found a major QTL, *qGMN7.1* explaining up to 23% PV for Mn concentration on LG 7 with a RIL population derived from 93-11 and PA64s. Further fine mapping of this QTL region uncovered a set of five genes *LOC_Os07g15350, LOC_Os07g15360, LOC_Os07g15390, LOC_Os07g15400* and *LOC_Os07g15370* within the target region of 49.3 kb. Subsequently, *LOC_Os07g15370* (*OsNRAMP5*) could be declared as the likely gene causing higher grain Mn accumulation (Liu et al. 2017). The authors also validated the findings using CSSL in the background of 93-11 containing *qGMN7.1* from PA64s. In wheat, a total of 16 QTLs contributing Se content at various stages viz., seedling, shoot and grain were discovered by Wang et al. (2017). Earlier, Pu et al. (2014) also reported five QTLs governing Se content in wheat.

Like cereal crops, grain legumes also remain important from the food security point due to higher content of protein and essential minerals. Analyses of mapping populations have provided a number of QTLs controlling mineral content in grain legumes (see Bohra et al. 2015). Ates et al. (2016) analysed the Se content in a RIL population developed from PI 320937 and Eston in lentil. The authors reported four QTLs, one on LG 2 and three on LG 5, with 16.9% being the highest PV explained by these QTLs. A non-exhaustive list of the QTLs available from latest studies is presented in Table 1.1.

1.3.2 Association Genetics to Discern Genomic Regions Linked with Mineral Traits

The ability of genome-wide association study (GWAS) to genetically dissect the trait-of-interest in a diverse panel (non-requirement of an artificially created population) with enhanced resolution makes this technique promising for associating nutrient content variation with genetic variants in crop plants (Diapari et al. 2014; Huang et al. 2015; Nawaz et al. 2015; Suwarno et al. 2015). A recent GWAS conducted on 378 brown rice accessions led authors to associate 20 QTLs with the variation in concentrations of five mineral elements, i.e. Fe, Zn, Se, Cd and Pb (Huang et al. 2015). More importantly, QTL colocalizations observed on chromosomes 5, 7 and 11 hold great potential in relation to biofortification breeding in rice.

In a similar manner, genetic basis of eight essential grain mineral contents was examined in USDA minicore collection of brown rice with GWAS, which enabled identification of 37 genomic regions controlling accumulation of minerals like Zn, Fe, Mn, Mg, K, etc. (Nawaz et al. 2015). In wheat, marker-trait association analysis (MTA) in 47 synthetic lines facilitated discovery of six QTLs, three each for grain Zn and Fe (Gorafi et al. 2016). Recently, a GWAS in a collection of 336 spring barley line for content of various minerals led identification of 11 QTLs for grain Fe content, 3 QTLs for grain Zn and 3 QTLs for grain Mn (Gyawali et al. 2017). In chickpea, association mapping in a set of 94 diverse genotypes suggested significant association of 8 markers with variation in contents of Zn and Fe, and the MTAs were detected on LGs 1, 4, 6 and 7. Concerning protein content, a total of 19 MTAs explaining variation for seed protein content were detected in chickpea on different five chromosomes following an association study involving 187 accessions (Jadhav et al. 2015). Upadhyaya et al. (2016b) performed GWAS in 336 chickpea genotypes using 16,376 SNP markers and reported seven significant MTAs for seed protein content explaining up to 41% PV. The authors also validated five genes in the parental lines and the derived RILs.

In sorghum, significant MTAs were detected for grain protein content on chromosome 2 and 4 based on a genome-wide scan of a global set of 265 lines (Rhodes et al. 2017). In a previous study, Owens et al. (2014) deciphered the role of carotenoid biosynthesis candidate gene in maize through GWAS. Similarly, in maize, screening of a large panel of 380 lines for genome-wide associations led authors to discover seven significant MTAs for β-carotene (*βC*) on LGs 1, 2, 8 and 10, with PV up to 16% (Suwarno et al. 2015). Additionally, significant MTAs were obtained for β-cryptoxanthin (βCX: 13 MTAs) and zeaxanthin (ZEA: 14 MTAs). A more recent association analysis in 233 tetraploid wheat accessions using a high-density SNP array delineated a set of 24 candidate genes, which could be related to carotenoid synthesis (Colasuonno et al. 2017).

1.4 Functional Genomics and Grain Micronutrients Accumulation in Crops

Unprecedented progress in plant functional genomics has shed new light on the candidate gene(s) and biosynthetic pathways associated with various complex traits including essential micronutrients of human importance. Several candidate gene(s) underlying grain Fe and Grain Zn content such as *OsYSL1*, *OsMTP1*, *OsNAS1*, *OsNAS3*, *OsNRAMP1*, heavy metal ion transporter, *OsAPRT* has been pinpointed in rice (Anuradha et al. 2012; Neelamraju et al. 2012). Based on association mapping and expression profiling, 16 genes involved in grain Fe and grain Zn accumulation were found in chickpea (Upadhyaya et al. 2016a). Concerning seed protein content, six candidate genes encoding ATP-dependent RNA helicase DEAD-box, cystathionine-beta synthase, CMP and dCMP deaminases, G10 and zinc finger protein were reported in chickpea (Upadhyaya et al. 2016b). Following cloning of two carotenoid cleavage dioxygenase (CCD) genes viz. *CCD1* and *CCD4*, differential expression profiles of metabolic gene and homoelogs including *PSY1*, *LCYe*, *HYD1/2* and *CCD1/4* established in wheat offered important insights into β-carotene enrichment in the endosperm of wheat grains (Qin et al. 2016). Rezaei et al. (2016a) reported a total of 32 candidate genes in chickpea participating in carotenoid synthesis pathways and examined their expression pattern at various seed developmental stages in five chickpea genotypes.

In recent years, increasing attention is being paid towards discovery of non-coding (nc) RNAs engaged in regulating important traits such as grain micronutrient content in plants. Among the different classes of ncRNAs discovered in plants, micro RNAs (miRNAs) are endogenous small non-coding riboregulators with their lengths varying from 20 to 24 nucleotides (see Mishra and Bohra 2018). Paul et al. (2016) discovered participation of some known as well novel miRNAs in Fe translocation through analysing the small RNA sequencing libraries constructed from the roots of soy*FER*1-overexpressing transgenic rice. Importantly, NRAMP4, coding for a metal transporter, was predicted as a target gene for the novel miRNAs (miR11, miR26, miR30 and miR31). The authors proposed activation of NRAMP4 as a result of the reduced expression of the above four novel miRNAs. The role of two genes GRMZM2G366919 and GRMZM2G178190 (members of NRAMP gene family) in grain Zn and Fe accumulation in maize was established through meta QTL analysis (Jin et al. 2015).

1.5 Transgenic Interventions for Enriching Grain Nutrient Content

Transgenic/genetic engineering (GE) is an alternative to conventional breeding enabling transfer of gene(s) across the species regardless of the sexual reproduction process. Improvement of grain micronutrient density through conventional breeding is greatly constrained by the limited variability for grain micronutrient available in the cultivated/crossable gene pool. In view of this, GE has been successfully employed for increasing micronutrients (Table 1.2), especially Zn, Fe, pro-vitamin A, etc. in grains of in staple crops (Aluru et al. 2008; Abid et al. 2017; Boonyaves et al. 2017; Masuda et al. 2012). Transgenic approach has yielded notable results in different crops; for instance several fold increment in Fe content in rice endosperm (Bashir et al. 2013; Zhang et al. 2012; Ogo et al. 2011). Similarly, fourfold increase in grain Fe content and two-fold increase in transgenic rice overexpressing nicotianamine synthase (*OsNAS*) genes is noteworthy (Johnson et al. 2011; Wirth et al. 2009). Likewise, a transgenic rice event overexpressing *OsNAS2* and soybean

Table 1.2 List of transgenes contributing to higher micronutrients in various crop plants

Crop	Micronutrient	Source	Gene	Reference
Rice	Fe	Rice and barley	*OsYSL2*, *HvNAS1*	Masuda et al. (2012)
Rice (Pusa-sugandhi II)	Fe	Rice	*Ferritin*	Paul et al. (2012)
Rice	Fe	Arabidopsis	*AtIRT1*, *AtNAS1*,	Boonyaves et al. (2017)
			PvFER	
Rice (IR64)	Fe	Soybean	*OsNAS2*	Trijatmiko et al. (2016)
			SferH-1	
Rice	Fe, Zn	Arabidopsis	*AtIRT1*	Boonyaves et al. (2016)
Rice (IR64)	Fe and Zn	Rice	*OsNAS*	Moreno-Moyano et al. (2016)
Rice (Nipponbare)	Vitamin A	Arabidopsis,	*AtNAS1*	Singh et al. (2017b)
		Maize, Common bean	*PvFERRITIN*	
		Bacteria	*CRTI*	
			ZmPSY	
Rice (Tsukinohikari and Tachisugata)	Fe and Zn	Soybean	*FER* 1	Paul et al. (2016)
Rice	Fe	Yeast	*refre1/372and OsIRO2*	Masuda et al. (2017)
Maize	Vitamin A	Bacterial	*crtB*, *crtI*	Aluru et al. (2008)
Wheat	Fe and Zn	*Aspergillus japonicus*	*phyA*	Abid et al. (2017)

ferritin (*SferH-1*) genes showed dramatic increase in contents of grain Fe (up to 15 μg/g) and Zn (45.7 μg/g) in the field trials conducted in two countries (Trijatmiko et al. 2016). The study offered further evidences confirming the endosperm enriched with Fe and Zn, and the bioavailability of the Fe. Enhanced concentration of grain Fe was demonstrated both in polished and unpolished rice through transgene expression of *AtIRT1, AtNAS1* and *PvFER* genes (Boonyaves et al. 2017). A previous study by the same group showed up to 9.6 μg/g DW increase in grain Fe in the polished rice grains consequent upon the expression of *AtIRT1* gene in association with *AtNAS1* and *PvFERRITIN* genes (Boonyaves et al. 2016). Recently, Abid et al. (2017) attempted to overcome the limitation posed by phytate, a chelating agent reducing the bioavailability of micronutrient viz., Fe, Zn in various crops. Transgenic expression of *Aspergillus japonicus* phytase gene (*phyA*) in wheat endosperm allowed higher bioavailability of grain Fe and Zn in wheat through enhancing the activity of phytase enzyme.

1.6 High-Throughput Ionome Profiling and Biofortification Breeding

With tremendously improving genotyping/sequencing platforms, large-scale and accurate phenotyping assumes greater significance with respect to improving micronutrient content in different crops. In this context, ionomics has been emerging as high-throughput "elemental profiling" approach that surveys the mineral nutrient of a living organism (Huang and Salt 2016; Baxter 2009). Precise and accurate measurement of existing grain micronutrients accelerates the progress of identification of genetic lines carrying high micronutrient content (Swamy et al. 2016). Several non-destructive high-throughput elemental analytical techniques viz., Atomic Absorption Spectrometry (AAS), inductively coupled plasma mass-spectrometry (ICP-MS), inductively coupled plasma-optical emission spectrometry (ICP-OES), synchrotron X-ray fluorescence microscopy (XFM), energy dispersive X-ray fluorescence spectrometry (Trijatmiko et al. 2016; Manickavelu et al. 2017; Khokhar et al. 2018) have been employed to measure nutrient density in plants. To this end, some community-oriented platforms have also been established rendering the ionomic data freely available to public. Examples include ionomic HUB or iHUB (http://www.ionomicshub.org/home/PiiMS) that allows researchers to access ionomic resources pertaining to Arabidopsis, rice, yeast and soybean (see Baxter et al. 2007). This international collaborative workspace supports tools enabling data annotation, data collection and workflow, and data sharing.

1.7 Whole-Genome Predictions for Improving Elemental Concentrations of Crops

Though the cost of genotyping is being reduced dramatically, the phenotyping bottlenecks still pose a big hurdle to crop breeding progress (see Bohra 2013). In such a scenario, new breeding methods like genomic selection (GS) holds great promise as GS accelerates selection cycles as well as selection gains per unit of time (http://genomics.cimmyt.org/). GS has been extensively used in livestock industry including sheep, dairy cattle, pig breeding, poultry (Meuwissen et al. 2016) and allowed selection of traits that are "hard-to-measure" and allowed increasing rate of genetic improvement in animal breeding programmes (van der Werf 2013).

In GS, prediction models are trained using a reference population or training set that is scored both genotypically and phenotypically (Meuwissen et al. 2016). Genomic estimated breeding values (GEBVs) are then calculated and based on these GEBVs selection is practiced in breeding population that is scored only at genotypic level, thus circumventing the need for costly and time-consuming phenotypic recordings. Lorenz et al. (2011) have described several methods that have been proposed to calculate genomic predictions such as random regression best linear unbiased prediction (RR-BLUP), least absolute shrinkage and selection operator (LASSO), reproducing kernel Hilbert spaces (RKHS) and support vector machine (SMV) regression, partial least squares (PLS) regression and principal component (PC) regression. Readers are encouraged to refer to other specialized reviews for greater details (Cabrera-Bosquet et al. 2012; Crossa et al. 2017; Meuwissen et al. 2001; Nakaya and Isobe 2012).

To date, limited studies have been performed in crops that assess the potential of GS *vis a vis* elemental content in crops. In wheat, Velu et al. (2016b) estimated genome-wide predictions for gain Zn and Fe contents in a panel of 330 diverse lines. Although the degree of accuracy varied between locations, moderate to high prediction accuracies (up to 0.69 for Zn and 0.73 for Fe) underlined the greater relevance of GS in improving Zn and Fe concentrations in wheat. Similarly, moderate to high prediction abilities were obtained for content of major (Mg, K and P) and minor (Mn, Fe and Zn) elements in wheat (Manickavelu et al. 2017). However, the authors reported higher prediction accuracies for major elements than that of minor elements, which could be explained on the basis of degree of heritability of the trait (element content) in the crop. Earlier, in maize Owens et al. (2014) reported prediction accuracies up to 0.71 with an average of 0.43 while analysing grain carotenoid traits.

As advocated by Manickavelu et al. (2017), tremendous opportunities exist to combine "Speed breeding" with GS, which brightens the scope of "Remote breeding" for improving nutrient density in crops based on the whole-genome predictions.

1.8 Conclusion and Perspectives

Advances in crop genomics have greatly assisted breeding of crop cultivars with enhanced nutrient density. The genetic nature of the mineral composition is increasingly resolved to empower researchers for subsequent trait manipulation. However, the low heritability of mineral content (especially minor elements) as reported by various researchers limits the prospects of conventional MAS for improving nutrient density in crop plants (Manickavelu et al. 2017). Given this, GWAS and GS are genome-scale methods to elucidate genetics of mineral accumulation, which is reported to be under the control of several gene(s)/QTLs exerting smaller effects on the phenotype (Owens et al. 2014). Both methods permit maximum benefits of radically decreasing cost and growing throughput of genotyping assays, in conjunction with considerably decreasing the count of plant phenotyping attempts. In parallel, transgenic systems allow researchers to exploit the variation lying outside crop's gene pool and provide a means to overcome crossability barriers. Evidences from latest studies involving cutting-edge omics techniques help better elucidate the phenomenon of nutrient metabolism in plants. Embracing these tools and technologies in biofortification breeding could improve the efficiency of breeding crops with improved levels of essential nutrients. We anticipate that the enhanced ability of biofortification breeding to efficiently produce nutrient-rich crop genotypes will be instrumental in addressing the burgeoning problem of hidden-hunger.

References

Abid N, Khatoon A, Maqbool A, Irfan M, Bashir A, Asif I, Shahid M, Saeed A, Brinch-Pedersen H, Malik KA (2017) Transgenic expression of phytase in wheat endosperm increases bioavailability of iron and zinc in grains. Transgenic Res 26:109–122

Adams ML, Lombi E, Zhao FJ, McGrath SP (2002) Evidence of low selenium concentrations in UK breadmaking wheat grain. J Sci Food Agric 82:1160–1165

Aluru M, Xu Y, Guo R, Wang Z, Li S, White W, Wang K, Rodermel S (2008) Generation of transgenic maize with enhanced provitamin a content. J Exp Bot 59:3551–3562

Amiri R, Bahraminejad S, Sasani S, Jalali-Honarmand S, Fakhri R (2015) Bread wheat genetic variation for grain's protein, iron and zinc concentrations as uptake by their genetic ability. Eur J Agron 67:20–26

Anuradha K, Agarwal S, Rao YV, Rao KV, Viraktamath BC, Sarla N (2012) Mapping QTLs and candidate genes for iron and zinc concentrations in unpolished rice of Madhukar×Swarna RILs. Gene 508:233–240

Anuradha N, Satyavathi CT, Bharadwaj C, Nepolean T, Sankar SM, Singh SP, Meena MC, Singhal T, Srivastava RK (2017) Deciphering genomic regions for high grain iron and zinc content using association mapping in pearl millet. Front Plant Sci 8:412

Ates D, Sever T, Aldemir S, Yagmur B, Temel HY, Kaya HB et al (2016) Identification QTLs controlling genes for Se uptake in lentil seeds. PLoS One 11:e0149210

Badigannavar A, Girish G, Ramachandran V, Ganapathi TR (2016) Genotypic variation for seed protein and mineral content among post-rainy season-grown sorghum genotypes. Crop J 4:61–67

Bänziger M, Long J (2000) The potential for increasing the iron and zinc density of maize through plant-breeding. Food Nutr Bull 21:397–400

Bashir K, Takahashi R, Akhtar S, Ishimaru Y, Nakanishi H, Nishizawa NK (2013) The knockdown of OsVIT2 and MIT affects iron localization in rice seed. Rice 6:31

Baxter I (2009) Ionomics: studying the social network of mineral nutrients. Curr Opin Plant Biol 12:381–386

Baxter I, Ouzzani M, Orcun S, Kennedy B, Jandhyala SS, Salt DE (2007) Purdue ionomics information management system. An integrated functional genomics platform. Plant Physiol 143:600–611

Blair MW, Wu X, Bhandari D, Astudillo C (2016) Genetic dissection of ICP-detected nutrient accumulation in the whole seed of common bean (*Phaseolus vulgaris* L.). Front Plant Sci 7:219

Blanco A, Simeone R, Gadaleta A (2006) Detection of QTLs for grain protein content in durum wheat. Theor Appl Genet 112:1195–1204

Bohra A (2013) Emerging paradigms in genomics-based crop improvement. Sci World J 585467:17

Bohra A, Sahrawat KL, Kumar S, Joshi R, Parihar AK, Singh U, Singh D, Singh NP (2015) Genetics- and genomics-based interventions for nutritional enhancement of grain legume crops: status and outlook. J Appl Genet 56:151–161

Bohra A, Jha UC, Kumar S (2016) Enriching nutrient density in staple crops using modern"-Omics" tools. In: Singh U, Praharaj CS, Singh SS, Singh NP (eds) Biofortification of food crops. Springer, New Delhi, pp 85–103

Boonyaves K, Gruissem W, Bhullar NK (2016) NOD promoter-controlled AtIRT1 expression functions synergistically with NAS and FERRITIN genes to increase iron in rice grains. Plant Mol Biol 90:207–215

Boonyaves K, Wu TY, Gruissem W, Bhullar NK (2017) Enhanced grain iron levels in rice expressing an iron-regulated metal transporter, nicotianamine synthase, and ferritin gene cassette. Front Plant Sci 8:130

Bouis HE, Saltzman A (2017) Improving nutrition through biofortification: a review of evidence from HarvestPlus, 2003 through 2016. Glob Food Sec 12:49–58

Boyles RE, Pfeiffer BK, Cooper EA, Rauh BL, Zielinski KJ, Myers MT, Brenton Z, Rooney WL, Kresovich S (2017) Genetic dissection of sorghum grain quality traits using diverse and segregating populations. Theor Appl Genet 130:697–716

Cabrera-Bosquet L, Crossa J, von Zitzewitz J, Dolors Serret M, Araus JL (2012) High-throughput phenotyping and genomic selection: the frontiers of crop breeding converge. J Integr Plant Bio 54:312–320

Cakmak I, Ozkan H, Braun HJ, Welch RM, Romheld V (2000) Zinc and iron concentrations in seeds of wild, primitive and modern wheats. Food Nutr Bull 21:401e403

Chao H, Wang H, Wang X, Guo L, Gu J, Zhao W, Li B, Chen D, Raboanatahiry N, Li M (2017) Genetic dissection of seed oil and protein content and identification of networks associated with oil content in Brassica napus. Sci Rep 7:46295

Colasuonno P, Lozito ML, Marcotuli I, Nigro D, Giancaspro A, Mangini G, De Vita P, Mastrangelo AM, Pecchioni N, Houston K, Simeone R, Gadaleta A, Blanco A (2017) The carotenoid biosynthetic and catabolic genes in wheat and their association with yellow pigments. BMC Genomics 18:122

Crespo-Herrera LA, Velu G, Singh RP (2016) Quantitative trait loci mapping reveals pleiotropic effect for grain iron and zinc concentrations in wheat. Ann Appl Biol 169:27–35

Crespo-Herrera LA, Govindan V, Stangoulis J, Hao Y, Singh RP (2017) QTL mapping of grain Zn and Fe concentrations in two Hexaploid wheat RIL populations with ample transgressive segregation. Front Plant Sci 8:1800

Crossa J, Pérez-Rodríguez P, Cuevas J, Montesinos-López O, Jarquín D, de los Campos G et al (2017) Genomic selection in plant breeding: methods, models, and perspectives. Trends Plant Sci 22:961–975

De Valença AW, Bake A, Brouwer ID, Giller KE (2017) Agronomic biofertilization of crops to fight hidden hunger in sub-Saharan Africa. Gobal Food Security 12:8–14

van der Werf J (2013) Genomic selection in animal breeding programs. In: Gondro C, van der Werf J, Hayes BJ (eds) Genome-wide association studies and genomic prediction. Springer, New York, NY, pp 543–561

Diapari M, Sindhu A, Bett K, Deokar A, Warkentin TD, Taran B (2014) Genetic diversity and association mapping of iron and zinc concentrations in chickpea (*Cicer arietinum* L.). Genome 57:1–10

Dwivedi SL, Sahrawat KL, Rai KN, Blair MW, Andersson M, Pfieffer W (2012) Nutritionally enhanced staple food crops. Plant Breed Rev 34:169–262

Eiche E, Bardelli F, Nothstein AK, Charlet L, Göttlicher J, Steininger R, Dhillon KS, Sadana US (2015) Selenium distribution and speciation in plant parts of wheat (*Triticum aestivum*) and Indian mustard (*Brassica juncea*) from a seleniferous area of Punjab. Sci Total Environ 505:952–961

Garcia-Oliveira AL, Tan L, Fu Y, Sun C (2009) Genetic identification of quantitative trait loci for contents of mineral nutrients in rice grain. J Integr Plant Biol 51:84–92

Giuliano G (2017) Provitamin A biofortification of crop plants: a gold rush with many miners. Curr Opin Biotechnol 44:169–180

Gomez-Becerra HF, Yazici A, Ozturk L, Budak H, Peleg Z, Morgounov A, Fahima T, Saranga Y, Cakmak I (2010) Genetic variation and environmental stability of grain mineral nutrient concentrations in Triticum dicoccoides under five environments. Euphytica 171(1):39–52

Gomez-Becerra HF, Erdem H, Yazici A, Tutus B, Torun L, Ozturk L, Cakmak I (2011) Grain concentrations of protein and mineral nutrients in a large collection of spelt wheat grown under different environments. J Cereal Sci 52:342–349

Gorafi YSA, Ishii T, Kim JS, Elbashir AAE, Tsujimoto H (2016) Genetic variation and association mapping of grain iron and zinc contents in synthetic hexaploid wheat germplasm. Plant Genet Resour. https://doi.org/10.1017/S1479262116000265

Graham RD, Welch RM, Bouis HE (2001) Addressing micronutrient malnutrition through enhancing the nutritional quality of staple foods: principles, perspectives and knowledge gaps. Adv Agron 70:77–142

Gregorio GB, Senadhira D, Htut H, Graham RD (2000) Breeding for trace mineral density in rice. Food Nutr Bull 21:382–386

Gu R, Chen F, Liu B, Wang X, Liu J, Li P, Pan Q, Pace J, Soomro AA, Lübberstedt T, Mi G, Yuan L (2015) Comprehensive phenotypic analysis and quantitative trait locus identification for grain mineral concentration, content, and yield in maize (Zea mays L.). Theor Appl Genet 128:1777–1789

Guerrero B, Llugany M, Palacios O, Valiente M (2014) Dual effects of different selenium species on wheat. Plant Physiol Biochem 83:300–307

Gyawali S, Otte ML, Chao S, Jilal A, Jacob DL, Amezrou R, Verma RPS (2017) Genome wide association studies (GWAS) of element contents in grain with a special focus on zinc and iron in a world collection of barley (*Hordeum vulgare* L.). J Cereal Sci 77:266–274

Harjes C, Rocheford T, Bai L, Brutnell T, Kandianis C, Sowinski S, Stapleton A, Vallabhaneni R, Williams M, Wurtzel E et al (2008) Natural genetic variation in lycopene epsilon cyclase tapped for maize biofortification. Science 319:330–333

Hirschi KD (2009) Nutrient biofortification of food crops. Annu Rev Nutr 29:401–421

Huang XY, Salt DE (2016) Plant ionomics: from elemental profiling to environmental adaptation. Mol Plant 9:787–789

Huang Y, Sun C, Min J, Chen Y, Tong C, Bao J (2015) Association mapping of quantitative trait loci for mineral element contents in whole grain rice (*Oryza sativa* L.). J Agric Food Chem 63:10885–10892

Ishikawa R, Iwata M, Taniko K, Monden G, Miyazaki N, Orn C et al (2017) Detection of quantitative trait loci controlling grain zinc concentration using Australian wild rice, Oryza meridionalis, a potential genetic resource for biofortification of rice. PLoS One 12(10):e0187224

Jadhav AA, Rayate SJ, Mhase LB, Thudi M, Chitikineni A, Harer PN, Jadhav AS, Varshney RK, Kulwal PL (2015) Marker-trait association study for protein content in chickpea (Cicer arietinum L.). J Genet 94:279–286

Jambunathan R (1980) Improvement of the nutritional quality of sorghum and pearl millet. Food Nutr Bull 2:11–16

Jin T, Chen J, Zhu L, Zhao Y, Guo J, Huang Y (2015) Comparative mapping combined with homology based cloning of the rice genome reveals candidate genes for grain zinc and iron concentration in maize. BMC Genet 16:17

Jittham O, Fu X, Xu J, Chander S, Li J, Yang X (2017) Genetic dissection of carotenoids in maize kernels using high-density single nucleotide polymorphism markers in a recombinant inbred line population. Crop J 5:63–72

Johnson AAT et al (2011) Constitutive overexpression of the OsNAS gene family reveals single-gene strategies for effective iron- and zinc-biofortification of rice endosperm. PLoS One 6:e24476

Kandianis CB, Stevens R, Liu WP, Palacios N, Montgomery K, Pixley K, White WS, Rocheford T (2013) Genetic architecture controlling variation in grain carotenoid composition and concentrations in two maize populations. Theor Appl Genet 126:2879–2895

Khazaei H, Podder R, Caron CT, Kundu SS, Diapari M, Vandenberg A, Bett KE (2017) Marker–trait association analysis of iron and zinc concentration in lentil (Lens culinaris Medik.) seeds. Plant Genome 10(2):28724070

Khokhar JS, Sareen S, Tyagi BS, Singh G, Wilson L, King IP et al (2018) Variation in grain Zn concentration, and the grain ionome, in field-grown Indian wheat. PLoS One 13:e0192026

Kondou Y, Manickavelu A, Komatsu K, Arifi M, Kawashima M, Ishii T, Hattori T, Iwata H, Tsujimoto H, Ban T, Matsui M (2016) Analysis of grain elements and identification of best genotypes for Fe and P in Afghan wheat landraces. Breed Sci 66:676–682

Krishnappa G, Singh AM, Chaudhary S, Ahlawat AK, Singh SK, Shukla RB et al (2017) Molecular mapping of the grain iron and zinc concentration, protein content and thousand kernel weight in wheat (*Triticum aestivum* L.). PLoS One 12:e0174972

Kumar S, Hash CT, Thirunavukkarasu N, Singh G, Rajaram V, Rathore A, Senapathy S, Mahendrakar MD, Yadav RS, Srivastava RK (2016) Mapping quantitative trait loci controlling high iron and zinc content in self and open pollinated grains of pearl millet [*Pennisetum glaucum* (L.) R. Br.]. Front Plant Sci 7:1636

Liu H, Wang ZH, Li F, Li K, Yang N, Yang Y, Huang D, Liang D, Zhao H, Mao H, Liu J, Qiu W (2014) Grain iron and zinc concentrations of wheat and their relationships to yield in major wheat production areas in China. Field Crop Res 156:151–160

Liu C, Chen G, Li Y, Peng Y, Zhang A, Hong K et al (2017) Characterization of a major QTL for manganese accumulation in rice grain. Sci Rep 7(1):17704

Lorenz AJ, Chao S, Asoro FG et al (2011) Genomic selection in plant breeding: knowledge and prospects. Adv Agron 110:77–123

Lung'aho MG, Mwaniki AM, Szalma SJ, Hart JJ, Rutzke MA, Kochian LV, Glahn RP, Hoekenga OA (2011) Genetic and physiological analysis of iron biofortification in maize kernels. PLoS One 6:e20429

Mahender A, Anandan A, Pradhan SK, Pandit E (2016) Rice grain nutritional traits and their enhancement using relevant genes and QTLs through advanced approaches. Springerplus 5:2086

Mallikarjuna MG, Thirunavukkarasu N, Hossain F, Bhat JS, Jha SK, Rathore A, Agrawal PK, Pattanayak A, Reddy SS, Gularia SK, Singh AM, Manjaiah KM, Gupta HS (2015) Stability performance of inductively coupled plasma mass spectrometry-phenotyped kernel minerals concentration and grain yield in maize in different agro-climatic zones. PLoS One 10:e0139067

Manickavelu A, Hattori T, Yamaoka S, Yoshimura K, Kondou Y, Onogi A et al (2017) Genetic nature of elemental contents in wheat grains and its genomic prediction: toward the effective use of wheat landraces from Afghanistan. PLoS One 12:e0169416

Masuda H, Ishimaru Y, Aung MS, Kobayashi T, Kakei Y, Takahashi M, Higuchi K, Nakanishi H, Nishizawa NK (2012) Iron biofortification in rice by the introduction of multiple genes involved in iron nutrition. Sci Rep 2:543

Masuda H, Shimochi E, Hamada T, Senoura T, Kobayashi T, Aung MS et al (2017) A new transgenic rice line exhibiting enhanced ferric iron reduction and phytosiderophore production confers tolerance to low iron availability in calcareous soil. PLoS One 12(3):e0173441

Meuwissen THE, Hayes BJ, Goddard ME (2016) Genomic selection: a paradigm shift in animal breeding. Anim Front 6:6–14

Meuwissen THE, Hayes BJ, Goddard ME (2001) Prediction of total genetic value using genome wide dense marker maps. Genetics 157:1819–1829

Mishra A, Bohra A (2018) Non-coding RNAs and plant male sterility: current knowledge and future prospects. Plant Cell Rep 37:177–191

Mohanty A, Marndi BC, Sharma S, Das A (2011) Biochemical characterization of two high protein rice cultivars from Assam rice collections. Oryza 48:171–174

Monsen ER (2000) Dietary reference intakes for the antioxidant nutrients: vitamin C, vitamin E, selenium, and carotenoids. J Am Diet Assoc 100:637–640

Moreno-Moyano LT, Bonneau JP, Sánchez-Palacios JT, Tohme J, Johnson AAT (2016) Association of increased grain iron and zinc concentrations with agro-morphological traits of biofortified rice. Front Plant Sci 7:1463

Morgounov A, Gómez-Becerra HF, Abugalieva A, Dzhunusova M, Yessimbekova M, Muminjanov H, Zelenskiy Y, Ozturk L, Cakmak I (2007) Iron and zinc grain density in common wheat grown in Central Asia. Euphytica 155:193–203

Muzhingi T, Palacios-Rojas N, Miranda A, Cabrera ML, Yeum KJ, Tang G (2017) Genetic variation of carotenoids, vitamin E and phenolic compounds in Provitamin A biofortified maize. J Sci Food Agric 97:793–801

Nakaya A, Isobe SN (2012) Will genomic selection be a practical method for plant breeding? Ann Bot 110:1303–1316

Nawaz Z, Kakar KU, Li XB, Li S, Zhang B, Shou HX, Shu QY (2015) Genome-wide association mapping of quantitative trait loci (QTLs) for contents of eight elements in brown rice (*Oryza sativa* L.). J Agric Food Chem 63:8008–8016

Neelamraju S, Mallikarjuna Swamy BP, Kaladhar K, Anuradha K, Venkateshwar Rao Y, Batchu AK, Agarwal S, Babu AP, Sudhakar T, Sreenu K, Longvah T, Surekha K, Rao KV, Ashoka Reddy G, Roja TV, Kiranmayi SL, Radhika K, Manorama K, Cheralu C, Viraktamath BC

(2012) Increasing iron and zinc in rice grains using deep water rices and wild species – identifying genomic segments and candidate genes. Q Assur Safety Crops Foods 4:138

Ogo Y, Itai RN, Kobayashi T, Aung MS, Nakanishi H, Nishizawa NK (2011) OsIRO2 is responsible for iron utilization in rice and improves growth and yield in calcareous soil. Plant Mol Biol 75:593–605

Ortiz-Monasterio JI, Palacios-Rojas N, Meng E, Pixley K, Trethowan R, Pena RJ (2007) Enhancing the mineral and vitamin content of wheat and maize through plant breeding. J Cereal Sci 46:293–307

Owens BF, Lipka AE, Magallanes-Lundback M, Tiede T, Diepenbrock CH, Kandianis CB, Kim E, Cepela J, Mateos-Hernandez M, Buell CR, Buckler ES, DellaPenna D, Gore MA, Rocheford T (2014) A foundation for provitamin A biofortification of maize: genome-wide association and genomic prediction models of carotenoid levels. Genetics 198:1699–1716

Pandey A, Khan MK, Hakki EE, Thomas G, Hamurcu M, Gezgin S, Gizlenci O, Akkaya MS (2016) Assessment of genetic variability for grain nutrients from diverse regions: potential for wheat improvement. Springerplus 5:1912

Paul S, Ali N, Gayen D, Datta SK, Datta K (2012) Molecular breeding of *Osfer*2 gene to increase iron nutrition in rice. GM Crops Food 3:310–316

Paul S, Gayen D, Datta SK, Dutta K (2016) Analysis of high iron rice lines reveals new miRNAs that target iron transporters in roots. J Expt Bot 67:5811–5824

Peleg Z, Saranga Y, Yazici A, Fahima T, Ozturk L, Cakmak I (2008) Grain zinc, iron and protein concentrations and zinc-efficiency in wild emmer wheat under contrasting irrigation regimes. Plant Soil 306:57–67

Phuke RM, Anuradha K, Radhika K, Jabeen F, Anuradha G, Ramesh T, Hariprasanna K, Mehtre SP, Deshpande SP, Anil G, Das RR, Rathore A, Hash T, Reddy BVS, Kumar AA (2017) Genetic variability, genotype × environment interaction, correlation, and GGE biplot analysis for grain iron and zinc concentration and other agronomic traits in RIL population of sorghum (*Sorghum bicolor* L. Moench). Front Plant Sci 8:712

Pinson SRM, Tarpley L, Yan W, Yeater K, Lahner B, Yakubova B et al (2015) Worldwide genetic diversity for mineral element concentrations in rice grain. Crop Sci 55:294–311

Poblaciones MJ, Rodrigo S, Santamaria O, Chen Y, McGrath SP (2014) Agronomic selenium biofortification in Triticum durum under Mediterranean conditions: from grain to cooked pasta. Food Chem 146:378–384

Pu ZE, Yu M, He QY, Chen GY, Wang JR, Liu YX et al (2014) Quantitative trait loci associated with micronutrient concentrations in two recombinant inbred wheat lines. J Integr Agric 13:2322–2329

Qin X, Zhang W, Dubcovsky J, Tian L (2012) Cloning and comparative analysis of carotenoid β-hydroxylase genes provides new insights into carotenoid metabolism in tetraploid (Triticum turgidum ssp. durum) and hexaploid (Triticum aestivum) wheat grains. Plant Mol Biol 80:631–646

Qin X, Fischer K, Yu S, Dubcovsky J, Tian L (2016) Distinct expression and function of carotenoid metabolic genes and homoeologs in developing wheat grains. BMC Plant Biol 16:155

Raboy V, Dickinson DB, Below FE (1984) Variation in seed total phosphorus, phytic acid, zinc, calcium, magnesium, and protein among lines of *Glycine max* and *G. soja*. Crop Sci 24:431–434

Rahman WMM, Zaman EMS, Thavarajah P, Thavarajah D, Siddique KHM (2013) Selenium biofortification in lentil (Lens culinaris Medikus subsp. culinaris): farmers' field survey and genotype × environment effect. Food Res Int 54:1596–1604

Renuka N, Mathure SV, Zanan RL, Thengane RJ, Nadaf AB (2016) Determination of some minerals and β-carotene contents in aromatic *indica* rice (*Oryza sativa* L.) germplasm. Food Chem 191:2–6

Rezaei MK, Deokar A, Tar'an B (2016) Identification and expression analysis of candidate genes involved in carotenoid biosynthesis in chickpea seeds. Front Plant Sci 7:1867

Rhodes DH, Hoffmann L Jr, Rooney WL, Herald TJ, Bean S, Boyles R, Brenton ZW, Kresovich S (2017) Genetic architecture of kernel composition in global sorghum germplasm. BMC Genomics 18:15

Roy SC, Sharma BD (2014) Assessment of genetic diversity in rice [*Oryza sativa* L.] germplasm based on agro-morphology traits and zinc-iron content for crop improvement. Physiol Mol Biol Plants 20:209–224

Ruel MT, Alderman H, The Maternal and Child Nutrition Study Group (2013) Nutrition-sensitive interventions and programmes: how can they help to accelerate progress in improving maternal and child nutrition? Lancet 382:536–551

Schroeder DG, Brown KH (1994) World health organ nutritional status as a predictor of child survival: summarizing the association and quantifying its global impact. Bull World Health Organ 72:569–579

Sing SP, Gruissem W, Bhullar NK (2017) Single genetic locus improvement of iron, zinc and β-carotene content in rice grains. Sci Rep 7:6883

Singh A, Sharma V, Dikshit HK, Aski M, Kumar H, Thirunavukkarasu N, Patil BS, Kumar S, Sarker A (2017a) Association mapping unveils favorable alleles for grain iron and zinc concentrations in lentil (Lens culinaris subsp. culinaris). PLoS One 12:e0188296

Singh A, Sharma V, Dikshit HK, Aski M, Kumar H, Thirunavukkarasu N, Patil BS, Kumar S, Sarker A (2017b) Association mapping unveils favorable alleles for grain iron and zinc concentrations in lentil(Lens culinaris subsp. culinaris). PLoS One 12:e0188296

Srinivasa J, Arun B, Mishra VK, Singh GP, Velu G, Babu R et al (2014) Zinc and iron concentration QTL mapped in a Triticum spelta × T. aestivum cross. Theor Appl Genet 127:1643–1651

Suwarno WB, Pixley KV, Palacios-Rojas N, Kaeppler SM, Babu R (2015) Genome-wide association analysis reveals new targets for carotenoid biofortification in maize. Theor Appl Genet 128:851–864

Swamy BPM, Rahman MA, Inabangan-Asilo MA, Amparado A, Manito C, Chadha-Mohanty P, Reinke R, Slamet-Loedin IH (2016) Advances in breeding for high grain zinc in rice. Rice (NY) 9:49

Terasawa Y, Ito M, Tabiki T, Nagasawa K, Hatta K, Nishio Z (2016) Mapping of a major QTL associated with protein content on chromosome 2B in hard red winter wheat (Triticum aestivum L.). Breed Sci 66:471–480

Tiwari C, Wallwork H, Balasubramaniam A, Mishra VK, Govindan V, Stangoulis J, Kumar U, Joshi AK (2016) Molecular mapping of quantitative trait loci for zinc, iron and protein content in the grains of hexaploid wheat. Euphytica 207:563–570

Trijatmiko KR, Dueñas C, Tsakirpaloglou N, Torrizo L, Arines FM, Adeva C et al (2016) Biofortified indica rice attains iron and zinc nutrition dietary targets in the field. Sci Rep 6:19792

Upadhyaya HD, Bajaj D, Das S, Kumar V, Gowda CL, Sharma S, Tyagi AK, Parida SK (2016a) Genetic dissection of seed-iron and zinc concentrations in chickpea. Sci Rep 6:24050

Upadhyaya HD, Bajaj D, Narnoliya L, Das S, Kumar V, Gowda CL, Sharma S, Tyagi AK, Parida SK (2016b) Genome-wide scans for delineation of candidate genes regulating seed-protein content in chickpea. Front Plant Sci 7:302

Velu G, Ortiz-Monasterio I, Singh R, Payne T (2011) Variation for grain micronutrients concentration in wheat core-collection accessions of diverse origin. Asian J Crop Sci 3:43–48

Velu G, Tutus Y, Becerra HFG, Hao Y, Demir L, Kara L, Crespo-Herrera L, Orhan S, Yazici A, Singh R, Cakmak I (2016a) QTL mapping for grain zinc and iron concentrations and zinc efficiency in a tetraploid and hexaploid wheat mapping populations. Plant Soil 411:81–99

Velu G, Crossa J, Singh RP, Hao Y, Dreisigacker S, Perez-Rodriguez P, Joshi AK, Chatrath R, Gupta V, Balasubramaniam A, Tiwari C, Mishra VK, Sohu VS, Mavi GS (2016b) Genomic prediction for grain zinc and iron concentrations in spring wheat. Theor Appl Genet 129:1595–1605

Velu G, Rai KN, Muralidharan V, Kulkarni VN, Longvah T, Raveendran TS (2007) Prospects of breeding biofortified pearl millet with high grain iron and zinc content. Plant Breed 126: 182–185

Wang P, Wang H, Liu Q, Tian X, Shi Y, Zhang X (2017) QTL mapping of selenium content using a RIL population in wheat. PLoS One 12:e0184351

White PJ, Broadley MR (2009) Biofortification of crops with seven mineral elements often lacking in human diets—iron, zinc, copper, calcium, magnesium, selenium and iodine. New Phytol 182:49–84

WHO (2017) https://www.who.int/nutrition/publications/globaltargets2025_policybrief_anaemia/en/

Wirth J et al (2009) Rice endosperm iron biofortification by targeted and synergistic action of nicotianamine synthase and ferritin. Plant Biotechnol J 7:631–644

Yamunarani R, Govind G, Ramegowda V, Vokkaliga H, Thammegowda HV, Guligowda SA (2016) Genetic diversity for grain Zn concentration in finger millet genotypes: potential for improving human Zn nutrition. Crop J 4:229–234

Yan J, Kandianis C, Harjes C, Bai L, Kim E, Yang X, Skinner D, Fu Z, Mitchell S, Li Q et al (2010) Rare genetic variation at Zea mays crtRB1 increases beta-carotene in maize grain. Nat Genet 42:322–327

Zhai SN, Xia XC, He ZH (2016) Carotenoids in staple cereals: metabolism, regulation, and genetic manipulation. Front Plant Sci 7:1197

Zhang Y, Xu YH, Yi HY, Gong JM (2012) Vacuolar membrane transporters OsVIT1 and OsVIT2 modulate iron translocation between flag leaves and seeds in rice. Plant J 72:400–410

Zhang M, Pinson SRM, Tarpley L, Huang X, Lahner B, Yakubova E et al (2014) Mapping and validation of quantitative trait loci associated with concentration of 16 elements in unmilled rice grain. Theor Appl Genet 127:137–165

Zhang H, Liu J, Jin T, Huang Y, Chen J, Zhu L, Zhao Y, Guo J (2017a) Identification of quantitative trait locus and prediction of candidate genes for grain mineral concentration in maize across multiple environments. Euphytica 213:90

Zhang D, Zhang H, Chu S, Li H, Chi Y, Triebwasser-Freese D, Lv H, Yu D (2017) Integrating QTL mapping and transcriptomics identifies candidate genes underlying QTLs associated with soybean tolerance to low-phosphorus stress. Plant Mol Biol 93:137–150

Zhao FJ, Su YH, Dunham SJ, Rakszegi M, Bedo Z, McGrath SP, Shewry PR (2009) Variation in mineral micronutrient concentrations in grain of wheat lines of diverse origin. J Cereal Sci 49:290–295

Zhou JF, Huang YQ, Liu ZZ, Chen JT, Zhu LY, Song ZQ, Zhao YF (2010) Genetic analysis and QTL mapping of zinc, iron, copper and manganese contents in maize seed. J Plant Genet Resour 11:593–595

Chapter 2
Marker-Assisted Breeding for Improving the Cooking and Eating Quality of Rice

Maxwell Darko Asante, Francis Osei Amoako-Andoh, Valentin Stanislas Edgar Traore, Matilda Ntowaa Bissah, and Joshua N. Cobb

2.1 Introduction

Rice is the most important food crop in the world. It is a major source of calories for over a half of the world's population. Unlike other cereals, rice is mainly eaten as whole grain, thus, cooking and eating quality is extremely important. The cooking and eating quality is measured by the easiness of cooking, texture, springiness, stickiness and chewiness of cooked rice (Champagne et al. 2010). These attributes are controlled by starch physicochemical properties, comprising of apparent amylose content (AAC), gelatinization temperature (GT), gel consistency (GC) as well as paste viscosity properties, which are popularly called RVA paste viscosity parameters because they are mainly measured using Rapid Visco Analyzer (Bao 2014).

These physicochemical properties of starch are measured using tedious and expensive biochemical methods. Assessment of these attributes is further complicated by the lack of discrete phenotypic classes in segregating populations, environmental effects and the triploid nature of the rice plant (He et al. 1999). It can therefore be challenging to breed for these traits in a timely manner. DNA markers

M. D. Asante (✉) · F. O. Amoako-Andoh
CSIR-Crops Research Institute, Kumasi, Ghana
e-mail: m.darkoasante@cropsresearch.org; f.amoako-andoh@cropsresearch.org

V. S. E. Traore
Institute of Environment and Agricultural Research (INERA), Ouagadougou, Burkina Faso
e-mail: inera.direction@fasonet.bf

M. N. Bissah
CSIR-Plant Genetic Resources Research Institute, Bunso, Ghana

J. N. Cobb
International Rice Research Institute, Los Baños, Philippines
e-mail: j.cobb@irri.org

A. M. I. Qureshi et al. (eds.), *Quality Breeding in Field Crops*,
https://doi.org/10.1007/978-3-030-04609-5_2

could be very useful in selecting for rice genotypes, with the desired cooking and eating qualities. Rice is the first crop plant to have its genome fully sequenced (Yu et al. 2002; Matsumoto et al. 2005). The sequencing of the japonica and indica genomes has helped rice Geneticist, Breeders and Cereal Chemists to identify useful molecular markers for marker-assisted selection for cooking and eating quality traits.

In this chapter, we discussed the various cooking and eating quality characteristics in rice comprising of apparent amylose content (AAC), gelatinization temperature (GT), gel consistency (GC), RVA pasting parameters (RVA), aroma and the use of molecular markers to facilitate identifying these traits for their selection in breeding programmes. Some successful examples in the use of DNA markers in breeding for cooking and eating quality in rice have been highlighted. Even though the scope and target of this chapter is MAS for single genes that enable breeders to manipulate key grain quality traits in rice, genomic selection and other more quantitative approaches are mentioned.

2.2 Cooking and Eating Properties of Rice

2.2.1 Apparent Amylose Content

The rice grain is composed of 80–90% starch. Starch consists of two glucose polymers—amylose and amylopectin. Amylose is largely a linear molecule with a degree of polymerization (DP) of 1000–5000 glucose units, whilst amylopectin is highly branched and contains relatively larger polymer units (Juliano 2003). Amylose content is typically referred to as apparent amylose content (AAC) because the iodine-based assay used for quantifying it often detects long chain amylopectin in addition to the true amylose (Takeda et al. 1987).

AAC is the most important determinant of the texture of cooked rice. Recently, long chain amylopectin rather than amylose content has been suggested as the major factor affecting the texture of cooked rice (Umemoto 2018). Thus, AAC which affects amylose content plus long chain amylopectin may be a better determinant of cooked rice texture compared to true amylose composition. AAC for rice ranges from 0% for waxy rice to approximately 30% for non-waxy types. Originally, rice varieties were classified as follows: waxy (0–5%), very low (5–12%), low (12–20%), intermediate (20–25%) or high (25–33%) AAC (Juliano and Villareal 1993).

As characterization of the *Wx* genetic locus has become better understood (discussed below), rice is now typically classified as waxy rice (0–2% AAC), very low AAC (3–9%), low AAC (10–18%), intermediate AAC (19–23%) and high AAC (>23%) (Bergman et al. 2004; Chen et al. 2008a, b). Cooked rice with high amylose content is flaky, dry, hard, and separates, while rice with low amylose content is sticky, moist, tender and glossy (Juliano 1979).

2.2.2 Genetics of Apparent Amylose Content

AAC is principally controlled by the *waxy* gene (*Wx*) on chromosome 6, which encodes the granule-bound starch synthase (GBSS) (Wang et al. 1995; Smith et al. 1997; He et al. 1999). The *waxy* gene has three alleles, *Wx*, Wx^a and Wx^b, which are found in waxy (sticky) rice, *indica* and *japonica* sub-species respectively (Sano 1984). The *indica* sub-species that harbours the Wx^a allele accumulates more GBSS protein in the endosperm than *japonica* sub-types that possess the Wx^b allele. Thus, *japonica* rice generally has a lower content of AAC than *indica* and thus is more sticky. Truly sticky rice is made up of approximately 100% amylopectin thus, has little or no AAC. The level of AAC is largely determined by the splicing pattern of the first intron in the *waxy* gene (Wang et al. 1995). When this intron is un-spliced it produces grains with no AAC. Partial and complete splicing in the intron leads to intermediate and high AAC, respectively.

Additional minor QTLs for AAC have been mapped at various positions within the rice genome (Tan et al. 1999; Lanceras et al. 2000; Aluko et al. 2004; Fan et al. 2005; Shi-yong et al. 2006; Wang et al. 2007; Zheng et al. 2008). The *waxy* gene is reported to act additively with five additional minor genes, namely: *AGPlar*, *PUL*, *SSI*, *SSII-3*, and *SSIII-2*, to affect AAC (Tian et al. 2009).

2.2.3 Gelatinization Temperature

Gelatinization temperature (GT) is the temperature at which 90% of starch granules swell irreversibly in hot water resulting in forfeiture of crystallinity and birefringence (Deja and Khush 2000). GT controls the cooking time of rice, and it ranges from 55 to 85 °C (Fitzgerald et al. 2009). Rice with high GT elongates less and will be generally undercooked when standard cooking methods are applied. GT is typically classified as low (55–69 °C), intermediate (70–74 °C) or high (75–79 °C) (Khush et al. 1979). GT has a strong negative correlation with alkaline spreading value (ASV), which measures the disintegration of rice grains in KOH. ASV is comparatively cheaper and faster to measure compared to the direct measurement of GT with instruments. Thus, ASV is normally used as a proxy to estimate GT. ASV scores of 2 (low), 3–5 (intermediate) and 6–7 (high) correspond to high (>74 °C), intermediate (70–74 °C) and low (<70 °C) GT, respectively (Bergman et al. 2004).

2.2.4 Genetics of Gelatinization Temperature

Gelatinization temperature is mainly controlled by the alkaline degeneration gene (*Alk* gene) on chromosome 6 of the rice genome, and it has been mapped at the same locus as *starch synthase II* (*SSIIa*), which encodes soluble starch synthase IIa protein (Umemoto et al. 2002; He et al. 2006) and is driven by two variants (GC/TT and G/A) of four nonsynonymous SNPs in the coding region of the *Alk* gene (Bao et al. 2006; Gao et al. 2011). The GC/TT polymorphism differentiates rice varieties with high or intermediate GT (possessing the GC allele) from those with low GT (possessing the TT allele) in 90% of a diverse set of 509 rice genotypes (Bao et al. 2006). Allelic variation at these two variants causes a substitution of leucine for phenylalanine, respectively (Waters et al. 2006).

The second variant (G/A) in exon 8 of the *Alk* gene classifies a diverse set of 70 genotypes into classes of high GT or low GT without distinguishing the intermediate phenotype (Waters et al. 2006) due to the substitution of methionine in the place of valine in the gene product. Rice genotypes with high GT have a combination of valine and leucine while those with low GT had a combination of either methionine and leucine or valine and phenylalanine at the same amino acid residues (Waters et al. 2006).

GT is also affected by additional minor QTLs, which can modify the effects of the major genes (Lanceras et al. 2000; Bao et al. 2002; Aluko et al. 2004; Fan et al. 2005; Shi-yong et al. 2006; Wang et al. 2007). Three genes, *SBE3*, *ISA* and *SSIV-2* as well as the *waxy* gene are reported to have minor effects on GT, and they act additively with the *Alk* gene to affect GT (Tian et al. 2009).

2.2.5 Gel Consistency

Gel consistency (GC) is a measure of how firm rice grain is after cooking. It helps to classify rice varieties of the same AAC, particularly in the high AAC class as hard, medium or soft texture (Paula and Fitzgerald 2012). GC is commonly determined by measuring the length of a cooled gel made from rice flour that has been cooked in 0.2 M KOH (Cagampang et al. 1973). GC is inversely related to length of the gel, and is classified as hard (length of gel <40 mm), medium (length of gel 41–60 mm) and soft (length of gel >61 mm) (Cagampang et al. 1973).

Sticky and low amylose rice are reported to have low GC (Paula and Fitzgerald 2012). However, rice with intermediate to high AAC could have high, intermediate or soft classifications. The correlation between AAC and GC in IRRI tropical indica material was found to be low (Paula and Fitzgerald 2012), implying that the traits are physiologically independent in the context of high or intermediate AAC.

2.2.6 Genetics of Gel Consistency

Due to the strong association of low AAC with low GC the *waxy* locus is reported to be the major determinant of GC (Tan et al. 1999; Lanceras et al. 2000; Tian et al. 2009). This was confirmed by the cloning of a major QTL for GC at the *waxy* locus (Su et al. 2011) where a SNP in exon 10 of the *waxy* gene was found to be highly associated with GC (Tran et al. 2011). Many minor QTLs for GC have been mapped

(He et al. 1999; Bao et al. 2002; Fan et al. 2005; Tian et al. 2005; Shi-yong et al. 2006; Wang et al. 2007) and include three additional genes in the starch biosynthesis pathway—*AGPiso*, *SBE3* and *ISA* (Tian et al. 2009).

2.2.7 Paste Viscosity Parameters

Rapid Visco Analyzer (RVA) is the preferred equipment for measuring paste viscosity parameters due to its low cost and high throughput. RVA profile mimics the cooking process of rice. The profile is generated by subjecting rice flour to a heat-hold-cool-hold temperature cycle (Juliano 1996). The primary RVA parameters include peak viscosity, PV (first peak viscosity after gelatinization), trough or hot paste viscosity, HPV (paste viscosity at the end of the 95 °C holding period) and final or cool paste viscosity, CPV (paste viscosity at the end of the test). These primary parameters are used to calculate secondary ones, including breakdown, BD (BD = PV − HPV), setback, SB (SB = CPV − PV) and consistency, CS (CS = CPV − HPV). Other RVA parameters include peak time and pasting temperature.

2.2.8 Genetics of Paste Viscosity Parameters

Most RVA parameters are reported to be highly associated with AAC and thus, mainly controlled by the *waxy* locus (Bao et al. 2000; Larkin et al. 2003) and as such many major QTLs for the various RVA parameters have been identified by many authors at that locus (Bao et al. 2000; Larkin et al. 2003; Wang et al. 2007; Zheng et al. 2012; Yao et al. 2017). QTLs for RVA parameters that influence eating quality including HPV, CPV, BD, SB and CS were found to co-locate the *waxy* locus, whilst those for RVA parameters that characterize the cooking process such as peak time and pasting temperature were found in the vicinity of the *Alk* gene (Wang et al. 2007).

Many minor QTLs on almost all the 12 chromosomes of rice have been identified (Bao et al. 2000; Wang et al. 2007; Zheng et al. 2012; Yao et al. 2017) for RVA parameters and in one experiment, as many as 106 QTLs for RVA parameters were detected across both *indica* and *japonica* sub-species of rice (Yao et al. 2017). For sticky rice, 10 out of 17 *starch synthesis*-related genes were reported to affect RVA parameters with *PUL* playing a dominant role (Yan et al. 2011).

2.2.9 Aroma

Aromatic rice also known as fragrant or perfumed rice is highly prized all over the world. The two most popular aromatic rice types are the Basmati cultivars from India and Pakistan and the Jasmines from Thailand. Other classical examples of fragrant rice are 'Dulhabhog' of Bangladesh, 'Azucena' and 'Milfor' of the

Philippines and 'Rojolele' of Indonesia (Khush et al. 1979). About 114 different volatile compounds have been linked to rice fragrance (Yajima et al. 1979). However, 2-acetyl-1-pyrroline or 2AP was reported as the main compound that differentiates fragrant rice from non-fragrant varieties (Buttery et al. 1983).

2.2.10 *Inheritance of Aroma*

Even though di-genic and multi-genic control of aroma have been reported by some authors (Pinson 1994; Lorieux et al. 1996), many studies appear to favour a monogenic recessive inheritance (Sood and Siddiq 1978; Berner and Hoff 1986; Ahn et al. 1992; Dong et al. 2001).

An aroma gene for rice was first mapped to chromosome 8, where it was linked with the RFLP marker, RG28 (Ahn et al. 1992). This gene, *BADH2*, was later cloned and reported to encode betaine aldehyde dehydrogenase 2 (Bradbury et al. 2005a). The *BADH2* gene codes for the biosynthesis of 2-acetyl-1-pyrroline (2AP) (Buttery et al. 1983). The accumulation of 2AP is reported to result from the absence of BADH2 activity, leading to increased levels of 4-aminobutyraldehyde/Δ^1-pyrroline, the immediate precursor of 2AP (Bradbury et al. 2008; Chen et al. 2008c).

An eight-base-pair deletion (8-bp) combined with three SNPs in exon 7 of the *BADH2* gene were found to be the cause of fragrance in both Basmati- and Jasmine-styled rice types where the non-fragrant rice varieties had a full copy of the gene. The 8-bp deletion in the *BADH2* gene was used to develop an allele-specific perfect marker for determining fragrance in rice (Bradbury et al. 2005b).

However, some rice types with fragrant phenotypes were found not to have the 8-bp deletion in exon 7 of the *BADH2* gene (Fitzgerald et al. 2008; Kovach et al. 2009). Rather, a 7-bp deletion in exon 2 of the *BADH2* gene was found to be the cause of fragrance in those varieties (Shi et al. 2008). Also, eight new alleles were discovered at the *BADH2* locus that conferred fragrance in 24 accessions that did not carry any of the previously identified alleles (Kovach et al. 2009). A new allele caused by an 803 bp deletion between exons 4 and 5 of the *BADH2* gene in rice variety 'Zaimiaoxiangnuo' was found, and a marker FMbadh2-E4-5 was developed for its detection (Sho et al. 2011). In another study, three genes located on chromosomes 3, 4 and 8 were reported to be the cause of fragrance in a rice variety called 'Pusa 1121' (Amarawathi et al. 2008). The QTL on chromosome 8 was mapped to the *BADH2* region, while that on chromosome 4 co-located with a *BADH1* gene. Fragrant Japanese rice landraces were grouped into two groups based on different SNP mutations in the *BADH2* gene (Ootsuka et al. 2014). The mutations were a known SNP located at exon 13 and a novel SNP at the exon 1–intron 1 junction.

Recently, a new allele of the *badh2* gene has been found in a fragrant *indica* rice accession called 'Velchi' (Khandagale et al. 2017). This new allele resulted from a 253-bp deletion in the promoter region of 'Velchi' as compared to non-fragrant rice accessions. Additionally, 'Velchi' had a 5-bp duplication that was used to design a new marker (velbadh2-p-UTR) to distinguish it from non-fragrant rice accessions.

These indicate the presence of allelic/genic diversity for fragrance in rice germplasm from various regions of world. These variations can be exploited by breeders to develop new varieties with increased levels of fragrance in rice.

2.3 DNA Markers Used for Predicting Cooking and Eating Quality in Rice

Molecular markers are useful when they are tightly linked to or within the gene of interest. In rice, genes for AAC, RVA, GT, GC and aroma have been cloned and functional markers have been developed to facilitate the selection for these traits.

2.3.1 Molecular Markers for AAC, RVA and GC

AAC and RVA are highly associated with three SNPs in the *waxy* gene, namely intron 1 (G → T), exon 6 (A → C) and exon 10 (C → T) substitutions (Larkin et al. 2003; Chen et al. 2008a, b). AAC is associated with two of the three SNPs—intron1 (G → T) and exon 6 (A → C) (Chen et al. 2008a, 2010; Asante et al. 2013). The third substitution in the *waxy* gene, exon 10 (C → T), differentiates high AAC rice with strong paste viscosity from those with weak paste viscosities (Chen et al. 2008a). Three allele-specific PCR markers targeting the three functional SNPs in the *waxy* gene were developed for genotyping in single PCR amplification (Chen et al. 2010). This marker is very efficient for predicting the AAC and RVA parameters of rice germplasm from various parts of the world (Chen et al. 2010; Asante et al. 2013). Using the three SNPs together, a collection of world germplasm had four *waxy* SNP haplotypes (Chen et al. 2008a, 2010; Asante et al. 2013). These waxy SNP haplotypes include TAC, GCC, GAC and GAT for low, intermediate, high and high AAC plus high RVA, respectively (Chen et al. 2008a, 2010; Asante et al. 2013). Recently, two new haplotypes of the SNPs within the *waxy* gene were identified in seven accessions of the USDA mini-core rice germplasm (Li et al. 2017).

A perfect marker for predicting intermediate amylose based on the A → C SNP in exon 6 has recently been developed and validated in a Chinese mini core germplasm collection (Zhou et al. 2018). Also, an SSR marker in intron 1 of the *waxy* gene (RM 190) was found to explain a large portion of the variation in AAC and most RVA parameters, and it is used for marker-assisted selection for cooking quality in rice (Bergman et al. 2001; Chen et al. 2008b; Li et al. 2017). A 23 bp sequence duplication (InDel) of exon 2 has also been found to be highly associated with AAC (Li et al. 2017). Combining all the SNPs in the *waxy* gene as haplotype has been found to be more predictive for AAC and RVA compared to the individual SNP markers (Chen et al. 2008a, 2010; Asante et al. 2013).

Markers developed specifically for GC are not common. However, a major QTL for GC located at the *waxy* locus (Su et al. 2011), and a SNP in exon 10 of the *waxy* gene, were found to be highly associated with GC (Tran et al. 2011). Accordingly, the markers developed to select for AAC can also be used to select for GC.

2.3.2 DNA Markers for GT

Molecular markers have been developed to predict GT using the two functional SNPs in exon 8 of the *ALK gene*—GC/TT and G/A (Umemoto and Aoki 2005; Bao et al. 2006; Gao et al. 2011). The GC and TT alleles are associated with intermediate/high and low GT, respectively. The G/A SNP differentiates genotypes with high GT from those with low GT. High and low GT are associated with the G and A alleles, respectively (Waters et al. 2006).

2.3.3 DNA Markers for Aroma in Rice

Many functional markers have been developed for predicting fragrance in rice. A perfect marker based on the 8-bp deletion in exon 7 of the BADH2 gene on chromosome 8 was developed (Bradbury et al. 2005b). This mutation is reported to be the cause of fragrance in a majority of rice genotypes (Kovach et al. 2009). This marker has thus been used extensively to predict the fragrance status of rice breeding lines (Asante et al. 2010; Jin et al. 2010; Bett-garber et al. 2015).

An indel marker targeting the 8-bp deletion and reported to be easier to use compared to the perfect marker (Bradbury et al. 2005b) was developed (Sakthivel et al. 2009). Another indel marker based on 8-bp deletion in the BADH2 gene called Aromarker has also been developed and successfully used for marker-assisted backcrossing (Yi et al. 2009; Jantaboon et al. 2011).

A fragrance allele caused by a deletion of 803 bp between exons 4 and 5 of the BADH2 gene in rice variety 'Zaimiaoxiangnuo' was also found, and a marker, FMbadh2-E4-5, was developed for its detection (Sho et al. 2011).

Another fragrance allele was due to a 7-bp deletion in exon 2 of the BADH2 gene. Functional markers, which can differentiate fragrance caused by the 8-bp deletion on exon 7 from that caused by the 7-bp deletion on exon 2 of chromosome 8, and non-fragrant from fragrant rice have been developed (Shi et al. 2008). Besides the 8-bp deletion, nine other mutations in the *BADH2* gene, which resulted in fragrance in various subpopulations of rice have been identified, and DNA markers that are associated with these mutations developed (Kovach et al. 2009).

2.4 Application of Molecular Markers for Rice Breeding in the Last Two Decades

SSR markers have been widely used for diversity studies, QTL mapping and marker-assisted selection for plant breeding in the last two decades. Molecular markers are especially useful in situations where they are closely linked or in linkage disequilibrium with a gene of interest (Dekkers 2004). Gene-based or functional markers have also been developed to facilitate selection for various traits in rice (Bradbury et al. 2005b; Mackill 2007; Masouleh et al. 2009; Sakthivel et al. 2009; Chen et al. 2010). The use of DNA markers in rice breeding has been occurring mostly for pyramiding disease resistance and grain quality genes using marker-assisted backcrossing (MAB) (Francia et al. 2005; Jena and Mackill 2008). The use of MAB has three components: use of a tightly linked marker or genic marker (also called a 'perfect' marker) for foreground selection, use of flanking markers for recombinant selection, and the use of background markers for selection against the donor genome at unlinked loci (Collard and Mackill 2008). One drawback to the above approach is the intensive amount of genotyping that must be done, primarily in the case of background selection.

Large backcross populations, usually on the order of 500 plants or more, must be produced to ensure that there are sufficient plants to carry out background selection after the foreground and recombinant selection have been performed. With flanking markers defining an interval of 3–5 cM, the foreground/recombinant selection will limit the population to less than 5% of the total backcross progeny. This process can get to be very expensive (Mackill 2007) and while SNPs are rapidly replacing SSRs and are relatively cheaper (McCouch et al. 2010), in many cases, it is worth considering the urgency associated with developing the stacked conversion. If time permits, making a BC3- or BC4-derived conversion can often take care of any meaningful residual background effects without the need for expensive fingerprinting of large backcross populations.

2.4.1 *Examples of Marker-Assisted Breeding for Cooking and Eating Quality in Rice*

1. In China, an elite restorer line, R8166, which is susceptible to blast and has poor grain quality (cooked rice retrogrades and becomes hard after cooling due to high AC and hard GC originating from its Wx^a genotype), was improved for the two traits by incorporating two *R* genes for blast and the Wx^b gene for AAC through marker-assisted backcrossing (MAB) (Wu-ming et al. 2017). During the MAB process, two lines, R163 and R167, with the lowest and highest percentage of the background of the recurrent parent were selected to generate hybrids with improved blast resistance and grain quality. R163, R167 and their resultant hybrids had significantly lower AAC, softer GC and lower grain chalkiness (Wu-ming et al. 2017).

2. In Thailand, scientists used MAS to develop new rice varieties for submergence tolerance and Jasmine-styled cooking quality for the rainfed and irrigated lowland ecosystems in the Mekong region of Southeast Asia (Jantaboon et al. 2011). The scientists made a cross between two rice varieties, IR57514 (widely adapted to the rainfed lowlands of the Mekong region, high yielding and submergence-tolerant with *Sub1* gene) and Kao Dawk Mali 105, popularly called KDML105 (fragrant, traditional Thai variety that has good cooking qualities). They developed a large population of recombinant inbred lines (RILs) using single seed descent (SSD). Preferred alleles of *Sub1*, *Wx*, *badh2* and *SSIIa* loci were selected using markers; R10783Indel, Waxy, Aromarker and GT11, respectively. MAS was done in a step-wise manner; Wx and Aromarker were used for genotyping of 2037 F7 RILs. RILS (341 lines) with the preferred Wx and Aromarker were selected and genotyped with the marker for submergence tolerance, R10783Indel, after which 97 RILs were kept. These 97 lines were genotyped with the GT11 marker, and two groups of ideal genotypes were selected for evaluation. These were 66 RILs with positive alleles for all of the four loci designated as ID1, whereas the remaining 31 RILs carrying a negative allele at *GT11* and positive alleles for the other three loci grouped as ID2. The DNA markers helped to develop ideal genotypes that have submergence tolerance, Jasmine-styled grain quality as well as superior yields compared to their parents (Jantaboon et al. 2011).
3. A group in India developed a Basmati quality rice variety that has resistance to bacterial blight through marker-assisted backcross-pedigree breeding strategy (Gopalakrishnan et al. 2008). They crossed a popular Basmati rice cultivar, 'Pusa Basmati 1' (PB 1) with a non-Basmati variety (IRBB55) which had two bacterial blight resistance genes, *xa13* and *Xa21*, and did their selection from a BC1F5 population.

   Markers linked to the bacteria blight resistance genes, *xa13* and *Xa21* from IRBB55, and the recurrent parent PB 1 allele for the *waxy* locus giving intermediate amylose content and a maintainer allele at fertility restorer locus were used for foreground selection. Background selection was done using microsatellite markers. An elite line that combines resistance to bacteria blight disease and Basmati-styled grain quality was commercially released as 'Improved Pusa Basmati 1' (Gopalakrishnan et al. 2008).
4. In China, another group developed rice lines that combine resistance to bacterial blight with high amylose content through marker-assisted selection (Ramalingam et al. 2002). They selected for Xa-21 bacterial blight resistance and high amylose content (Wx^b allele) (Ramalingam et al. 2002). Selection was done from four F2 populations involving *indica* × *indica*, *indica* × intermediate and *japonica* × *indica* crosses. Eighty-six F2 lines with the *Xa-21* gene were selected from all four F2 generations using a PCR marker which is linked to the gene. Subsequently, lines that were both homozygous and heterozygous for the Wx^b allele were selected as a *waxy* gene microsatellite marker. Finally, 20 true breeding lines that combine homozygous for the Wx^b allele high amylose and having the *Xa-21* gene were selected for field trials.

5. The most cultivated rice variety in Myanmar, 'Manawthukha' was improved for its fragrance and amylose content (AAC) using marker-assisted backcrossing procedure (Yi et al. 2009). 'Manawthukha' is non-fragrant and has high AAC, intermediate GT and GC. It was crossed to Basmati 370 to introgress the Basmati grain qualities—fragrance and intermediate AAC. The preferred alleles of the *BADH2* and *Wx* genes were introgressed from Basmati 370 into 'Manawthukha' by making four backcrosses and allowing the BC4F1 to self to obtain a BC4F2 population. Twelve BC4F2 lines that carried a large percentage of the background of 'Manawthukha' and the homozygous Basmati-styled grain quality alleles were selected and planted in multi-location trials (four in Myanmar, and one in Thailand). The agronomic performances of the lines were similar to that of 'Manawthukha', and they had the intermediate amylose and fragrance of the donor Basmati 370 parent (Yi et al. 2009).
6. In China, a key maintainer line, II-32B, used for breeding hybrid rice has poor grain quality because it is non-fragrant and has high AAC and GT (Jin et al. 2010). The grain quality of II-32B was improved using marker-assisted selection. Functional markers for *Wx*, *SSIIa* and *BADH2* genes for AAC, GT and fragrance, respectively, were used. II-32B was crossed to 'Yixiang B', a fragrant maintainer line that has low AAC and low GT. The F1s were allowed to self to obtain an F2 population. A total of 300 F2 lines were genotyped with the functional markers for *Wx*, *SSIIa* and *BADH2* genes. F2 lines homozygous for the grain quality alleles of the donor parent Yixiang B (*Wx*-(CT)17 microsatellite, *SSIIa*-TT SNP and 8-bp deletion allele in the *BADH2* gene) were selected and backcrossed to II-32B as a recurrent parent. The BC1F1 generation was advanced to BC2F2 population using MAS. BC2F2 lines homozygous for *Wx*-(CT)17, *SSIIa*-TT and 8-bp *BADH2* deletion alleles were selected and allowed to self to BC2F4 generation. Agronomic evaluation was done on 5–10 plants selected randomly from each BC2F4 line. Eventually, 17 lines which were homozygous for *Wx*-(CT)17, *SSIIa*-TT and *fgr* gene markers were selected. Improved II-32B lines with fragrance and lower AAC and GT were obtained (Jin et al. 2010).
7. In Ghana, genotyping-by-sequencing markers (GBS) were used to facilitate selection of lines that combine good grain quality with resistance to blast and rice yellow mottle virus (RYMV) diseases (Asante 2012). A rice cultivar, 'Digang', resistant to RYMV and blast but with poor grain quality, non-fragrant and with high AAC was crossed to fragrant Jasmine 85, which is susceptible to the two diseases but has low AAC. The grain quality loci of Jasmine 85 and the background of 'Digang' were selected simultaneously using the GBS data. Fragrant lines with low AAC are being evaluated for their reaction to blast and RYMV as well as yield.

Diverse rice germplasm from Africa and other parts of the world have been genotyped using functional markers for fragrance, GT, AAC and RVA paste viscosity to identify alleles of interest to facilitate the selection of parents for breeding for cooking and eating quality in Ghana and elsewhere (Asante et al. 2010, 2013).

2.4.2 *Modern Markers Technologies for Breeding for Rice Grain Quality and Other Traits*

Rice is the first crop plant to have its genome fully sequenced. The ever-declining cost for sequencing, facilitating the partial and complete re-sequencing of genomes of various accessions, has led to the identification of arrays of high-density SNPs. Popular SNPs technologies include LGC, Illumina, Axiom (Affymetrix) and Douglas Scientific. A review of the various SNP genotyping platforms has recently been done (Barabaschi et al. 2016).

The rice community has taken advantage of these SNP technologies to create platforms capable of genotyping thousands of SNPs. Medium and high-resolution SNP arrays deployed for genotyping rice include a 44 K SNP chip (Zhao et al. 2011), 50 K SNP chips (Chen et al. 2014; Singh et al. 2015), 600 K high-density rice array (Thomson et al. 2017) and the 700 K high-density rice array (McCouch et al. 2016).

In addition to the fixed SNP arrays, Next Generation Sequencing (NGS) methods, such as genotyping-by-sequencing (GBS), provides high-density SNPs. GBS is based on high-throughput, next generation sequencing of genomic fragments generated by restriction enzymes (REs), and the cost per sample is relatively low (Elshire et al. 2011; Poland and Rife 2012). It enables the discovery of SNPs whilst genotyping (Elshire et al. 2011).

The availability of millions of genome-wide SNPs from both fixed arrays and NGS allows breeders to do genomics-assisted breeding—MAS and genomic selection (GS). The high density of SNPs across all chromosomes in the genome allows diversity analysis, genome-wide association studies, gene cloning, QTL mapping and DNA fingerprinting to be done more effectively with higher reliability of results.

The deployment of SNP-based genomics-assisted breeding and in combination with appropriate breeding schemes, is drastically improving the way rice breeding is being done around the world. Many rice breeding programmes have shifted from pedigree selection to single seed descent, otherwise known as rapid regeneration advance method (RGA). RGA is orders of magnitude cheaper than pedigree selection and helps to generate promising lines within a shorter periods and is particularly suited for rice because the plants can grow in small containers and forced to flower early. The combination of breeding methods such as RGA, automated phenotyping and availability of high-throughput SNP platforms for rice will improve selection for grain quality, yield and stress-related traits.

The on-going genomic revolution is expected to immensely benefit breeders, especially those who work with model crops such as rice. Genomic selection using NGS markers is likely to take over rice breeding in the near future, and it will help to accelerate the rate of genetic gain within rice breeding programmes.

2.5 Conclusions

Cooking and eating quality of rice is determined by fragrance and starch physicochemical properties, comprising of AAC, GT, GC and paste viscosity. Phenotyping for these traits is tedious, time-consuming and expensive. In addition, phenotyping for the starch properties of individuals in a population can only be done on the grain after harvesting, which extends the time required to collect the relevant data and thus negatively affect genetic gain. Breeding for cooking and eating quality in rice is thus a major candidate for marker-assisted breeding along with breeding for disease resistance. Major genes for fragrance, AAC and GT have been cloned, and the mutations in these genes are used to develop functional markers for selection. These markers have been successfully applied to improve the cooking and eating qualities of both inbred and hybrid rice in various countries. The advent of high-throughput SNP genotyping arrays and NGS techniques has made genomics-assisted breeding for yield, biotic and abiotic stresses, and grain quality the most efficient ways of developing improved rice varieties.

Acknowledgements The Alliance for a Green Revolution in Africa (AGRA) funded the lead author's research on rice grain quality.

References

Ahn SN, Bollich CN, Tanksley SD (1992) RFLP tagging of a gene for aroma in rice. Theor Appl Genet 84:825–828. https://doi.org/10.1007/BF00227391

Aluko G, Martinez C, Tohme J et al (2004) QTL mapping of grain quality traits from the interspecific cross Oryza sativa X O. glaberrima. Theor Appl Genet 109:630–639. https://doi.org/10.1007/s00122-004-1668-y

Amarawathi Y, Singh R, Singh AK et al (2008) Mapping of quantitative trait loci for basmati quality traits in rice (Oryza sativa L.). Mol Breed 21:49–65. https://doi.org/10.1007/s11032-007-9108-8

Asante MD (2012) Genetic analysis of grain quality traits in rice. Univeristy of Ghana, Accra

Asante MD, Kovach MJ, Huang L et al (2010) The genetic origin of fragrance in NERICA1. Mol Breed 26:419–424. https://doi.org/10.1007/s11032-009-9382-8

Asante MD, Offei SK, Gracen V et al (2013) Starch physicochemical properties of rice accessions and their association with molecular markers. Starch/Staerke 65:1022–1028. https://doi.org/10.1002/star.201300058

Bao J (2014) Genes and QTLs for rice grain quality improvement. In: Yan W, Bao J (eds) Rice – germplasm, genetics and improvement. IntechOpen, London, pp 239–278

Bao JS, Zheng XW, Xia YW et al (2000) QTL mapping for the paste viscosity characteristics in rice (Oryza sativa L.). Theor Appl Genet 100:280–284. https://doi.org/10.1007/s001220050037

Bao JS, Wu YR, Hu B et al (2002) QTL for rice grain quality based on a DH population derived from parents with similar apparent amylose content. Euphytica 128:317–324. https://doi.org/10.1023/A:1021262926145

Bao JS, Corke H, Sun M (2006) Nucleotide diversity in starch synthase IIa and validation of single nucleotide polymorphisms in relation to starch gelatinization temperature and other physicochemical properties in rice (*Oryza sativa* L.). Theor Appl Genet 113:1171–1183. https://doi.org/10.1007/s00122-006-0355-6

Barabaschi D, Tondelli A, Desiderio F et al (2016) Next generation breeding. Plant Sci 242:3–13. https://doi.org/10.1016/j.plantsci.2015.07.010
Bergman CJ, Delgado JT, McClung AM, Fjellstrom RG (2001) An improved method for using a microsatellite in the rice waxy gene to determine amylose class. Cereal Chem 78:257–260
Bergman CJ, Bhattcharya K, Ohtsubo K (2004) Rice end-use quality analysis. In: Champagne E (ed) Rice chemsitry and technology, 3rd edn. AACC, St. Paul, MN, pp 415–472
Berner D, Hoff B (1986) Inheritance of scent in American long grain rice. Crop Sci 26:876–878
Bett-garber KL, Bryant RJ, Grimm CC et al (2015) Physicochemical and sensory analysis of U.S. rice varieties developed for the basmati and jasmine markets. Cereal Chem 94:602–610
Bradbury LMT, Fitzgerald TL, Henry RJ et al (2005a) The gene for fragrance in rice. Plant Biotechnol J 3:363–370. https://doi.org/10.1111/j.1467-7652.2005.00131.x
Bradbury LMT, Henry RJ, Jin Q et al (2005b) A perfect marker for fragrance genotyping in rice. Mol Breed 16:279–283. https://doi.org/10.1007/s11032-005-0776-y
Bradbury LMT, Gillies SA, Brushett DJ et al (2008) Inactivation of an aminoaldehyde dehydrogenase is responsible for fragrance in rice. Plant Mol Biol 68:439–449. https://doi.org/10.1007/s11103-008-9381-x
Buttery RG, Ling LC, Juliano B, Turnbaugh JG (1983) Cooked rice aroma and 2-acetyl-1-pyrroline. J Agric Food Chem 31:823–826. https://doi.org/10.1021/jf00118a036
Cagampang GB, Perez CM, Juliano BO (1973) A gel consistency test for eating quality of rice. J Sci Food Agric 24:1589–1594. https://doi.org/10.1002/jsfa.2740241214
Champagne ET, Bett-Garber KL, Fitzgerald MA et al (2010) Important sensory properties differentiating premium rice varieties. Rice 3:270–281. https://doi.org/10.1007/s12284-010-9057-4
Chen MH, Bergman CJ, Pinson SRM, Fjellstrom RG (2008a) Waxy gene haplotypes: associations with pasting properties in an international rice germplasm collection. J Cereal Sci 48:781–788. https://doi.org/10.1016/j.jcs.2008.05.004
Chen MH, Bergman CJ, Pinson SRM, Fjellstrom RG (2008b) Waxy gene haplotypes: Associations with apparent amylose content and the effect by the environment in an international rice germplasm collection. J Cereal Sci 47:536–545. https://doi.org/10.1016/j.jcs.2007.06.013
Chen S, Yang Y, Shi W et al (2008c) Badh2, encoding betaine aldehyde dehydrogenase, inhibits the biosynthesis of 2-acetyl-1-pyrroline, a major component in rice fragrance. Plant Cell 20:1850–1861. https://doi.org/10.1105/tpc.108.058917
Chen MH, Fjellstrom RG, Christensen EF, Bergman CJ (2010) Development of three allele-specific codominant rice Waxy gene PCR markers suitable for marker-assisted selection of amylose content and paste viscosity. Mol Breed 26:513–523. https://doi.org/10.1007/s11032-010-9419-z
Chen H, Xie W, He H et al (2014) A high-density snp genotyping array for rice biology and molecular breeding. Mol Plant 7:541–553. https://doi.org/10.1093/mp/sst135
Collard BC, Mackill DJ (2008) Marker-assisted selection: an approach for precision plant breeding in the twenty-first century. Philos Trans R Soc B Biol Sci 363:557–572. https://doi.org/10.1098/rstb.2007.2170
Deja NC, Khush GS (2000) Rice grain quality evaluation procedures. In: Singh RK, Singh US, Khush GS (eds) Aromatic rices. International Rice Research Institute, Manila, pp 15–28
Dekkers JC (2004) Commercial application of marker- and gene-assisted selection in livestock: strategies and lessons. J Anim Sci 82:E313–E328. https://doi.org/10.2527/2004.8213_supplE313x
Dong Y, Tsuzuki E, Terao H (2001) Trisomic genetic analysis of aroma in three Japanese native rice varieties (Oryza sativa L.). Euphytica 117:191–196. https://doi.org/10.1023/A:1026502115436
Elshire RJ, Glaubitz JC, Sun Q et al (2011) A robust, simple genotyping-by-sequencing (GBS) approach for high diversity species. PLoS One 6:1–10. https://doi.org/10.1371/journal.pone.0019379
Fan CC, Yu XQ, Xing YZ et al (2005) The main effects, epistatic effects and environmental interactions of QTLs on the cooking and eating quality of rice in a doubled-haploid line population. Theor Appl Genet 110:1445–1452. https://doi.org/10.1007/s00122-005-1975-y
Fitzgerald MA, Sackville Hamilton NR, Calingacion MN et al (2008) Is there a second fragrance gene in rice? Plant Biotechnol J 6:416–423. https://doi.org/10.1111/j.1467-7652.2008.00327.x

Fitzgerald MA, McCouch SR, Hall RD (2009) Not just a grain of rice: the quest for quality. Trends Plant Sci 14:133–139. https://doi.org/10.1016/j.tplants.2008.12.004

Francia E, Tacconi G, Crosatti C et al (2005) Marker assisted selection in crop plants. Plant Cell Tissue Organ Cult 82:317–342. https://doi.org/10.1007/s11240-005-2387-z

Gao Z, Zeng D, Cheng F et al (2011) ALK, the key gene for gelatinization temperature, is a modifier gene for gel consistency in rice. J Integr Plant Biol 53:756–765. https://doi.org/10.1111/j.1744-7909.2011.01065.x

Gopalakrishnan S, Sharma RK, Anand Rajkumar K et al (2008) Integrating marker assisted background analysis with foreground selection for identification of superior bacterial blight resistant recombinants in Basmati rice. Plant Breed 127:131–139. https://doi.org/10.1111/j.1439-0523.2007.01458.x

He P, Li SG, Qian Q et al (1999) Genetic analysis of rice grain quality. Theor Appl Genet 98:502–508. https://doi.org/10.1007/s001220051098

He Y, Han Y, Jiang L et al (2006) Functional analysis of starch-synthesis genes in determining rice eating and cooking qualities. Mol Breed 18:277–290. https://doi.org/10.1007/s11032-006-5505-7

Jantaboon J, Siangliw M, Im-mark S et al (2011) Ideotype breeding for submergence tolerance and cooking quality by marker-assisted selection in rice. Field Crop Res 123:206–213. https://doi.org/10.1016/j.fcr.2011.05.001

Jena KK, Mackill DJ (2008) Molecular markers and their use in marker-assisted selection in rice. Crop Sci 48:1266–1276. https://doi.org/10.2135/cropsci2008.02.0082

Jin L, Lu Y, Shao Y et al (2010) Molecular marker assisted selection for improvement of the eating, cooking and sensory quality of rice (Oryza sativa L.). J Cereal Sci 51:159–164. https://doi.org/10.1016/j.jcs.2009.11.007

Juliano BO (1979) Amylose analysis in rice—a review. In: Proceedings of the workshop on chemical aspects of rice grain quality. International Rice Research Institute, Los Banos, Laguna, pp 251–260

Juliano BO (1996) Juliano, B. O. In: Walker Ce HJE (ed) Applications of the rapid visco analyser. Newport Scientific, Sydney, NSW, pp 19–24

Juliano BO (2003) Rice chemistry and quality. PhilRice, Nueva Ecija

Juliano BO, Villareal CP (1993) Grain quality evaluation of world rices. International Rice Research Institute, Manila

Khandagale KS, Zanan RL, Mathure SV, Nadaf AB (2017) Haplotype variation of Badh2 gene, unearthing of a new fragrance allele and marker development for non-basmati fragrant rice "Velchi" (Oryza sativa L.). Agric Gene 6:40–46. https://doi.org/10.1016/j.aggene.2017.09.003

Khush GS, Paule CM, De La Cruz MN (1979) Rice quality evaluation and improvement at IRRI. In: Proceedings of the workshop on chemical aspects of rice grain quality. International Rice Research Institute, Manila, pp 21–32

Kovach MJ, Calingacion MN, Fitzgerald MA, Mccouch SR (2009) The origin and evolution of fragrance in rice (Oryza sativa L.). PNAS 10:14444–14449. https://doi.org/10.1073/pnas.0904077106

Lanceras JC, Huang ZL, Naivikul O et al (2000) Mapping of genes for cooking and eating qualities in Thai jasmine rice (KDML105). DNA Res 7:93–101. https://doi.org/10.1093/dnares/7.2.93

Larkin PD, McClung AM, Ayres NM, Park WD (2003) The effect of the Waxy locus (granule bound starch synthase) on pasting curve characteristics in specialty rices (Oryza sativa L.). Euphytica 131:243–253. https://doi.org/10.1023/A:1023962406605

Li K, Bao J, Corke H, Sun M (2017) Association analysis of markers derived from starch biosynthesis related genes with starch physicochemical properties in the USDA rice mini-core collection. Front Plant Sci 8:424. https://doi.org/10.3389/fpls.2017.00424

Lorieux M, Petrov M, Huang N et al (1996) Aroma in rice: genetic analysis of a quantitative trait. Theor Appl Genet 93:1145–1151. https://doi.org/10.1007/BF00230138

Mackill DJ (2007) Molecular markers and marker-assisted selection in rice genomics-assisted crop improvement. In: Varshney RK, Tuberosa R (eds) Genomics assisted crop improvement: vol. 2: genomics applications in crops. Springer, New York, NY, pp 147–168

Masouleh AK, Waters DLE, Reinke RF, Henry RJ (2009) A high-throughput assay for rapid and simultaneous analysis of perfect markers for important quality and agronomic traits in rice using multiplexed MALDI-TOF mass spectrometry. Plant Biotechnol J 7:355–363. https://doi.org/10.1111/j.1467-7652.2009.00411.x

Matsumoto T, Wu J, Kanamori H et al (2005) The map-based sequence of the rice genome. Nature 436:793–800. https://doi.org/10.1038/nature03895

McCouch SR, Zhao K, Wright M et al (2010) Development of genome-wide SNP assays for rice. Breed Sci 60:524–535. https://doi.org/10.1270/jsbbs.60.524

McCouch SR, Wright MH, Tung C-W et al (2016) Open access resources for genome-wide association mapping in rice. Nat Commun 7:10532. https://doi.org/10.1038/ncomms10532

Ootsuka K, Takahashi I, Tanaka K et al (2014) Genetic polymorphisms in Japanese fragrant landraces and novel fragrant allele domesticated in northern Japan. Breed Sci 64:115–124. https://doi.org/10.1270/jsbbs.64.115

Paula R, Fitzgerald M (2012) Genetic diversity of rice grain quality. Genet Divers Plants 2012: 285–310. https://doi.org/10.5772/35119

Pinson SRM (1994) Inheritance of aroma in six rice cultivars. Crop Sci 34:1151. https://doi.org/10.2135/cropsci1994.0011183X003400050002x

Poland JA, Rife TW (2012) Genotyping-by-sequencing for plant breeding and genetics. Plant Genome J 5:92. https://doi.org/10.3835/plantgenome2012.05.0005

Ramalingam J, Basharat HS, Zhang G (2002) STS and microsatellite marker-assisted selection for bacterial blight resistance and waxy genes in rice, *Oryza sativa* L. Euphytica 127:255–260

Sakthivel K, Rani NS, Pandey MK et al (2009) Development of a simple functional marker for fragrance in rice and its validation in Indian basmati and non-basmati fragrant rice varieties. Mol Breed 24:185–190. https://doi.org/10.1007/s11032-009-9283-x

Sano Y (1984) Theor Appl Genet 68:467

Shao GN, Tang A, Tang SQ et al (2011) A new deletion mutation of fragrant gene and the development of three molecular markers for fragrance in rice. Plant Breed 130:172–176. https://doi.org/10.1111/j.1439-0523.2009.01764.x

Shi W, Yang Y, Chen S, Xu M (2008) Discovery of a new fragrance allele and the development of functional markers for the breeding of fragrant rice varieties. Mol Breed 22:185–192. https://doi.org/10.1007/s11032-008-9165-7

Shi-yong SUN, Wei HAO, Hong-xuan LIN (2006) Identification of QTLs for cooking and eating quality of rice grain. Rice Sci 13:161–169

Singh N, Jayaswal PK, Panda K et al (2015) Single-copy gene based 50 K SNP chip for genetic studies and molecular breeding in rice. Sci Rep 5. https://doi.org/10.1038/srep11600

Smith AM, Denyer K, Martin C (1997) The synthesis of the starch granule. Annu Rev Plant Physiol Plant Mol Biol 48:67–87

Sood B, Siddiq E (1978) A rapid technique for scent determination in rice. Indian J Genet Plant Breed 38:268–271

Su Y, Rao Y, Hu S et al (2011) Map-based cloning proves qGC-6, a major QTL for gel consistency of japonica/indica cross, responds by waxy in rice (Oryza sativa L.). Theor Appl Genet 123:859–867. https://doi.org/10.1007/s00122-011-1632-6

Takeda Y, Hizukuri S, Juliano BO (1987) Structures of rice amylopectins with low and high affinities for iodine. Carbohydr Res 168:79–88. https://doi.org/10.1016/0008-6215(87)80008-3

Tan YF, Li JX, Yu SB et al (1999) The three important traits for cooking and eating qualities of rice grain are controlled by a single locus. Theor Appl Genet 99:642–648. https://doi.org/10.1007/s001220051279

Thomson MJ, Singh N, Dwiyanti MS et al (2017) Large-scale deployment of a rice 6 K SNP array for genetics and breeding applications. Rice 10:1–13. https://doi.org/10.1186/s12284-017-0181-2

Tian R, Jiang GH, Shen LH et al (2005) Mapping quantitative trait loci underlying the cooking and eating quality of rice using a DH population. Mol Breed 15:117–124. https://doi.org/10.1007/s11032-004-3270-z

Tian Z, Qian Q, Liu Q et al (2009) Allelic diversities in rice starch biosynthesis lead to a diverse array of rice eating and cooking qualities. Proc Natl Acad Sci 106:21760–21765. https://doi.org/10.1073/pnas.0912396106

Tran NA, Daygon VD, Resurreccion AP et al (2011) A single nucleotide polymorphism in the Waxy gene explains a significant component of gel consistency. Theor Appl Genet 123:519–525. https://doi.org/10.1007/s00122-011-1604-x

Umemoto T (2018) Genes affecting eating and processing qualities. In: Sasaki T, Ashikari M (eds) Rice genomics, genetics and breeding. Springer, Singapore, pp 417–434

Umemoto T, Aoki N (2005) Single-nucleotide polymorphisms in rice starch synthase IIa that alter starch gelatinisation and starch association of the enzyme. Funct Plant Biol 32:763–768

Umemoto T, Yano M, Satoh H et al (2002) Mapping of a gene responsible for the difference in amylopectin structure between japonica-type and indica-type rice varieties. Theor Appl Genet 104:1–8. https://doi.org/10.1007/s001220200000

Wang Z, Zheng F, Shen G et al (1995) Post-transcriptional regulation of the waxy gene. Plant J 7:613–622. https://doi.org/10.1046/j.1365-313X.1995.7040613.x

Wang LQ, Liu WJ, Xu Y et al (2007) Genetic basis of 17 traits and viscosity parameters characterizing the eating and cooking quality of rice grain. Theor Appl Genet 115:463–476. https://doi.org/10.1007/s00122-007-0580-7

Waters DLE, Henry RJ, Reinke RF, Fitzgerald MA (2006) Gelatinization temperature of rice explained by polymorphisms in starch synthase. Plant Biotechnol J 4:115–122. https://doi.org/10.1111/j.1467-7652.2005.00162.x

Wu-ming X, Xin P, Li-xin LUO et al (2017) Development of elite restoring lines by integrating blast resistance and low amylose content using MAS. J Integr Agric 17(1):16–27. https://doi.org/10.1016/S2095-3119(17)61684-8

Yajima I, Yanal T, Nakamura M et al (1979) Volatile flavor components of cooked kaorimai (Scented rice, o. sativa japonica). Agric Biol Chem 43:2425–2429. https://doi.org/10.1080/00021369.1979.10863850

Yan CJ, Tian ZX, Fang YW et al (2011) Genetic analysis of starch paste viscosity parameters in glutinous rice (*Oryza sativa* L.). Theor Appl Genet 122:63–76. https://doi.org/10.1007/s00122-010-1423-5

Yao X, Wang J, Liu J et al (2017) Mapping quantitative trait loci associated with starch paste viscosity in rice (*Oryza sativa* L.) under different environmental conditions. Plant Breed 136:591–602. https://doi.org/10.1111/pbr.12505

Yi M, Nwe KT, Vanavichit A et al (2009) Marker assisted backcross breeding to improve cooking quality traits in Myanmar rice cultivar Manawthukha. Field Crop Res 113:178–186. https://doi.org/10.1016/j.fcr.2009.05.006

Yu J, Hu S, Wang J et al (2002) A draft sequence of the rice genome (Oryza sativa L. ssp.). Science 296:79–92

Zhao K, Tung C-W, Eizenga GC et al (2011) Genome-wide association mapping reveals a rich genetic architecture of complex traits in Oryza sativa. Nat Commun 2:467. https://doi.org/10.1038/ncomms1467

Zheng X, Wu JG, Lou XY et al (2008) The QTL analysis on maternal and endosperm genome and their environmental interactions for characters of cooking quality in rice (*Oryza sativa* L.). Theor Appl Genet 116:335–342. https://doi.org/10.1007/s00122-007-0671-5

Zheng L, Zhang W, Liu S et al (2012) Genetic relationship between grain chalkiness, protein content, and paste viscosity properties in a backcross inbred population of rice. J Cereal Sci 56:153–160. https://doi.org/10.1016/j.jcs.2012.05.003

Zhou L, Chen S, Yang G, et al (2018) A perfect functional marker for the gene of intermediate amylose content Wx-in in rice (*Oryza sativa* L.). Breed Appl Biotechnol 18:103–109

Chapter 3
Improving the Nutritional Value of Potatoes by Conventional Breeding and Genetic Modification

John E. Bradshaw

3.1 Introduction

The potato (*Solanum tuberosum*) is the world's fourth most important food crop after maize, rice and wheat with 382 million tonnes fresh-weight (FW) of tubers produced in 2014 from 19.1 million hectares of land, in 164 countries, giving a global average yield of 20.0 t ha^{-1} (http://faostat.fao.org). About 60% of production (231 million tonnes) was in Asia (187), Africa (26) and Latin America (18) as a result of steady increases in recent years, particularly in China and India. Indeed, China (96 million tonnes) is now the number one potato producer in the world and India (46) is second, with the Russian Federation (32) third, Ukraine (24) fourth and the USA (20) fifth. As a staple food and vegetable, the potato is a major supplier of carbohydrate (starch) in the human diet, and is even being considered for human life support in space (Wheeler 2009). The potato tuber also provides significant amounts of protein, with a good amino acid balance, minerals, vitamins, micronutrients and phytonutrients which include antioxidants. It is high in dietary fibre, especially when eaten unpeeled with its skin. Fresh potatoes are virtually free of fat and cholesterol and have a water content of about 80%. Details of chemical composition can be found in the book *Advances in Potato Chemistry and Technology second Edition* (Singh and Kaur 2016), along with analytical techniques.

As a major food crop, the potato has an important role to play in the United Nations "2030 Agenda for Sustainable Development" which started on 1 January 2016 (http://faostat.fao.org). The agenda includes 17 goals, the second of which is to end hunger, achieve food security and improved nutrition, and promote sustainable agriculture. The background to the agenda is as follows. In 2015, out of a world population of 7.3 billion, around 800 million people suffered from hunger and more than two billion from one or more micronutrient deficiencies, known as "hidden

J. E. Bradshaw (✉)
Retired Plant Breeder and Geneticist, James Hutton Institute, Dundee, UK

A. M. I. Qureshi et al. (eds.), *Quality Breeding in Field Crops*,
https://doi.org/10.1007/978-3-030-04609-5_3

hunger", especially in women, infants and children. Ironically about 1.9 billion people were overweight, of whom 600 million were obese. Furthermore, the world population is expected to reach 8.5 billion by 2030 and 9.7 billion in 2050. By 2030 the aim of the agenda is to "ensure access by all people, in particular the poor and people in vulnerable situations, including infants, to safe, nutritious and sufficient food all year round". For potatoes, the need is to increase production and improve nutritional value.

In 2014 (http://faostat.fao.org) there was still a large difference in average yields between Africa (13.65 t ha^{-1}) and North America (43.68), whereas the average difference between Asia (18.82 t ha^{-1}) and Europe (22.17) was not so great. Bradshaw (2009) reviewed prospects for increasing yield per hectare, particularly in Asia, Africa and Latin America where increases in food production are required to match population growth, but new land is not readily available.

In this chapter, we are going to explore improving the nutritional value and health properties of potatoes by conventional breeding and genetic modification. Prerequisites are genetic variation and rapid and reliable screens for the target traits, as well as evidence that alteration of the chosen traits will improve nutritional value and human health. Potatoes need to be cooked because of the indigestibility of their ungelatinized starch (Burton 1989). Cooking is frequently by baking, boiling, steaming, roasting, deep-oil frying or microwave cooking, although in their native Andes a broad diversity of additional preparation methods is employed (Bradshaw and Bonierbale 2010). Cooking results in physical, chemical and enzymic changes in the potato tuber and these affect its nutritional value to different degrees (Tian et al. 2016). Furthermore, following food digestion, the nutrients must be taken up by the human body (i.e. bioavailable) in order to be utilized. Similar considerations apply to processing which increases the commercial value of potatoes. Since the 1950s processing has grown into a global industry which is still expanding. In North America and some European countries between 50% and 60% of the crop is processed (Li et al. 2006; Kirkman 2007), and processors are building factories in countries where the potato is primarily grown as a staple food, a trend that is likely to continue. The major processed products are potato crisps (chips), French fries and other frozen products, followed by dehydrated products, chilled-peeled potatoes and canned potatoes. The uses of potatoes as a source of starch and for molecular farming are beyond the scope of this chapter.

We are not going to consider more general aspects of cooking and processing quality such as improved flavour and texture, dry-matter (DM) and starch content, freedom from after-cooking blackening and enzymic browning, light-coloured fry-products post-harvest and post-storage, and novel starches for countries with a starch industry. However, it is important to remember that improving nutritional value and health properties needs to be done in the context of wider breeding objectives including those for "Sustainable Development". The relative priority to be given to nutritional value is a difficult question with no easy answer.

It will prove useful to start with a brief overview of the evolution of modern potatoes and methods available for their improvement by conventional breeding and genetic modification.

3.2 Evolution of Modern Potatoes and Breeding Methods

3.2.1 Domestication in South America

Wild tuber-bearing *Solanum* species are distributed from the southwestern USA (38°N), through the Andes, to central Argentina and adjacent Chile (41°S), and cover a great ecogeographical range (Spooner and Hijmans 2001). Their taxonomy is complicated and has undergone major revision. Hawkes (1990) recognized 219 wild tuber-bearing species and arranged them into 19 series of subsection *Potatoe* of section *Petota* of subgenus *Potatoe* of genus *Solanum*. Spooner et al. (2014) reduced this number of species to 107, now all classified as *Solanum* section *Petota* (tuber-bearing species) and partitioned into three nuclear clades rather than 19 series. Spooner (2016) explained how these changes came about. Wild relatives of cultivated potatoes are held in a number of genebanks around the world and are a valuable resource for potato breeding (Huaman et al. 2000a), but have not been used to any appreciable extent for improving the nutritional value of cultivated potatoes. Utilization of wild species has been reviewed by Jansky (2009) and Bradshaw and Bonierbale (2010).

The journey from gathering wild tubers to cultivation and domestication started early in the human colonization of the Americas, as wild potato remains have been found in a late Pleistocene settlement in south central Chile dated to around 12,500 years before present (Ugent et al. 1987; Moseley 2001). Interestingly, Louderback and Pavlik (2017) have just reported the earliest evidence of wild potato use in North America, at 10,900–10,100 years before present, in the form of well-preserved starch granules extracted from ground stone tools at North Creek Shelter, southern Utah. The granules were identified as those of *S. jamesii*, which is known to be highly nutritious, having twice the protein, zinc and manganese content of cultivated *S. tuberosum* and three times the calcium and iron content. Thus, a summer-active and highly productive herbaceous perennial would have provided a reliable, yearlong source of carbohydrate and minerals that significantly improved dietary quality.

Spooner et al. (2005) provided molecular taxonomic evidence for a single domestication in the highlands of southern Peru from the northern group of members of the *S. brevicaule* complex of diploid species typified by *S. bukasovii* (now *S. candolleanum*: Spooner et al. 2014). The result of domestication over 7000 years ago (Simmonds 1995) was a diploid cultigen *S. tuberosum* Stenotomum Group (Dodds 1962) from which all other cultivated potatoes were derived (Fig. 3.1). Dodds (1962) classified cultivated potatoes into five informal groups within one species (*S. tuberosum*) in which Andigena (tetraploid), Chaucha (triploid), Phureja (diploid) and Tuberosum (tetraploid) groups were derived from Stenotomum (diploid). Phureja was selected from Stenotomum by Andean farmers for lack of tuber dormancy and faster tuber development so that they could grow up to three crops a year in the lower, warmer, eastern valleys of the Andes. Phureja potatoes were therefore able to spread into northern Ecuador, Colombia and Venezuela and are the

Dodds 1962:

S. brevicaule complex of diploid species e.g.

S. bukasovii (2*x*)

(wild) ↓ (domestication in southern Peru)

S. acaule (4*x*) × *S. tuberosum* Stenotomum Gp (2*x*) × *S. megistacrolobum* (2*x*)

(wild) ↓ (cultivated) | | ↓ (wild)

S. juzepczukii (3*x*) | ↓ *S. ajanhuiri* (2*x*)

↓ ↓ Phureja Gp (mostly 2*x*)

unreduced gamete(s)

× ← Andigena Gp (4*x*) × Stenotomum Gp (2*x*)

↓ Wild species? →↓ ↓

S. curtilobum (5*x*) Tuberosum Gp (4*x*) Chaucha Gp (3*x*)

Spooner et al. 2007:

S. tuberosum Andigenum Gp (Andigena, Chaucha, Phureja, Stenotomum)

S. tuberosum Chilotanum Gp (Chilean Tuberosum)

Fig. 3.1 Origin of cultivated groups (Gp = Group) of *S. tuberosum* (Dodds 1962; Spooner et al. 2007) and cultivated species with bitter taste and frost tolerance (*S. ajanhuiri*, *S. juzepczukii* and *S. curtilobum*) (2, 3, 4 and 5*x* = diploid, triploid, tetraploid and pentaploid)

second most widely cultivated type in South America after Andigena. The latter is grown throughout the upland Andes of South America, presumably because farmers found the tetraploid superior to the diploids for yield and other traits. Tuberosum potatoes were selected from Andigena types for tuber production in long days in coastal Chile and are referred to as Chilean Tuberosum. They are a genetically distinct group of potatoes with a different cytoplasm to Andigena potatoes (Raker and Spooner 2002; Hosaka 2004).

Spooner et al. (2007) also regard Andigena, Chaucha, Phureja, Stenotomum and Tuberosum as a single species *S. tuberosum*, but now divided into just two Cultivar Groups. These are the Andigenum Group of upland Andean landraces containing diploids, triploids and tetraploids, and the Chilotanum Group of lowland tetraploid Chilean landraces. Three frost tolerant species are also cultivated in the Andes: *S. ajanhuiri* (diploid), *S. juzepczukii* (triploid) and *S. curtilobum* (pentaploid) (Fig. 3.1). In practice, tetraploid *S. tuberosum* can be regarded as the autotetraploid of diploid Stenotomum. The gametophytic self-incompatibility of the diploid does not operate in the autotetraploid so that natural pollination results in a mixture of self- and cross-pollination (Brown 1993). Goniocalyx (diploid) has been recognized as a northern subspecies/subgroup of Stenotomum, noted for tubers with a bright yellow flesh (Hawkes 1990).

The cultivated potatoes of South America are a valuable resource for breeding, including improvement of nutritional value. They are also a testament to the achievements of farmers as selectors and maintainers of vegetatively propagated landraces which arose through naturally occurring sexual reproduction. Following its creation in Lima, Peru in 1970, The International Potato Centre (CIP) (http://www.cipotato.org) assembled a collection of potato cultivars (landraces) native to nine countries in Latin America. Following elimination of duplicate accessions, the collection comprised 3527 landraces of which 552 were diploids, 128 triploids, 2836 tetraploids (2644 Andigena, 144 Chilean Tuberosum and 48 hybrids) and 11 pentaploids (Huaman et al. 1997). A core set of 306 Andigena accessions was established to aid utilization (Huaman et al. 2000b, c). They were chosen to represent the widest morphological diversity and to maximize geographical representation, and hence should be valuable for future breeding both in South America and worldwide.

3.2.2 South America to the World

Potatoes (tetraploid *S. tuberosum*) were introduced into Europe in the 1570s but initially remained a botanic curiosity, being grown and studied in physic gardens for interest and medicinal purposes. Their potential as a food crop was first seen in Ireland at the end of the seventeenth century; but it was food shortages that proved the stimulus to potato cultivation throughout Europe during the eighteenth century, because military and economic strength depended upon adequately fed manpower (Burton 1989; Reader 2008). Thus the eighteenth century saw potatoes accepted as a food throughout Europe and the nineteenth century saw their ascendancy as a major food crop (Burton 1989). It is likely that the early introductions of potatoes to Europe came from both the Andes and coastal Chile. Analysis of DNA from 49 herbarium specimens has confirmed the presence in Europe of Andean potatoes from around 1700 and Chilean potatoes from 1811 (Ames and Spooner 2008). Rough Purple Chili was an important Chilean Tuberosum introduction into the USA in 1851 (Goodrich 1863) because its descendants were widely used as female parents in crosses with European Tuberosum at the end of the nineteenth century (Ortiz 2001). This is one reason why Chilean Tuberosum cytoplasm predominates in modern cultivars.

During the seventeenth and eighteenth centuries many European countries developed widespread political and commercial interests in the rest of the world, and their colonists and missionaries took with them their common crops including potatoes (Burton 1989). Potato production expanded worldwide during the nineteenth century, but it was the expansion in China and India during the second half of the twentieth century that led to these countries becoming the two most important producers in the world. Today the potato is grown as a summer crop in the tropical highlands of Central and South America and in the lowland temperate regions of the world, as a winter crop in the lowland subtropics, and as a spring and an autumn crop in the Mediterranean. It can be grown all year round in some parts of the world such as the equatorial highlands of South America and East Africa, and parts of China and Brazil.

Named potato cultivars appeared in Europe and North America from the 1730s (Reader 2008). They were distinct clones which had been selected, multiplied and marketed from seedlings raised from seeds in the berries from natural open-pollination. The use of artificial hybridization was advocated as early as 1807 (Knight 1807), but selection of clones from natural open-pollination remained a common practice into the early twentieth century. Since then more scientific breeding methods have been developed (Bradshaw and Bonierbale 2010). The extent of progress since 1807 in potato breeding worldwide can be judged from the World Catalogue of Potato Varieties 2009/10 (Pieterse and Hils 2009), which lists more than 4500 cultivars from 102 countries covering all potato growing regions in the world. This is a remarkable achievement for a crop which was unknown outside of South America until almost 500 years ago, and which was derived from a narrow genetic base. It represents adaptation of the potato to the wide range of daylengths, environments and end uses already mentioned. Modern cultivars are also the building blocks for breeding new cultivars with improved characteristics.

Most potato cultivars are clonally (vegetatively) propagated through seed (daughter) tubers and are genetically uniform. However, in the torrid zones of the lowland tropics and subtropics, cultivars based on current methods of true potato seed (TPS) propagation are an attractive proposition despite being genetically variable (Bradshaw and Bonierbale 2010). In future, uniform diploid F1 hybrids for TPS propagation should be possible. Lindhout et al. (2011) have demonstrated the feasibility of F1 hybrid breeding through the introduction of a gene (*Sli*) that inhibits gametophytic self-incompatibility. Such hybrids will probably replace the genetically heterogeneous tetraploid cultivars currently available from $4x \times 4x$, $4x \times 2x$ and $2x \times 2x$ crosses in which the $2x$ parents produce unreduced $2n$ gametes (Bradshaw and Bonierbale 2010).

3.2.3 Conventional Breeding Methods

The germplasm available for improving the nutritional value of potatoes is modern cultivars, the cultivated landraces of South America and the wild relatives of potatoes. Breeding methods and techniques have been reviewed by Bradshaw and Bonierbale (2010). A brief summary of key features is as follows.

Worldwide, breeding cultivars at the tetraploid level for clonal propagation has traditionally involved making hybridizations between pairs of parents with complementary phenotypic features, followed by multi-stage, multi-trait selection over a number of clonal generations. Since the 1990s, a number of ways have become available to improve the efficiency and effectiveness of this conventional breeding, such as progeny testing, family selection, marker-assisted selection and genomic selection (Bradshaw 2017). The parents used will be modern cultivars or similar elite germplasm. Increasingly they will have genes for new traits introgressed from landraces and wild relatives, particularly genes for disease and pest resistance. Introgression is essentially backcrossing, as explained and reviewed by Bradshaw and Bonierbale (2010). It also often involves ploidy manipulation including the production of haploids (also called dihaploids) of tetraploid *S. tuberosum* from 1958 onwards (Hougas et al. 1958). Parents can also come from population improvement schemes aimed at providing tetraploid or diploid parents suitable for crossing with adapted tetraploid parents (often cultivars) in the breeding of finished cultivars. The starting populations usually comprise cultivated landraces alone or together with wild relatives. Some of the schemes involved selection for tuber production in long days, again as explained and reviewed by Bradshaw and Bonierbale (2010). One example will be relevant later in this chapter. Carroll (1982), in Scotland, produced a population of diploid Phureja/Stenotomum adapted to long-day conditions in Europe. Direct hybridization of the improved diploid population with tetraploid potato cultivars via unreduced pollen grains ($4x \times 2x$ crosses) resulted in tetraploid hybrids, some of which were superior to standard tetraploid cultivars in both total and marketable yield (Carroll and De Maine 1989). Furthermore, diploid cultivars, such as Inca Dawn, Inca Sun and Mayan Gold, were produced from the population, but have been targeted at niche markets because their yield is only two-thirds that of Tuberosum potatoes.

3.2.4 Genetic Modification Methods

Major improvements to already successful potato cultivars can be made by various types of *Agrobacterium*-mediated genetic transformation; namely cisgenesis, intragenesis, gene silencing by RNA interference and transgenesis, as discussed with potato examples by Bradshaw (2016, 2017). We shall see the importance of understanding the biochemistry of the trait of interest, and the control of gene expression for the genes that encode the key enzymes for delivering the required biochemistry. Although the CaMV 35S promoter has frequently been used for constitutive expression, we shall also see that tuber-specific expression is often important to avoid adverse effects on plant growth and tuber yield. Indeed, selecting the best genetically modified (GM) potatoes for commercialization requires a lot of work; namely seeking the best constructs and promoters and eliminating any undesirable transformants and somaclonal variants that arise from the transformation per se and in vitro regeneration, respectively (Davies 2002). Combinations of traits can be achieved by stacking genes through co-transformation with separate constructs, transformation with a multi-gene cassette (4–6 is current limit) or re-transformation with further genes. In future, the use of site-directed transformation (rather than random insertion) should be possible, the potato genome sequence having been published in

Nature on 14 July 2011 (Potato Genome Sequencing Consortium 2011). Sites of gene insertion can be chosen that are conducive to high levels of gene expression and multiple transgenes stacked at the same site, making it easy to move them into other germplasm by crossing (Mahfouz et al. 2016). Gene editing is also possible by site-directed DNA sequence modifications. Despite the potential of genetic transformation, the period 1995–2013 saw a reluctance to accept the first GM potatoes in Europe and North America, and the same may be true of the new GM potatoes now available for human consumption in the USA (Bradshaw 2016).

3.3 Problems Ancient and Modern

3.3.1 Domestication and Glycoalkaloids

Potato tubers contain small quantities of steroidal glycoalkaloids which are usually regarded as undesirable toxic compounds, despite having some anticarcinogenic properties (Friedman and McDonald 1997; Friedman and Levin 2016). Other plant parts have higher concentrations and it is generally thought that glycoalkaloids confer some protection against various pathogens. Tuber concentrations above 15 mg 100 g^{-1} FW confer bitterness and above 20 mg are considered unsafe for human consumption, resulting in symptoms typically associated with food poisoning. Factors which can affect tuber concentrations are cultivar, climate, storage environment, maturity, damage, temperature and exposure to light (Sinden et al. 1984; Dale and Mackay 1994). Hence most causes of excessive glycoalkaloid levels are under the control of the breeder, the grower or those involved in processing and marketing. Potato cultivars vary in content, but commercial potatoes rarely exceed 20 mg 100 g^{-1} FW. However, some wild species used in breeding have higher contents.

The domestication of the potato is assumed to have involved selection for less bitter and hence less toxic tubers. However, Johns and Alonso (1990) found that some genebank accessions of *S. bukasovii* (now *S. candolleanum*: Spooner et al. 2014) had tuber glycoalkaloid levels which were consistently close to the levels found in many clones of *S. tuberosum* Stenotomum Group. They concluded that exploitation and domestication of this species would have required little or no selection for lower glycolkaloid content, unlike their samples of other candidates for domestication, namely *S. canasense*, *S. leptophyes* and *S. sparsipilum*. Nevertheless, it seems fair to credit Andean farmers with making the potato an edible crop.

Van Gelder et al. (1988) determined the compositions of glycosidic-bound steroidal alkaloids of *Solanum* species used in potato breeding by capillary gas chromatography using simultaneous nitrogen-specific (NPD) and flame ionization detection (FID). Total glycoalkaloid contents varied from 12.3 to 734.8 mg 100 g^{-1} FW. Similar small-sized tubers of cultivars grown under the same conditions showed lower contents of 12.6–72.1 mg 100 g^{-1} FW, with values of 4–36 mg for normal-sized field grown tubers. The highest contents were found in *S. vernei*, a source of resistance to potato cyst nematodes.

Today there is concern that introgression breeding from wild species may inadvertently result in unacceptably high concentrations of tuber glycoalkaloids, and hence new cultivars are commonly tested to ensure that their contents are below the accepted safe limit of 20 mg 100 g^{-1} FW. American cultivar Lenape was removed from commerce in 1970 due to its high glycoalkaloid levels which averaged 29 mg 100 g^{-1} FW over 39 locations and rose to 65 mg in stressed conditions. It is thought that the high levels were due to *S. chacoense* in its immediate ancestry (Zitnak and Johnston 1970; Sinden et al. 1984). Nevertheless, glycoalkaloid contents are highly heritable and the breeder can control the levels in new cultivars by appropriate selection of wild species and parental cultivars and by testing selected hybrids and potential cultivars (Sinden et al. 1984). It may also be necessary to check wild species for unusual forms of glycoalkaloids when used for the first time.

Genetic modification can also be used to control glycoalkaloid concentrations. About 95% of total content is accounted for by α-solanine and α-chaconine, which are differently glycosylated forms (galactose and glucose, respectively) of the aglycone, solanidine. Although not all of the steps in their biosynthesis are known, several intermediates have been identified (Itkin et al. 2013; Kaminski et al. 2016), and virtual elimination of tuber glycoalkaloids by gene silencing is feasible (Fig. 3.2). Cholesterol is synthesized from acetyl-CoA, converted to solanidine, and then by two separate pathways to α-solanine and α-chaconine. Altered glycoalkaloid content has been associated with a number of candidate genes. Kaminski et al. (2016) used next generation sequencing, bulk segregant analysis to identify a genomic region (chromosome 1) with candidate genes responsible for differential glycoalkaloid content in a diploid mapping population of *S. tuberosum* crossed with *S. sparsipilum* and backcrossed to *S. tuberosum*. The differences in total glycoalkaloid levels came equally from α-solanine and α-chaconine. Genes were found encoding critical enzymes in the synthesis of cholesterol: sterol 24-C-methyltransferase (*SMT1*), sterol desaturase (*SD*) and C-4 sterol methyl oxidase (*SMO*). Sawai et al. (2014) used RNA interference to silence the *SSR-2* gene encoding sterol side chain reductase 2 which reduces cycloartenol to cycloartanol. The resulting tubers had significantly lower levels of cholesterol and glycoalkaloids (about 10% of controls) without affecting plant growth.

Umemoto et al. (2016) focused on the biosynthetic steps from cholesterol to solanidine. They identified two cytochrome P450 monooxygenase genes, *PGA1* (chromosome 6) and *PGA2* (chromosome 7), that encode enzymes mediating two oxidation steps. Silencing *PGA1* or *PGA2* by RNA interference resulted in a large reduction in glycoalkaloid content and the creation of novel phenotypes, including the suppression of flower development and tuber sprouting. In other respects, the transgenic plants grew normally in the vegetative stages and produced tuber yields in the greenhouse similar to control plants. It was inferred from the metabolites that accumulated in the transgenic plants, and were produced by in vitro enzyme assays, that *PGA1* and *PGA2,* respectively, encoded enzymes that catalysed the 26- and 22-hydroxylation steps in the biosynthetic pathway. Itkin et al. (2013) also

Fig. 3.2 Genes mentioned in text for biosynthesis of α-solanine and α-chaconine

Acetyl-CoA

↓

↓

Cycloartenol

SSR2 ↓ ↓ *SMT1*

Cycloartanol 24-Methylenecycloartanol

SMO ↓

SD ↓

Cholesterol

PGA2 ↓

PGA1 ↓

Furostanol-intermediate

GAME4 ↓

Solanidine

SGT1 ↓ ↓ *SGT2*

γ-Solanine γ-Chaconine

β-Solanine β-Chaconine

↓ *SGT3* ↓

α-Solanine α-Chaconine

investigated a gene (*GAME4*) encoding an enzyme involved in the conversion of cholesterol to glycoalkaloids by oxidation of a furostanol-intermediate to its 26-aldehyde. Again, silencing the gene by RNA interference showed a large reduction to very low levels (< 0.5 mg 100 g^{-1}) of α-solanine and α-chaconine without adverse effects.

Finally, Shepherd et al. (2015) focused on the genes that encode the enzymes for the biosynthesis of γ-solanine from UDP-galactose and solanidine (*SGT1*), γ-chaconine from UDP-glucose and solanidine (*SGT2*), and both α-solanine and α-chaconine from UDP-rhamnose and β-solanine and β-chaconine, respectively (*SGT3*). They used antisense fragments of cDNA to down-regulate *SGT1*, *SGT2* or *SGT3* in cultivar Desiree. Down-regulation of *SGT1* reduced the concentration of α-solanine, but had no effect on α-chaconine (previous work had found an increase in α-chaconine); down-regulation of *SGT2* reduced the concentration of α-chaconine, but increased the concentration of α-solanine; and down-regulation of *SGT3* reduced the concentrations of both. By using metabolite profiling, the researchers not only revealed expected changes in specific glycoalkaloid levels, but also significant changes in several other metabolites, some of which could be explained in terms of known pathways.

3.3.2 Frying and Acrylamide

Today, processors are under pressure to reduce acrylamide formation in crisps (chips) and French fries because of concerns about its effects on human health (Amrein et al. 2003; Pedreschi 2009; Pinhero et al. 2009). Acrylamide is classified as a Group 2A (probably carcinogenic to humans) carcinogen. Mandatory mitigation measures for reducing acrylamide in food are expected to come into force in the European Union by spring 2018 with benchmark levels for potato crisps and French fries of 750 and 500 μg kg^{-1}, respectively (European Commission).

Acrylamide is formed during high-temperature cooking (frying, baking and roasting) and food processing by Maillard browning reactions between reducing sugars and free asparagine. Lower levels of these compounds are obvious targets for conventional breeding and genetic engineering. Muttucumaru et al. (2017) analysed the acrylamide-forming potential of twenty potato cultivars grown at two sites (Doncaster and Woburn) in the United Kingdom (UK). They modelled the relationship between the ratio of free asparagine to reducing sugars and the levels of acrylamide. They identified a value of 2.26 as the tipping point in the ratio below which free asparagine concentration could affect acrylamide formation. This is an important result as reducing sugars have been the main determinant of acrylamide-forming potential in most datasets, but with free asparagine sometimes contributing to the variation. In contrast, acrylamide formation in heated wheat and rye flour is determined by the free asparagine concentration unless the flour is derived from sulphur-deprived wheat (Muttucumaru et al. 2017).

The J R Simplot Company in the USA is taking the genetic transformation route to improving processing cultivars being grown in North America, but it is not yet clear if they will be acceptable to companies such as McDonald's. Simplot's first Innate™ potatoes (www.innatepotatoes.com) were deregulated by the US Department of Agriculture (USDA) on 7 November 2014, and then voluntarily reviewed by the US Food and Drug Administration (FDA) who concluded on 20 March 2015 that they were as safe and nutritious as their conventional counterparts.

Chawla et al. (2012), of Simplot Plant Sciences, have shown that tuber-specific silencing of *asparagine synthetase 1* (by RNA interference) reduces the acrylamide-forming potential of potatoes grown in the field without affecting tuber shape and yield. Initially, Rommens et al. (2008) found that simultaneous silencing of *asparagine synthetase-1* and *-2* (*Ast1* and *Ast2*) limited the formation of asparagine and reduced the acrylamide-forming potential of tubers. However, although the phenotype of silenced lines appeared normal in the greenhouse, unexpectedly, field-grown tubers were small and cracked. This prompted a more thorough analysis of the role of the two asparagine genes in potato. Chawla et al. (2012) showed that tuber-specific silencing of *Ast1* reduced asparagine levels by 60–80% in processing cultivars Russet Ranger and Russet Burbank. This was sufficient to reduce acrylamide levels by up to 70% compared with the controls and hence is one possible solution to the acrylamide problem. Simplot have also been able to reduce the tuber formation of reducing sugars (glucose and fructose) from starch by lowering transcript levels for genes *PhL* (amyloplast-targeted phosphorylase-L) and *R1* (water dikinase) (Rommens et al. 2006).

Acrylamide producing potential can also be lowered by conventional breeding. Shepherd et al. (2010) examined the genetic variation released by hybridization of a processing clone (12601ab1) and a table cultivar (Stirling), both from the breeding programme at the Scottish Crop Research Institute (now the James Hutton Institute). The processing clone was known to be resistant to cold-temperature sweetening (Bradshaw et al. 2008), a desirable trait for cold-storage without using a sprout suppressant. A subset of 43 offspring clones from a replicated yield trial in 2007 were stored at 4 °C for 4 months. Fry colour and acrylamide content of their crisps were determined after frying at 180 °C, as well as the concentrations of reducing sugars and asparagine in their tubers. Most of the variation (>88%) between the 43 clones and 5 controls was genetical. Multiple regression analysis revealed that fry colour was determined by reducing sugar content (71.4% of variation), whereas acrylamide content was determined by both reducing sugars and asparagine (53.4% of variation). There was no correlation between reducing sugars and asparagine. The clone with the lowest acrylamide content was third lowest for reducing sugars and fourth lowest for asparagine, and was also second best for fry colour (light colour). The researchers concluded that both reducing sugars and asparagine should be targeted in conventional breeding.

Novy et al. (2017) reported the release in the USA in 2015 of cultivar Payette Russet, which was notable for its cold-sweetening resistance and associated low acrylamide formation, making it ideally suited for processing into French fries and other potato products. Low asparagine and reducing sugar concentrations contributed to an 81% reduction in acrylamide content in French fries relative to cultivars Ranger Russet and Russet Burbank following 8 months storage at 9 °C.

Finally, it is important to consider the best agronomic practice to keep acrylamide concentrations as low as possible, particularly the effect of fertilizer-application on the acrylamide-forming potential of the harvested tubers. In the UK up to 240 kg N ha^{-1} of nitrogen (N) fertilizer is applied, whereas potatoes are usually grown without sulphur (S) fertilization. Muttucumaru et al. (2013) applied different combinations of N and S to 13 cultivars (French fry, chipping and boiling types), grown in a field trial in the UK (Woburn) in 2010. The N levels were 0, 100 or 200 kg ha^{-1} and the S ones were 0, 15 or 40 kg ha^{-1}. Potatoes were analysed immediately after harvest for free asparagine, other free amino acids, sugars, and acrylamide-forming potential at 177 °C. The research gave complicated results but showed that N can affect acrylamide-forming potential but that the effect is type- and cultivar-dependent, with most cultivars showing an increase in acrylamide formation in response to increased N, but two showing a decrease. S application reduced glucose concentrations and mitigated the effect of high N application on the acrylamide-forming potential of some of the French fry type of potatoes.

3.4 Resistant Starch and Glycaemic Index

The form of energy storage in potato tubers is almost entirely starch, which is made up of two polymers: amylose (20–30%), which is a linear molecule, and amylopectin (70–80%), which is branched. In considering the nutritional value of potatoes we need to consider their glycaemic index (GI) (Monro and Mishra 2009). This is a ranking system that allows carbohydrate foods to be classified on a scale of 0–100 as either low (<55), medium (56–69) or high (>70), where glucose is used as the reference food and assigned a score of 100. GI measures the food's effects on post-prandial blood glucose levels over two hours. High GI foods result in a rapid increase and corresponding insulin responses. They can therefore contribute to obesity, glucose intolerance, type-2 diabetes and cardiovascular disease, but could be considered beneficial to sports persons after exercise because they produce a rapid supply of muscular glycogen. Interestingly, when Borch et al. (2016) did a systematic review of published observational studies, they did not find convincing evidence to suggest an association between intake of potatoes and risks of obesity, type-2 diabetes and cardiovascular disease. Furthermore, Akilen et al. (2016) found that the physiological functions of carbohydrate foods (potato, rice and pasta) consumed ad libitum at meal time on food intake, appetite, blood glucose, insulin and gut hormone responses in children, are not predicted by their GI. In other words, advice based on the GI of a fixed amount (50 g) of available carbohydrate may not be representative of post-prandial satiety and glycaemia when carbohydrates are consumed freely in a mixed meal. With these reservations in mind, we will now look at the GI of potatoes.

GI values reported for potatoes by Foster-Powell et al. (2002) range from 23 to 111, where the glucose control is 100. In Great Britain (GB), Henry et al. (2005) reported GI values from 56 to 94 for eight potato cultivars. Potatoes with waxy textures produced medium GI values whilst floury potatoes had high GI values. There is also some evidence that coloured potatoes have a lower GI than white ones (78.7 vs. 93.0), and that this is possibly due to an inhibitory effect of anthocyanins (polyphenols) on intestinal α-glucosidase (Ramdath et al. 2014). Despite the wide range of values, potatoes are usually considered a high GI food and this has raised concerns about their nutritional value and led to the search for low-GI potatoes (Pinhero et al. 2016). However, it must be remembered that GI is affected by cooking method and potatoes are usually eaten in mixed meals. Compared to boiling or mashing, for example, some cooking methods (e.g., frying, microwaving and baking) reduce the postprandial glycaemic response significantly, and a cooling step or co-digestion with protein, lipid and vinegar are effective ways to decrease the GI of potato (Tian et al. 2016). More detailed information on the effect of cooking methods on the glycaemic index of potatoes can be found in the reviews by Nayak et al. (2014) and Tian et al. (2016). Here, however, we are concerned with genetic variation in GI of potato.

Clinical evaluation of GI is expensive. Hence effective, quick and low-cost alternatives have been sought and validated, such as in vitro starch digestibility as determined by the rate and extent of glucose release from enzymatic digestion under controlled conditions. Working with seven potato cultivars (GI 53–103), Ek et al. (2014a) found a strong positive correlation ($r = 0.91$) between in vivo and in vitro GI methods, particularly in vitro starch hydrolysis of cooked potatoes at 120 min. In contrast, amylose, dietary fibre and total starch content were not correlated with either in vitro starch digestibility or GI. The researchers concluded that low-GI potato cultivars can be identified by screening using a high-throughput in vitro digestion procedure. They classified cultivar Carisma™ (a waxy type with GI = 53) as low-GI, cultivar Nicola (GI = 69) as medium-GI, and the other five cultivars as high-GI. They explained the low GI of cultivar Carisma in terms of the physical and chemical properties of its starch (Ek et al. 2014b). Cultivar Carisma™ was launched as a low-GI potato in Australia in 2010 and is now marketed and grown in North America. Another new conventionally bred cultivar (currently unnamed) developed by Agriculture and Agri-Food Canada (www.agr.gc.ca/potato-cultivars) is being trialled as a low-GI potato. It is reported to have a moderate index of 65 when served hot and a very low one of 34 when served cold.

Total starch can be classified by its digestibility into nutritionally important fractions: rapidly digestible starch (RDS), slowly digestible starch (SDS) and resistant starch (RS) (Pinhero et al. 2016). Rapidly digestible starch is the amount of starch hydrolysed in the first 20 min of in vitro incubation with a-amylase, SDS is the amount hydrolysed between 20 and 120 min, and RS is the starch left after 120 min, calculated by subtracting RDS and SDS from the total. Resistant starch reaches the large intestine essentially intact and has similar physiological effects and health benefits as fibre (Birt et al. 2013; Lockyer and Nugent 2017). Hence RS is generally included in total dietary fibre along with soluble and insoluble fibre. Interestingly, baked white potatoes in their skin compare favourably with other vegetables by having 2.2 g dietary fibre 100 g^{-1} (1.5 without skin) (Storey and Anderson 2014). As amylose is digested more slowly than amylopectin, RS is generally positively associated with the amount of amylose present in starch; but other factors are involved (Lockyer and Nugent 2017). The amount of RS also depends on preparation methods; for example, Englyst et al. (1992) found an increase from 7% to 13% on cooling after cooking.

Pinhero et al. (2016) evaluated 14 early potato cultivars grown in Canada for dietary fibre, total starch, RDS, SDS and RS, as well as for estimated GI (eGI). They found a strong positive correlation ($r > 0.96$) between eGI and RDS and a strong negative one ($r < -0.96$) between eGI and RS, irrespective of whether the potatoes were uncooked, cooked or retrograded (boiled and cooled for 48 h at 4 °C). Hence RDS and RS were major factors contributing to eGI. Monro et al. (2009) also found genetic variation for starch digestibility among 37 New Zealand potato breeding lines. Slowly digestible starch ranged from 7% to 37% of total starch and RS from 12% to 27% after a post-cooking cooling treatment that increased SDS and RS. The researchers concluded that there was sufficient genetic variation for selection in conventional breeding, and that the glycaemic impact of some potatoes may be substantially reduced by cool-storing after cooking. Raatz et al. (2016) found that RS content varied with cooking method and serving temperature but not between three North American cultivars, Yukon Gold, Dark Red Norland and Russet Burbank. Baked potatoes had higher RS contents than boiled ones, and chilled potatoes (4 °C) had more RS than either hot (65 °C) or reheated ones (4 °C for 6 days, reheated to 65 °C).

The overall conclusion from the studies reported in this section is that low-GI potato clones can be identified in a conventional breeding programme by screening using a high-throughput in vitro digestion procedure.

Finally, it is worth mentioning that the introduction of inulin to potato is another way to improve its nutritional value by reducing its energy density through increased dietary fibre. Hellwege et al. (2000) have developed transgenic potato tubers that synthesize the full range of inulin molecules naturally occurring in globe artichoke (*Cynara scolymus*).

3.5 Protein Content and Amino Acid Balance

Potatoes are commonly perceived as a source of carbohydrate, but on a dry-weight (DW) basis they contain about 10% protein (2% on FW basis), equal to that of most cereals such as rice or wheat. They are a good source of lysine, but low contents of sulphur amino acids (methionine and cysteine) limit their nutritive value (Friedman 1996). This was confirmed by Bártová et al. (2015) in glasshouse produced tubers of four Andigenum accessions and one Tuberosum cultivar (Desiree): protein content was 5.2–8.0% DW, patatin (the major storage protein) content was 17.1–41.7% of protein and methionine and cysteine were the amino acids with the lowest concentrations in all five genotypes.

Where potatoes are an important dietary constituent, breeding for increased protein quantity and quality can be justified, but it is difficult to find good examples in the scientific literature. In contrast, differences between cultivars have often been reported. Pęksa et al. (2013), for example, found differences between 13 cultivars in protein content and quality, and in concentrations of amino acids, that were not related to flesh colour (seven purple-fleshed, four red-fleshed and two yellow-fleshed). Galdón et al. (2010) also found differences in protein content and quality, and in concentrations of amino acids, between ten traditional cultivars from Tenerife.

Protein quantity and quality can also be improved by genetic modification. Chakraborty et al. (2010) developed transgenic potatoes with enhanced nutritive value by tuber-specific expression of a seed protein, Amaranth Albumin 1 (AmA1), from *Amaranthus hypochondriacus*, in seven Indian cultivars suitable for cultivation in different agro-climatic regions. A construct having a full-length *AmA1* gene, under the control of a tuber-specific promoter (GBSS: granule-bound starch synthase), was used for large-scale *Agrobacterium tumefaciens*-mediated transformation. Analyses of the transgenic tubers revealed up to 60% increase in total protein content and a significant increase in lysine, tyrosine and sulphur amino acids. The transgenics also exhibited enhanced photosynthetic activity with a concomitant increase in total biomass and a moderate increase in tuber yield. The data on field performance and safety evaluation indicated that the transgenic potatoes are suitable for commercial cultivation. Furthermore, in vitro and in vivo studies on experimental animals demonstrated that the transgenic tubers are also safe for human consumption. As AmA1 is a nonallergenic protein that originated from an edible crop, transgenic potato crops expressing AmA1 should receive public acceptability.

3.6 Mineral Biofortification

In their review of relationships between yield and mineral concentrations in potato tubers, White et al. (2009) covered the mineral elements required by humans, the mineral composition of potatoes and the effects of fertilizer application and genetic variation on tuber mineral concentrations. The review identified considerable genetic variation between modern cultivars for N, P, K, S, Ca, Mg, Cu, Fe, Mn and Zn, and also between Andean landraces for Ca, Fe and Zn and between wild species for Ca. The researchers were quietly optimistic that it should be possible to increase tuber mineral concentrations (biofortification) by combining genotypes that have naturally higher tuber mineral concentrations (genetic biofortification) with appropriate fertilization strategies (agronomic biofortification) to deliver greater quantities of essential minerals to the diet without compromising yield. Furthermore, the bioavailability of mineral elements in potatoes is generally high. For example, Macdonald-Clarke et al. (2016) found that the bioavailability of potassium (K) is as high from potatoes as from potassium gluconate supplements.

We are now going to look at a few recent examples of variation in mineral content among potato cultivars and in germplasm collections before considering iron (Fe) and zinc (Zn) in more detail, as well as the significance of calcium (Ca).

3.6.1 Variation in Mineral Content Among Potato Cultivars and in Germplasm Collections

Brown et al. (2010, 2011, 2012b, 2013, 2014) looked at the mineral content of potential (tetraploid) cultivars being trialled in the North West of the USA. They concluded that genetic variation for tuber Fe content exists and that breeding for enhanced Fe content would be feasible. In contrast, their genepool was not a good source of Zn for biofortification through conventional breeding. They found genetic variation for Ca and Mg but pointed out that potato is not a rich source of the former. They also found genetic variation for K and P in their red/specialty market class and thought that an increase in K content was a worthwhile objective. Finally, if required, genetic improvements could be made in Cu and S content.

Nassar et al. (2012) used inductively coupled plasma-optical emission spectroscopy to determine the mineral content of 16 cultivars, grown at five locations, on a per serving basis for per cent recommended daily intake (% RDI). Discriminant analysis showed that both genotype and growing location were important. They found differences between sites and between cultivars, with Freedom, Yukon Gold and Russet Burbank contributing most to % RDI. They concluded that one serving per day of these cultivars makes a significant contribution to the % RDI for the macro-minerals K, Mg and P and the trace minerals Cu, Fe, Se and Zn, for which global deficiencies are common.

Haynes et al. (2012) looked at the Cu, Fe, Mn and Zn contents of cultivar Atlantic and 17 hybrids between tetraploid Tuberosum and diploid hybrids of Phureja-Stenotomum, grown for 2 years at three locations in the East of the USA. Genetic variation was large for all four micronutrients, suggesting that contents could be improved through breeding.

Subramanian et al. (2017) evaluated the variation in tuber mineral concentrations among 49 accessions of wild and cultivated tuber-bearing *Solanum* species in the Commonwealth Potato Collection (CPC). Selected CPC accessions, representing the eco-geographical distribution of wild potatoes, were grown to maturity in peat-based compost under controlled greenhouse conditions. Tubers from five plants of each accession were harvested, bulked and their mineral composition determined on acid-digested material using inductively coupled plasma mass spectrometry (ICP-MS). Among the germplasm investigated, there was a greater range in concentrations of Ca, Cu, Fe, Mn and Zn (6.7-fold, 4.0-fold, 3.6-fold, 3.0-fold and 4.5-fold, respectively) than in concentrations of K, Mg, P and S (all less than 3.0-fold). Breeding for high tuber concentrations of Ca, Fe and Zn were thought worthwhile.

3.6.2 Genetic Biofortification of Iron and Zinc

Kromann et al. (2017) summarized the need for Fe and Zn biofortification as follows. It has been estimated that more than 60% of the world's population is dietary deficient in Fe and more than 30% in Zn. Iron deficiency leads to anaemia, reduces physical growth and cognitive development in young children, and is estimated to be responsible for a large proportion of the world's maternal deaths. Symptoms of Zn deficiency include stunting, diarrhoea and pneumonia in children, and Zn deficiency contributes significantly to recurrent infections and infant mortality. In contrast to Fe, Zn is not significantly stored in the human body and must be ingested daily.

In 2004, CIP chose Fe and Zn as potato targets in the HarvestPlus Biofortification Challenge Program (Paget et al. 2014). Although the concentration of Fe (and Zn) in potato is low compared with cereals and legumes, its bioavailability is greater due to the presence of high levels of ascorbic acid (promoter of Fe absorption) and low levels of phytic acid (inhibitor of Fe absorption) (Fairweather-Tait 1983). Andre et al. (2015) estimated the bioaccessibility (potential bioavailability) and bioavailability of Fe from 12 Andean potato clones using an in vitro gastrointestinal digestion procedure. They found that between 64% and 79% of the Fe is released from the tuber matrix during in vitro gastrointestinal digestion and is therefore available at the intestinal level. Furthermore, on average, 32% and 24.5% of the ascorbic acid and chlorogenic acid (the principal phenolic acid), respectively, were also bioaccessible from boiled tubers. There were negative relationships between bioavailable Fe values and phenolic concentrations, whereas ascorbic acid concentrations were positively associated with the ferritin values. However, intestinal absorption of intrinsic Fe from potato tubers could not be detected by the method used, so further work is required to conclusively determine the bioavailability of intrinsic Fe from potato tubers. More information on bioaccessibility and bioavailability can be found in the review by Andre et al. (2014).

Burgos et al. (2007) determined the Fe and Zn concentrations in 37 native Andean potato accessions grown in two highland locations and found significant genotypic variation, as well as significant variation due to environments and genotype × environment interaction. Concentrations in raw, peeled tubers ranged from 9 to 37 mg Fe kg^{-1} DW and from 8 to 20 mg Zn kg^{-1} DW, with diploid Phureja accession 703274 and diploid Stenotomum accession 701165 showing the highest levels of Fe and Zn, respectively, in both locations. Fe and Zn concentrations were positively correlated on a FW basis in each site and no losses occurred on boiling. The only significant differences in Fe content of peeled compared with unpeeled potatoes could be attributed to contamination with soil iron, as confirmed by elevated levels of aluminium in the samples. The variation found was ample for breeding to increase Fe and Zn levels. Furthermore, a large genetic diversity was confirmed from screening 579 native Andean landraces and improved potato cultivars held at CIP (https://research.cip.cgiar.org/).

Paget et al. (2014) made a genetic evaluation of Fe and Zn (and also Ca) in diploid potatoes, produced from a base population of Stenotomum, Goniocalyx and Phureja accessions from the study of Burgos et al. (2007). All analyses were conducted on peeled tubers. Mineral content was determined by inductively coupled plasma–optical emission spectrophotometry (ICP). The first generation (G1) comprised 17 full-sib families (16 different female parents) in four half-sib groups (four male parents), with family sizes of 23–36 genotypes. It was grown at Huanuco, Peru, at an altitude of 3800 m. In total, there were 487 analyses for mineral and DM content. Forty individuals (genotypes) with high Fe and Zn contents and desirable agronomic characteristics were selected as potential parents of the next generation (G2). In practice, 32 full-sib families were produced from two sets of four female and four male parents crossed in a factorial design. Seedlings were transplanted into the field in Huancayo to produce tuber families for assessment of clones in three-plant plots at Ayacucho, Peru, at an altitude of 2761 m. In total, 1329 clones were analysed for mineral and DM content, with family size ranging from 19 to 74 genotypes. The maximum mineral contents in G2 were 42.5 mg Fe kg^{-1} DW, 38.9 mg Zn kg^{-1} DW and 689.7 mg Ca kg^{-1} DW. Additive, but not non-additive, genetic variance was detected for Fe, Zn and Ca content of tubers and narrow-sense heritability estimates for individual plants on a DW basis were moderate (0.41 and 0.43

for Fe in G1 and G2, 0.38 and 0.36 for Zn in G1 and G2, and 0.42 and 0.57 for Ca in G1 and G2). Genetic correlations between minerals were all positive when analysed on a DW basis (0.04–0.72), and also small and positive (0.05–0.18) between tuber DM and minerals on a FW basis but negative (−0.14 to −0.38) on a DW basis. The researchers concluded that recurrent individual plant selection should be effective in raising Fe and Zn contents. Furthermore, the level of within-family variation was such that superior individuals should be identified from within a number of families. The researchers also concluded that the improved material should provide a useful source of dietary Fe and Zn. They estimated that in Peru, the best G1 material would provide 56% of the estimated average requirement (EAR) of Fe for children aged 4–8 years and 28% for female adults aged 19–30. Similarly, consumption would provide 37% of the EAR of Zn for children aged 4–8 years and 22% for female adults aged 19–30.

3.6.3 Agronomic Biofortification of Zinc

Genetic biofortification of Zn, but not Fe, can be complemented with agronomic biofortification (Kromann et al. 2017). Although the details are beyond the scope of this chapter, it deserves a brief mention. White et al. (2012, 2017) studied the effect of foliar fertilization of Zn on a number of cultivars in the UK where the Zn concentration in the flesh of maincrop potatoes, currently grown without foliar Zn fertilization, is on average 14.2 mg kg^{-1} DW. In the first field study, on cultivar Maris Piper, a maximum tuber concentration of 30 mg Zn kg^{-1} DW was achieved at a foliar Zn application rate of 5.4 g per plot (1.08 g per plant). Tuber yields were unaffected by foliar applications of up to 3.6 g Zn per plot, at which level tuber concentrations were about 25–30 mg Zn kg^{-1} DW. In the second field study, foliar Zn fertilizers increased tuber Zn concentrations of cultivar Saxon from 17.0 to 49.5 mg kg^{-1} DW, but excessive applications reduced tuber yield. With zinc oxide, a 10% loss of yield occurred with a tuber Zn concentration of about 20 mg kg^{-1} DW, which was lower than in the first study. The researchers concluded that Zn-biofortified Saxon potatoes could increase the bioavailable Zn in a meal by about 11.4%.

Kromann et al. (2017) conducted greenhouse and field experiments in the Ecuadorian Andes, to investigate the potential to increase tuber Fe and Zn concentrations with various rates of Fe and Zn applied as foliar and soil fertilizers. Samples were analysed for Fe and Zn by inductively coupled plasma optical emission spectrometry (ICPOES). The germplasm comprised diploid, triploid and tetraploid cultivars. They found that enhanced Zn supply to potato plants can increase tuber Zn concentrations without an adverse effect (toxicity) on yield, and that the increase was similar across cultivars. High rates of foliar Zn application gave a 2.51-fold increase in tuber Zn and high rates of soil Zn application a 1.91-fold increase. In contrast, the experiments showed no positive correlation between Fe fertilization and Fe concentration in tubers. The highest tuber concentrations of both Fe and Zn were found in the diploid cultivars Chaucha amarilla and Chaucha roja. The rankings of cultivars by tuber concentrations of the two minerals were identical, indicating a positive correlation between the genotypic concentrations of the two minerals. Finally, Kromann et al. (2017) reported that CIP has more than 50 genetically biofortified potato clones with the capacity to absorb and contain more than 35 mg Fe and 35 mg Zn kg^{-1} tuber flesh DW.

3.6.4 Significance of Calcium

The potato is not a rich source of Ca, but high tuber Ca content from Ca fertilization is associated with reduced incidences of disease (common scab) and physiological disorders (hollow heart) and hence is of interest to breeders (Zorrilla et al. 2014; Chung et al. 2016).

Zorrilla et al. (2014) wanted to combine the high yield and specific gravity of chipping (crisping) cultivar Atlantic with the internal tuber quality and moderate resistance to pitted (common) scab of cultivar Superior, which also has a high tuber Ca content compared with Atlantic. After hybridization of these cultivars, they were able to identify eight promising offspring clones that had similar chipping quality to Atlantic, but higher tuber Ca content and lower incidences of hollow heart and pitted scab.

In contrast, Chung et al. (2016) studied the genetics of tuber Ca concentration in wild *Solanum* species. They crossed high-Ca-accumulating *S. microdontum* (clone M15) and low-Ca-accumulating *S. kurtzianum* (clone K12) and then intercrossed 12 F_1 individuals to create a segregating F_2 population (123 clones) with significant variation in tuber Ca content and tuber size. Plants were grown in a greenhouse under controlled conditions to ensure tuber formation. Calcium content was determined using an atomic absorption spectrophotometer. Large tuber size was associated with low tuber Ca concentration. Twelve out of 42 SSR markers were associated with tuber Ca content, including one (SSR4743) on chromosome 7 which was linked to a cation exchanger-like (*CAX3-like*) gene known to be involved with Ca uptake in plants. The allele corresponding to high tuber Ca in the F_2 population matched that of the *S. microdontum* grandparent (high Ca). The researchers thought that SSR4743 could be used to develop a molecular marker to identify tubers with high Ca levels. Hybrids between *S. microdontum* and cultivated potatoes would be expected to produce tubers in the field in long days and hence could be assessed for yield and Ca content.

Finally, it will be recalled that Paget et al. (2014) found genetic variation for calcium content in their diploid population derived from Stenotomum, Goniocalyx and Phureja accessions studied by Burgos et al. (2007). However, they did not consider Ca a high priority compared with Fe and Zn.

3.7 Vitamin Biofortification

3.7.1 Vitamins B9 (Folate) and B1 (Thiamin)

Folates (vitamin B9: tetrahydrofolate and its derivatives) are essential micronutrients in the human diet whose deficiency is associated with increased risk of neural tube defects, cardiovascular diseases, anaemia and some cancers (Goyer and Navarre 2007; Robinson et al. 2015). Current folate intake is suboptimal in most of the world's populations, even in developed countries. Although folic acid supplements and food fortification have proved efficient, they are expensive and hard to implement in developing countries. However, in the USA, for example, a small portion of russet baked potato contains about 6% of the Recommended Daily Allowance (RDA) of 400 μg day^{-1} for an adult. Hence increasing the folate concentration in potatoes by breeding should have a positive benefit (Goyer and Navarre 2007; Robinson et al. 2015). Similar arguments apply to thiamin (vitamin B1) (Goyer and Sweek 2011; Navarre et al. 2016).

Folate concentration varies among potato germplasm. Goyer and Navarre (2007) analysed 67 genotypes including white-, purple-, red-, blue- and yellow-fleshed potatoes and several wild species. Folate concentrations varied from 521 to 1373 ng g^{-1} DW, with white-fleshed cultivar Ranger Russet at 1037 ng g^{-1} DW. The five genotypes with the highest folate levels were colour-fleshed, of which four were yellow. Skin contained about 30% higher folate concentrations than flesh. Cold-storage of tubers for 7 months generally increased folate contents, but results on the effect of cooking were inconsistent. Goyer and Navarre (2009) subsequently found that folate concentrations were 2.6- to 3.4-fold higher in young tubers (baby new potatoes) than in mature ones at harvest.

Goyer and Sweek (2011) examined a wider range of germplasm for both folate and thiamin concentrations, namely 54 field-grown clones from 33 accessions of primitive Andigenum cultivars and three modern potato cultivars in 2009 and 2010, together with 64 glasshouse-grown accessions of 25 wild potato species in 2010. Thiamin concentrations ranged from 490 to 2325 ng g^{-1} FW (2908–16,401 ng g^{-1} DW in 2010) and folate concentrations from 15 to 416 ng g^{-1} FW (388–2098 ng g^{-1} DW in 2010). The majority of clones in both years had thiamin concentrations of 1200–1600 ng g^{-1} FW and folate concentrations of 100–200 ng g^{-1} FW. The researchers identified Andigenum accession PI 320377 (a Phureja from Colombia) as promising material to integrate in breeding programmes for thiamin enhancement, and likewise clone RN 104.02 (from accession PI 473260) for folate enhancement. Among the wild species, *S. vernei* subsp. *vernei* and *S. acaule* subsp. *aemulans* had the highest thiamin concentrations (≥8633 ng g^{-1} DW) and *S. boliviense* accession 597736 had the highest folate concentration (3031 ng g^{-1} DW), but the other *S. boliviense* accession had a much lower amount, suggesting variability within species. *S. vernei* subsp. *vernei* also had relatively high folate concentrations.

Robinson et al. (2015) continued the search for high folate potato germplasm for breeding purposes. They screened 250 individual plants from 77 accessions and 10

Solanum species (*S. stipuloideum*, *S. chacoense* subsp. *chacoense*, *S. candolleanum*, *S. acaule*, *S. demissum*, *S. microdontum*, *S. okadae*, *S. tuberosum* Andigenum Group, *S. boliviense*, *S. vernei*) and found a 10.5-fold range of folate concentrations (221–2336 ng g^{-1} DW). Some individuals from *S. boliviense* (PI 597736), *S. tuberosum* Andigenum Group (PI 225710 and PI 320377, both Phureja from Colombia) and *S. vernei* (PI 558149 and PI 230468) had double the folate concentrations of commercial cultivars such as Russet Burbank (1040 ng g^{-1} DW).

Currently screening germplasm for folate (and thiamin) is a time consuming and tedious analytical process that involves a tri-enzyme extraction and microbial assay. Furthermore, wild species may produce small or even no tubers in the field so that they need to be grown in winter in greenhouses or crossed with adapted cultivated potatoes. Hence Robinson et al. (2015) are seeking single nucleotide polymorphisms (SNPs) linked to folate in a segregating population from a cross between the high folate individual from diploid *S. boliviense* (PI 597736) and a diploid *S. tuberosum* clone. They will then be able to practise marker-assisted selection for high folate on a large number of individuals in subsequent breeding work. The same could be done with the high folate individuals from *S. vernei*. Robinson et al. (2015) have also secured for field evaluation, hybrids from a cross between the high folate individual from Phureja accession PI 225710 and a diploid *S. tuberosum* clone. If the hybrids produce $2n$ gametes by first division restitution (FDR), the genes for high folate can be efficiently transferred to the tetraploid cultivar genepool through $4x$ (cultivar) × $2x$ (hybrid) crosses, as explained by Jansky (2009). The other possibility is to use colchicine for mitotic chromosome doubling of the diploids so that crossing can be done at the tetraploid level. Similar approaches could be used in breeding for higher thiamine contents.

Robinson et al. (2015) thought that doubling the folate content of modern cultivars was a realistic target. Based on the current per capita consumption of 137 g day^{-1} in the United States, such a potato would provide around 11% of the recommended daily need of 400 μg, assuming 20% DM and 80% retention during cooking. They also pointed out that further increases could be achieved by optimizing the time of harvest and storage conditions, as young tubers contain up to two-fold the amount found in mature tubers and folates accumulate up to two-fold in tubers stored at cold temperature.

3.7.2 Vitamins C (Ascorbic Acid), B3 (Niacin), B5 (Pantothenic Acid) and B6 (Pyridoxine)

Vitamins C (L-ascorbic acid), B3 (niacin), B5 (pantothenic acid) and B6 (pyridoxine) are essential micronutrients in the human diet with scurvy a well-known symptom of vitamin C deficiency. A summary of their importance can be found in a book chapter by Navarre et al. (2016). Although potatoes offer a good supply of vitamin B3, it will not be considered further as very little research has been done on it in potato.

Juhaszi et al. (2014) developed a novel method for simultaneous detection of vitamins C, B5 and B6 by GC-MS (gas chromatograph–mass spectrometer) that could handle 30–40 samples per day. It is faster than high-performance liquid chromatography (HPLC) and insensitive to tuber colour, unlike some spectrophotometric methods. Juhaszi et al. (2014) found variation for vitamin content of tubers in two diploid populations. Furthermore, the sink leaves of greenhouse-grown plants reflected the vitamin content of field-grown tubers. However, the researchers did not report the heritability of the variation and it is difficult to infer the prospects for breeding new cultivars with higher concentrations of the vitamins than currently available.

We are going to concentrate on ascorbic acid as potatoes are an important worldwide source of this vitamin, contributing, for example, about 20% of the dietary intake in Europe. As a staple food, the potato can be a more important source of vitamin C than fruits and vegetables that have higher contents but are only complementary components of the diet. Love and Pavek (2008) argued that a three-pronged approach should be taken to increase the ascorbic acid content of potatoes, involving breeding, improved crop management and modification of cooking processes. They reported tuber ascorbic acid concentrations as high as 29.8 mg 100 g^{-1} FW and as low as 11.5 mg 100 g^{-1} FW among parental clones in North American breeding programmes. They concluded that adequate variability exists within adapted, high-quality potato germplasm for rapid progress in breeding for higher ascorbic acid concentration of tubers. They also concluded that crop management research needs to define practices that will slow the natural decline that occurs near the end of field growth and during storage, a response partially conditioned by plant stress. Furthermore, research into cooking procedures might help reduce the oxidative and enzymatic degradation of ascorbic acid that results from exposure to moisture, heat and air.

Burgos et al. (2009b) identified considerable genetic variation in ascorbic acid content among 25 Andean potato cultivars grown in three environments, despite some genotype × environment interaction. Concentrations were evaluated by a spectrophotometric method based on the ability of ascorbic acid to reduce the dye 2,6-dichloroindophenol. They ranged in freshly harvested, raw, peeled tubers from 22.2 to 121.4 mg 100 g^{-1} DW (6.5–36.9 mg 100 g^{-1} FW), with diploid cultivar (CIP) 704393 (Goniocalyx Group) showing the highest levels in all three locations (105.1–121.4 mg 100 g^{-1} DW or 33.6–34.0 mg 100 g^{-1} FW). Burgos et al. (2009b) point out that this value is higher than those reported for Indian, Canadian and Norwegian potato cultivars but lower than the highest value reported for Korean cultivars, albeit in the central part of the tuber. Cultivar 704393 also had the highest level after cooking (54% retention on boiling) and after storage for 26 weeks under farmers' conditions (34% retention). The range of percentages for retention after storage (22.0–61.6%) was similar to those reported for potatoes from Tenerife and for European and Indian cultivars. Unpeeled tubers retained more ascorbic acid on boiling than baked or microwaved tubers. In conclusion, 100 g of fresh-harvested and boiled potatoes of cultivar 704393 could provide adults with 17–20% of their RDA (100–120 mg day^{-1}) of ascorbic acid.

Cho et al. (2013) identified 10 out of 268 advanced clones in their Korean programme as having high ascorbic acid contents of 32.3–38.9 mg 100 g^{-1} FW. Among them, four double-cropping clones with short dormancy were selected for further analysis, with cultivars Dejima and Chubaek as controls. Ascorbic acid content was higher in younger tubers than in mature ones, and its content was highest in the selected clones 90 days after planting. During storage, ascorbic acid content significantly declined in all clones. The decrease was highest in H06035-4 and H06085-2 compared to cultivars Chubaek and Dejima, so the differences between selected clones and control cultivars narrowed after 4 months of storage. Nevertheless, H06035-4 was selected as a promising high ascorbic acid clone because of its agronomic characteristics and consistency of high ascorbic content.

3.8 Phenolic Acids, Anthocyanins and Carotenoids

Potatoes contain a number of beneficial phytochemicals, including phenolic compounds and carotenoids, many of which are antioxidants (Brown 2005; Ezekiel et al. 2013). Phenolic compounds present in potatoes include phenolic acids and anthocyanins. Akyol et al. (2016) reviewed their health-promoting effects, along with the effects of growing conditions, post-harvest storage and cooking processes on phenolic compounds. All potatoes contain chlorogenic acid as the principal phenolic acid. The pink, red, blue and purple colours in potato tubers are due to anthocyanins whose antioxidant properties have been demonstrated in red- and purple-fleshed potatoes. Use of oxygen radical absorbance capacity and ferric reducing ability of plasma, revealed antioxidant levels in red- and purple-fleshed potatoes that were 2–3 times higher than in white-fleshed potatoes (Brown et al. 2003). Reyes et al. (2005) concluded from high positive correlations that total anthocyanins and total phenolics in purple- and red-fleshed potatoes are mainly responsible for their antioxidant capacity.

Yellow and orange flesh are due to carotenoids, which are also antioxidants with health-promoting properties such as protection against certain cancers (Colditz et al. 1985), cardiovascular diseases (Gaziano et al. 1995) and particularly macular eye degeneration (Snodderly 1995). Short-term feeding of foods rich in the retinal carotenoids lutein and zeaxanthin can substantially increase macula pigment density in human subjects (Hammond et al. 1997). Lachman et al. (2016) discuss the health-promoting properties of carotenoids, their concentrations in potatoes with different flesh colours, and the factors which affect individual and total carotenoid levels, such as genotype and environment, tuber development, storage, cooking and heat processing.

Kaspar et al. (2011) confirmed that yellow-fleshed potatoes had higher concentrations of phenolic acids and carotenoids than white-fleshed ones and that purple-fleshed potatoes had higher concentrations of phenolic acids and anthocyanins. Furthermore, consumption of pigmented potatoes decreased oxidative damage and inflammatory cytokine concentrations in healthy men. Tierno et al. (2015) found that losses of total phenolics, anthocyanins and carotenoids occurred on boiling peeled tubers of highly pigmented potatoes, with big differences between genotypes in retention percentages: 45–74% for phenolics, 27–82% for anthocyanins and 32–100% for carotenoids. Hence the enhancement of anthocyanins and carotenoids, by conventional breeding and genetic modification, provides an opportunity to improve the nutritional value of potatoes and their processed products. But it will be important to assess new cultivars for their total and individual phytochemicals after cooking and processing.

3.8.1 Anthocyanins

The genetics and biochemistry of anthocyanin biosynthesis were reviewed by Holton and Cornish (1995) for maize (*Zea mays*), snapdragon (*Antirrhinum majus*) and petunia (*Petunia hybrida*). Colourless dihydroflavonols represent a branch point in flavonoid biosynthesis as they are the intermediates in the production of both coloured anthocyanins and colourless flavonols. Lewis et al. (1998a, b) used HPLC to quantify the anthocyanins, flavonoids and phenolic acids in tetraploid cultivars of *S. tuberosum* with coloured tubers, and in wild, tuberous *Solanum* species, the tubers of which were mostly white or light purple. When the flesh was also pigmented, the same anthocyanins were found as in the skin, but at much lower concentrations. The flesh tissues in which anthocyanin pigmentation can be observed are the vascular bundle, the cortex, the peri-medullary zone and the central pith (van Eck 2007). Combinations of different pigments and pigment patterns occur. Whereas red- and purple-skinned potatoes are relatively common, potatoes with varying distributions of anthocyanins in the flesh (rings, arcs and radiating stars) are rare outside the native cultivars of the Andes. Solidly coloured red or purple tuber flesh is unusual throughout the world, but occurs as rare segregants in certain red- and purple-skinned breeding populations, as found by Brown et al. (2003). Total anthocyanin content ranged from 6.9 to 35 mg 100 g^{-1} FW in the red-fleshed clones and from 5.5 to 17.1 in the purple-fleshed ones. A few pigmented flesh types are available commercially as specialty potatoes, for example cultivars Cranberry Red and All Blue (Brown et al. 2003). Interest now centres on increasing the concentrations of anthocyanins as a way of improving the levels of antioxidants in potato tubers, but first it is useful to consider some genetics.

The inheritance of anthocyanin pigmentation in cultivated potatoes (diploid and tetraploid) was reviewed by van Eck (2007). Key genes/alleles are at locus *R* (chromosome 2) for the production of red anthocyanins, locus *P* (chromosome 11) for the production of purple (and blue) anthocyanins, and locus *I* (chromosome 10) which determines whether or not *R* and *P* are expressed in the tubers (both skin and flesh). Two alleles are recognized at the *I* locus: *Iep* giving pigmentation in the epidermis of the tuber and *Ico* in the outer cortex. Three alleles are recognized at the *R* locus: *R*, R^{pw} and *r*, with R^{pw} giving pink tubers. De Jong et al. (2003) were able to show that although the dosage of *R* in tetraploids contributes to the intensity of skin colour, it is not a major factor. White potatoes lack either dominant allele *I* or both *R* and *P*; pink and red potatoes lack *P* but have *I* with R^{pw} and *R*, respectively; and purple potatoes have *P* and *I*. The *I* locus is in fact part of the *B-I-F* linkage group whose alleles determine the tissue-specific regulation of anthocyanin biosynthesis (van Eck 2007). In tetraploid potatoes, the *I* locus was originally referred to as *D* for developer by Salaman (1910). Candidate genes for the loci were proposed by De Jong et al. (2004) and subsequently confirmed (Fig. 3.3). The *P* locus on chromosome 11 codes for flavonoid 3′, 5′-hydroxylase (Jung et al. 2005); the *R* locus (*drf*) on chromosome 2 codes for dihydroflavonol 4-reductase (Zhang et al. 2009a); and the *D* (*I*) locus on chromosome 10 encodes an R2R3 MYB transcription factor that regulates expression of multiple anthocyanin structural genes in tuber skin (Jung et al. 2009). As purple-skinned potatoes can have genotype *R*--- or *rrrr*, both the 'red' and 'not-red' alleles of *drf* must code for catalytically active enzymes. The enzyme encoded by the red allele appears capable of reducing dihydromyricetin as

naringenin

↓ F3H

← dihydrokaempferol →

↓ ↓ F3'H ↓ F3'5'H

↓ dihydroquercetin → dihydromyricetin

↓ F3'5'H

↓ DFR ↓ DFR ↓ DFR

leucopelargonidin leucocyanidin leucodelphinidin

↓ ↓ ↓

pelargonidin-3-glucoside cyanidin-3-glucoside delphinidin-3-glucoside

(red) (pink) (purple)

↓ ↓ ↓

F3H	flavanone 3–hydroxylase
F3'H	flavonoid 3'–hydroxylase
F3'5'H	flavonoid 3', 5'–hydroxylase
DFR	dihydroflavonol 4-reductase

Fig. 3.3 Part of anthocyanin biosynthetic pathway, simplified (Holton and Cornish 1995)

well as dihydrokaempferol (and dihydroquercetin), whereas the non-red allele(s) appears capable of reducing the former, but not the latter. It is assumed that similar arguments apply to the pink allele and dihydroquercetin. In contrast, the *pppp* genotype is a loss of function of enzyme activity. The dominant R2R3 MYB domain allele is transcribed in red and purple tubers, but the recessive allele is not transcribed in white ones with genotype iiii (Jung et al. 2009). Zhang et al. (2009b) detected three QTLs (Quantitative Trait Loci) on chromosomes 5, 8 and 9 that influenced the extent (none to complete) of flesh pigmentation (purple) in the progeny of a diploid crosssegregating for the trait.

Let us now turn to quantitative variation in the concentrations of anthocyanins. Proof of the existence of such variation has come from the breeding and release in North America of specialty potatoes such as AmaRosa (Brown et al. 2012a) and Purple Pelisse (Vales et al. 2012), which are remarkable for their unusual colours. Cultivar AmaRosa resulted from a seedling from a cross made in 2000 between a red-fleshed clone and a bulk of pollen from other red-fleshed clones. The red-skinned, red-fleshed seedling was multiplied and evaluated over a number of years, including extensive trials in 2006 and 2007. AmaRosa tubers had higher total anthocyanin and hydrophilic oxygen radical absorption capacity (H-ORAC) than the control cultivar All Blue, while total carotenoids and lipophylic oxygen radical absorption capacity (L-ORAC) were equivalent. Hence overall antioxidant capacity was greater. Cultivar Purple Pelisse resulted from a cross made in 2000 between a red-skinned, red-fleshed clone and a bulk of pollen from red-fleshed potatoes. It was multiplied from a seedling identified in 2001 as having purple-skinned, purple-fleshed tubers. It was released as a cultivar in 2009 after extensive trials in 2006 and 2007. Purple Pelisse tubers had higher total anthocyanin and hydrophilic oxygen radical absorption capacity (H-ORAC) than the control cultivar All Blue, and higher total antioxidant values than cultivars Russet Burbank and Snowden.

3.8.2 Carotenoids

Structurally, carotenoids are terpenoids, derived by condensation of prenyl pyrophosphates (Sandmann 2001), whose conjugated double-bond system determines their colour and is responsible for their biological actions as antioxidants. The conversion of geranylgeranyl pyrophosphate to phytoene by phytoene synthase is the first dedicated step in the formation of carotenoids (Fig. 3.4). We shall see that the main carotenoids present in cultivated potatoes are lutein, lutein-5,6-epoxide, zeaxanthin, antheraxanthin, violaxanthin and neoxanthin, and that beta-carotene, the precursor of vitamin A, is either absent or present in low concentrations. Orozco et al. (2013) confirmed that carotenoids can be present in potatoes in esterified as well as free form, and in their profiles of 60 cultivars found a direct correlation between the carotenoid content and the esterified fraction. They thought that esterification facilitated accumulation of carotenoids and hence the presence of esters should be included as a breeding trait. Sulli et al. (2017) confirmed the presence of

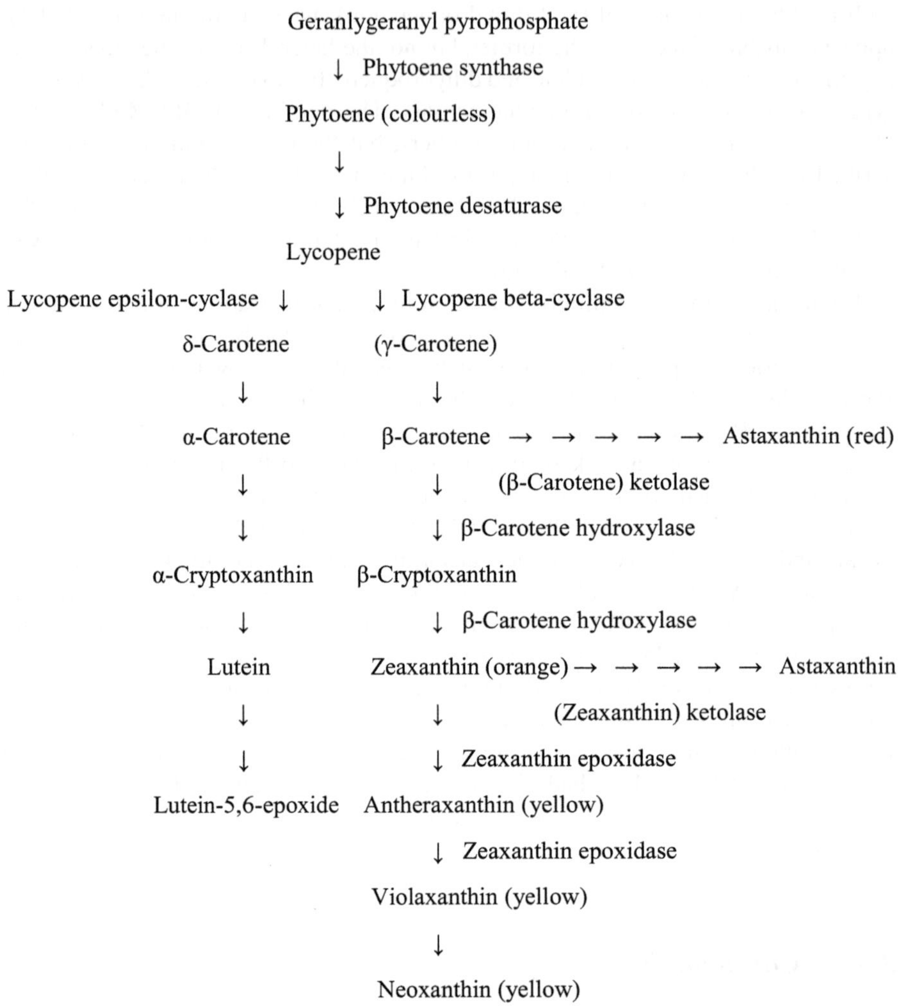

Fig. 3.4 Carotenoid biosynthetic pathway (simplified)

esters in their core set of ten cultivars and commented on their positive contribution to carotenoid stability, but also their negative effect on bioavailability.

Burgos et al. (2012) found that changes in concentrations of individual carotenoids due to boiling varied significantly among accessions of native Andean potatoes with diverse intensities of yellow flesh. Accessions (CIP) 705821 (Phureja) and 705172 (Phureja) were light yellow, 704393 (Goniocalix) and 701862 (Goniocalix) were intermediate yellow, and 702472 (Goniocalix), 705799 (Phureja) and 704218 (Phureja) were deep yellow. Boiling significantly reduced the violaxanthin and antheraxanthin concentrations of all accessions but lutein and zeaxanthin concentrations were not affected or were higher than in raw tubers. Burgos et al.

(2013) then estimated the lutein and zeaxanthin concentrations in boiled, freeze dried and milled samples of seven yellow-fleshed accessions (including four from previous study) before and after different steps (gastric, duodenal and micellar phase) of in vitro digestion. For all accessions, the amounts of lutein and zeaxanthin in the micelles after micellarization (i.e. the bioaccessible amounts) were significantly lower than the original amounts found in the boiled samples. The accession 701862 showed the highest bioaccessible lutein concentration (281 out of 396 μg 100 g^{-1} FW) and the accessions 703566 and 704218 showed the highest bioaccessible zeaxanthin concentrations (665 out of 1196 and 608 out of 1106 μg 100 g^{-1} FW, respectively). Considering the mean potato intake in the Andes (500 g day^{-1}), the accession 701862 would provide 14% of the lutein intake suggested for health benefits and the accessions 703566 and 704218 would provide 50% more than the suggested zeaxanthin intake. Hence higher levels of lutein and zeaxanthin are sensible targets when breeding new cultivars with higher levels of beneficial carotenoids, given their role in protection against macular eye degeneration (Snodderly 1995). Furthermore, there are relatively few dietary sources of zeaxanthin (Morris et al. 2004).

Bonierbale et al. (2009) confirmed that diploid Phureja is a good source of yellow-fleshed potatoes with high levels of lutein or zeaxanthin. They developed, validated and applied near-infrared reflectance spectroscopy (NIRS) to the characterization of 152 Phureja accessions (colour range: cream, light yellow and deep yellow) for total and individual carotenoid contents of raw tubers. NIRS estimated total carotenoids (103–2135 μg 100 g^{-1} FW) and zeaxanthin concentrations (trace to 1290 μg 100 g^{-1} FW) with high accuracy and could differentiate accessions with low, medium and high concentrations of violaxanthin (trace to 278 μg 100 g^{-1} FW), antheraxanthin (3–354 μg 100 g^{-1} FW), lutein (55–189 μg 100 g^{-1} FW) and β-carotene (trace to 18 μg 100 g^{-1} FW). NIRS can therefore be used to assess large numbers of genotypes (clones) in a breeding programme, in preference to costlier and more time consuming HPLC. Bonierbale et al. (2009) found that cluster analysis partitioned the 152 accessions into six groups. Groups 1 and 2 showed the highest mean total carotenoid, antheraxanthin and zeaxanthin concentrations and the lowest mean β-carotene concentration; groups 3 and 4 showed medium mean total carotenoid, antheraxanthin, zeaxanthin and β-carotene concentrations; and groups 5 and 6 showed the lowest mean total carotenoid, antheraxanthin and zeaxanthin concentrations and the highest mean β-carotene concentration. Accessions 701025 and 703566 (Group 1) showed the highest estimated zeaxanthin concentrations (above 1000 μg 100 g^{-1} FW). Forty-three accessions showed relatively high concentrations of β-carotene (above 10 μg 100 g^{-1} FW) with 703552, 706782, 701570, 706787, 706759, 706767, 706763, 706829, 704213, 704227 and 703581 showing the highest values. Burgos et al. (2009a) took a more detailed look at total and individual carotenoid concentrations in 23 of the 152 accessions, using spectrophotometry and HPLC. They confirmed the three main sets of groups formed by cluster analysis and that total carotenoid concentration was positively correlated with antheraxanthin and zeaxanthin concentrations, but negatively correlated with β-carotene concentration. Burgos et al. (2009a) also pointed out that the concentrations of carotenoids in the diploids they studied were much higher than those reported for tetraploid cultivars. Breithaupt and Bamedi (2002), for example, reported total carotenoid contents of 58–175 μg 100 g^{-1} FW in four yellow-fleshed cultivars and 38–62 μg 100 g^{-1} FW in four white-fleshed ones.

High heritabilities for flesh colour intensity have been reported for tetraploid breeding clones (0.93 on clonal mean basis) (Haynes et al. 1996) and for long-day-adapted, diploid breeding families (0.99 for half-sib families) (Haynes 2000), as well as a negative correlation with tuber size, as measured by weight (Haynes et al. 1994). Hence, in a practical breeding programme, selection for intense yellow flesh in similar sized tubers should result in higher contents of total carotenoids without a reduction in tuber weight. If necessary, the different degrees of yellowness can be quantified by spectrophotometry (Burgos et al. 2012). However, the breeder may also be interested in the concentrations of individual carotenoids, so let us look at some more results.

Individual carotenoids can be separated and quantified by HPLC with photodiode array (PDA) detection, the eluting compounds being identified by their absorbance spectra and co-elution with commercially available authentic standards (Lu et al. 2001; Wolters et al. 2010). Eleven clones from the Phureja/Stenotomum (diploid) breeding population of Haynes (2000) were analysed by Lu et al. (2001) along with two tetraploid controls. They detected six major carotenoids in all of the material: neoxanthin, violaxanthin, lutein-5,6-epoxide, lutein, zeaxanthin and an unknown carotenoid. Total carotenoid content in the diploid clones ranged from 136 (white flesh) to 1435 μg 100 g^{-1} FW (intense yellow flesh), and the main carotenoids were lutein-5,6-epoxide (34%), violaxanthin (33%) and lutein (26%). In contrast, tetraploid cultivars Superior (white flesh) and Yukon Gold (yellow flesh) contained 64 and 111 μg 100 g^{-1} FW of total carotenoids, and the main components were violaxanthin (40%) and lutein (22%). Both total and individual carotenoid contents were positively correlated with intensity of yellow. Again, this suggests that selecting for more intense yellow flesh will result in higher levels of carotenoids. The clones analysed by Lu et al. (2001) were chosen because they produced at least 5% of $2n$ pollen, and hence could be used to improve the nutritional value of tetraploid potatoes through $4x$ by $2x$ hybridizations, as done with cultivar Yukon Gold. It was this Canadian cultivar that opened the way to acceptance of yellow-fleshed potatoes by American consumers.

Haynes et al. (2011) investigated the inheritance of carotenoid content in $4x$ by $2x$ hybridizations. They crossed three diploid clones, that produced $2n$ pollen and had high (dark yellow flesh), medium (moderate yellow flesh) and low (white/cream flesh) carotenoid levels, to a light yellow-fleshed tetraploid breeding clone. In both of the yellow-fleshed diploids, the proportion of zeaxanthin was low at 3.1% or less of the total carotenoid content. After harvest, in each of two years, carotenoids were extracted and quantified by HPLC in 13 or 14 randomly selected clones from each family, as well as cultivar Yukon Gold as control. A continuous distribution of carotenoid concentration with high- and low-segregants was observed in all three families. There were significant differences among clones and high broad-sense heritabilities for zeaxanthin, total LCY-b pathway carotenoids (zeaxanthin + antheraxanthin + violaxanthin + neoxanthin), lutein and total carotenoids (LCY-b + lutein), despite clone × year interactions. Average contents in the clones ranged from 57 to 280 μg zeaxanthin 100 g^{-1} FW (Yukon Gold 103), 177–548 μg LCY-b 100 g^{-1} FW (Yukon Gold 262) and 62–169 μg lutein 100 g^{-1} FW (Yukon Gold 81). Based on flesh colour segregation, the tetraploid parent was duplex (*Chy2Chy2chy2chy2*) for the *Chy2* allele 3 governing yellow-flesh, the white-fleshed diploid was homozygous recessive (*chy2chy2*), and the two yellow-fleshed diploid parents were heterozygous (*Chy2chy2*) and produced 2*n* gametes by a second division restitution mechanism. The researchers concluded that tetraploid clones can be bred with higher levels of individual and total carotenoids than Yukon Gold.

Brown et al. (1993) studied orange-fleshed potatoes from the diploid breeding population of Haynes (2000) and showed that this colour was associated with large amounts of zeaxanthin. When averaged over a number of backcrosses, zeaxanthin levels were much higher in orange than white segregants (1426 compared with 304 μg 100 g^{-1} FW), whereas lutein levels were identical (146 μg 100 g^{-1} FW). Zeaxanthin also accounted for slightly more than half (51%) of the carotenoid content of a deep yellow, long-day-adapted Phureja cultivar, Inca Dawn (DB375/1), bred at the Scottish Crop Research Institute (Morris et al. 2004) (Fig. 3.5). Antheraxanthin (25%), violaxanthin (11%), an unidentified carotenoid (8%) and lutein (3%) were the other major carotenoids present. About 15% of the carotenoids were esterified. During tuber storage at 4 °C for 9 months in the dark, esterified carotenoids increased by 73%, zeaxanthin and antheraxanthin levels decreased by 11% and 7%, respectively, and the levels of lutein and an unidentified carotenoid both increased by 6%. Morris et al. (2004) compared cultivar Inca Dawn with two tetraploid cultivars, Desiree (cream/yellow flesh) and Pentland Javelin (white flesh). Total carotenoid contents were 36.3 (Inca Dawn), 4.91 (Desiree) and 1.63 μg g^{-1} DW (Pentland Javelin) in peeled, mature, whole tuber samples from glasshouse grown plants. The major carotenoids in cultivar Desiree were violaxanthin (51%), lutein (20%), neoxanthin (11%) and antheraxanthin (8%) with 23% present in the esterified form. Morris et al. (2004) found that there was an inverse trend between the level of zeaxanthin epoxidase transcript level and tuber carotenoid content in a range of potato germplasm, with implications for the regulation of carotenogenesis in potato tubers.

Römer et al. (2002) inhibited zeaxanthin conversion to violaxanthin in cultivars Baltica and Freya by tuber-specific down-regulation of the *zeaxanthin epoxidase* gene, using transformation with both sense and antisense constructs encoding the gene. As a consequence, zeaxanthin content was elevated 4- to 130-fold, reaching values up to 40 μg g^{-1} DW, while the amount of violaxanthin was dramatically reduced. In some transformants the monoepoxy intermediate antheraxanthin accumulated. Surprisingly, total carotenoid levels increased by up to 5.7-fold compared with controls. The transformed potatoes had dark yellow to orange flesh. Wolters et al. (2010) subsequently demonstrated that only genotypes combining the presence of dominant *β-carotene hydroxylase 2* (*Chy2*) allele 3 with homozygosity for recessive *zeaxanthin epoxidase* (*zep*) allele 1 (locus on chromosome 2) produced orange-fleshed tubers that accumulated large amounts of zeaxanthin (>250 μg 100 g^{-1}

Fig. 3.5 Long-day-adapted Phureja Group potatoes from the population of Carroll (1982). Source: SCRI and Fig. 1.3 in Handbook of Plant Breeding 7, Springer (with permission)

FW). They performed association analysis between SNP haplotypes and flesh colour phenotypes in a wide range of diploid (including orange) and tetraploid (none were orange) potato genotypes. Just one out of eleven *Chy2* alleles (3) changed white into yellow flesh (in diploids and tetraploids), whereas none of the *lycopene epsilon cyclase* (*Lcye*) alleles (locus on chromosome 12) had a large effect on flesh colour. The *Lcye* gene product is required for the synthesis of α-carotene, the precursor of lutein. The *Chy2* allele is the most likely candidate for the yellow flesh allele (*Y*) at the locus on chromosome 3 (Bonierbale et al. 1988; Brown et al. 2006). Genes other than *Chy2* and *Lcye* determine the intensity of the yellow flesh colour. The *zep* allele contained a non-LTR retrotransposon sequence in intron 1 and had a lower expression level (lower steady state mRNA level). Only 5 out of 230 tetraploid cultivars contained *zep* allele 1, all in simplex (single copy), thus confirming that this allele is rare in the tetraploid potato gene pool. More recently, Sulli et al. (2017) were able to divide a collection of five tetraploid (*S. tuberosum*) and five diploid (four Phureja and one *S. chacoense*) genotypes into three groups based on *Chy2* allele 3 and *zep* allele 1. The first group comprised four tetraploids and Phureja cultivar Mayan Gold, heterozygous *Chy2*---- and *Chy2chy2*, respectively, with yellow flesh and high levels of xanthophylls (antheraxanthin, violaxanthin, neoxanthin and lutein) and xanthophyll esters (23–41% of xanthophylls). The second group comprised one tetraploid and *S. chacoense*, homozygous recessive *chy-2chy2chy2chy2* and *chy2chy2*, respectively, with white-flesh and low levels of carotenoids. Finally, the third group comprised three Phureja, heterozygous

Chy2chy2 and homozygous recessive *zepzep*, with orange flesh and high levels of non-esterified zeaxanthin.

It might be concluded from the previous paragraph that conventional breeding for high zeaxanthin levels and orange flesh would prove difficult. Nevertheless, in Japan, Sakamoto et al. (2017) have reported breeding a new tetraploid potato cultivar Nagasaki Kogane with good eating quality, high carotenoid content (yellowish/orange flesh) and resistance to pests (cyst nematodes) and diseases (bacterial wilt and potato virus Y (PVY)). It was selected from a cross between cultivars Saikai 35 (Phureja cytoplasm) and Saikai 33, where the former was a cross between chromosome-doubled (tuber-disc method) cultivar Inca-no-mezame and cultivar Sakurafubuki. Diploid cultivar Inca-no-mezame was derived from crosses involving a Phureja clone and has a high carotenoid content. Cultivar Nagasaki Kogane contained lutein and zeaxanthin, but at lower concentrations than cultivar Inca-no-mezame (43.8 compared with 79.7 μg lutein 100 g^{-1} FW and 801.5 compared with 1131.7 μg zeaxanthin 100 g^{-1} FW). By way of comparison, the popular double-cropping tetraploid cultivar Dejima contained 48.1 μg lutein 100 g^{-1} FW and no zeaxanthin. Finally, diploid cultivar Inca-no-hitomi was selected from the open-pollinated progeny of cultivar Inca-no-mezame for its yellowish-orange flesh colour and other desirable agronomic characteristics (Kobayashi et al. 2008). They had similar total carotenoid contents which were four times higher than yellow-fleshed tetraploid cultivar Kita-akari, which in turn was ten times higher than white-fleshed tetraploid cultivar Irish Cobbler.

3.8.3 Genetic Engineering of Carotenoids

Vitamin A deficiency remains a major health problem in some parts of the world where it is the major cause of blindness in children. Beta-carotene (provitamin A), the precursor of vitamin A, is therefore a target for genetic engineering of carotenoids in potatoes as it is either absent or present only in low concentrations in *S. tuberosum* and closely related *Solanum* species.

Ducreux et al. (2005) produced greenhouse grown transgenic plants of tetraploid Tuberosum cultivar Desiree (light yellow flesh) and diploid Phureja cultivar Mayan Gold (deep yellow flesh) which expressed an *Erwinia uredovora CrtB* gene encoding phytoene synthase specifically in the tubers. In some transgenic lines of Desiree, total carotenoid levels of developing tubers had increased from 5.6 to 35.5 μg g^{-1} DW, β-carotene from negligible to 10.30 μg g^{-1} DW, lutein from 0.73 to 11.01 μg g^{-1} DW and violaxanthin from 2.18 to 8.52 μg g^{-1} DW. Likewise, in Mayan Gold, total carotenoid levels had increased from 24.4 to 78.1 μg g^{-1} DW, β-carotene from negligible to 6.51 μg g^{-1} DW, lutein from 2.36 to 18.48 μg g^{-1} DW, antheraxanthin from 4.23 to 12.50 μg g^{-1} DW and violaxanthin from 11.71 to 29.16 μg g^{-1} DW.

Diretto et al. (2007a) went further and transformed potato cultivar Desirée with a mini-pathway of key genes (enzymes) for the synthesis of β-carotene from geranylgeranyl diphosphate. Expression of three bacterial genes from *Erwinia herbicola* encoding phytoene synthase (*CrtB*), phytoene desaturase (*CrtI*) and lycopene beta-cyclase (*CrtY*), under tuber-specific (patatin) promoter control, resulted in tubers with deep yellow flesh without any adverse leaf phenotypes. Carotenoid levels in one transformant had increased from 5.8 to 114.4 μg g^{-1} DW and β-carotene from 0.013 to 47.4 μg g^{-1} DW (and α-carotene 0–6.2, lutein 1.0–23.1, violaxanthin 0.7–5.6), compared with "Golden Rice 2" at 31 μg g^{-1} DW. Assuming a β-carotene to retinol conversion of 6 : 1, this is sufficient to provide 50% of the RDA of vitamin A from 250 g FW of "Golden Potatoes" (GP) (Diretto et al. 2007a). More recently, Chitchumroonchokchai et al. (2017) provided data on bioavailability from transgenic lines GP17 and GP30. Beta-carotene (>36 μg g^{-1} DW) and α-carotene (>31 μg g^{-1} DW) were the most abundant carotenoids in GP17 and GP30 with total concentrations of these provitamin A carotenoids (68 and 91 μg g^{-1} DW in GP17 and GP30, respectively) being more than 100-fold greater than that in cultivar Desiree. Golden Potatoes also contained 2–3 times more lutein (>19 μg g^{-1} DW). Furthermore, α-tocopherol, an isomer of vitamin E, was unexpectedly detected in high concentrations in GP17 and GP30 (>78 μg g^{-1} DW), which were 15 times greater than in cultivar Desiree. Retention of all of these compounds on boiling was greater than 82%. Boiled potatoes were subjected to in vitro digestion (simulated oral, gastric and small intestinal phases of digestion) to determine the extent of transfer of carotenoids and α-tocopherol from the food matrix to micelles (i.e. their bioaccessibility). The average figures for GP17 and GP30 were 14% for the provitamin A carotenoids, 31% for lutein and 24% for α-tocopherol. The micelles were then added to monolayers of Caco-2 human intestinal epithelial cells to investigate apical uptake (i.e., bioavailability). Apical uptake during incubation for 4 h was around 15% for provitamin A carotenoids, 20% for lutein and 53% for α-tocopherol. The researchers concluded that for children and women of reproductive age, a 150-g serving of boiled Golden Potatoes has the potential to contribute 42% and 23%, respectively, of their daily requirement for provitamin A (measured as retinol activity equivalents), as well as 34% and 17%, respectively, of their daily vitamin E requirement.

Diretto et al. (2007b) also produced more modest increases in total carotenoid and β-carotene levels by the tuber-specific (patatin promoter) silencing of the genes encoding β-carotene hydroxylases CHY1 and CHY2 in cultivar Desiree. Beta-carotene levels increased from 0.00225 up to 0.085 μg g^{-1} DW and total carotenoids from 4.9 up to 14.3 μg g^{-1} DW. These changes were accompanied by a decrease in the immediate product of β-carotene hydroxylation, zeaxanthin, but not of downstream violaxanthin and neoxanthin.

Finally, Diretto et al. (2006) found that tuber-specific (patatin promoter) silencing in cultivar Desiree of the gene encoding the enzyme for the first dedicated step in lutein biosynthesis, *lycopene epsilon cyclase* (*LCY-e*), also resulted in modest increases of β-carotene from 0.00317 up to 0.044 μg g^{-1} DW and of total carotenoids from 4.68 up to 9.98 μg g^{-1} DW. These changes were not accompanied by a decrease in lutein, suggesting that the enzyme lycopene epsilon-cyclase is not rate-limiting for lutein accumulation.

There is also interest in engineering potatoes to accumulate ketocarotenoids. Astaxanthin, for example, has been associated with a range of health benefits and is used extensively as a feed additive in aquaculture (Morris et al. 2006). In nature, the main sources of astaxanthin are marine bacteria and microalgae. The conversion of β-carotene to astaxanthin requires the action of a hydroxylase as well as a ketolase. Morris et al. (2006) described the levels and types of carotenoid that accumulate in both tetraploid Tuberosum cultivar Desiree and diploid Phureja cultivar Mayan Gold on transformation with an algal *β-carotene oxygenase* gene (*bkt1* encoding a β-ketolase from *Haematococcus pluvialis*). A patatin promoter was used for tuber-specific expression. Two major ketocarotenoids, ketolutein and astaxanthin, absent in untransformed controls, were detected and gave tubers a reddish colour. The level of unesterified astaxanthin reached 13.9 μg g^{-1} DW and ketolutein reached 9.8 μg g^{-1} DW in some *bkt1* expressing Phureja lines but were much lower (0.6 and 0.5 μg g^{-1} DW, respectively) in the Tuberosum background. As astaxanthin biosynthesis requires the concerted action of both a carotene 4,4′ oxygenase (ketolase) and a 3,3′ hydroxylase, the researchers concluded that high activity of the 3,3′ hydroxylase occurred naturally in potato tubers. Subsequently, Campbell et al. (2015) investigated approaches to optimizing tuber astaxanthin content. They achieved a nutritionally significant concentration of 77 μg g^{-1} DW, which has the potential to provide the recommended 5 mg daily intake of astaxanthin from 55 g DW of potato (ca. 220 g FW). They selected a clone (01H15.57) from diploid Phureja cross 01H15 [Inca Sun (DB378(1), deep yellow) × Yema de Huevo (deep yellow)] with a high zeaxanthin content of 41.6 μg g^{-1} DW (total carotenoid content 53.9). They then managed to increase its astaxanthin content from 0 to 77.1 μg g^{-1} DW (total carotenoid content 93.5) by co-expressing the cauliflower *Or* transgene (tuber-specific GBSS promoter) and the *Brevundimonas crtZ* (hydroxylase) and *crtW* (ketolase) transgenes in tandem (*crtZW*) (constitutive 35S CaMV promoter). The *crtW* gene came from the marine bacterium *Brevundimonas* sp., strain SD212 and encodes an enzyme (ketolase) which accepts both β-carotene and zeaxanthin as substrates. Co-expression with the CrtZ enzyme from the same species results in efficient conversion of β-carotene to astaxanthin. The cauliflower *Or* (orange) gain of function gene mutation encodes a DnaJ cysteine-rich domain-containing protein whose transgene had been shown to increase total carotenoid levels in cultivar Desiree, from about 4 to 24 μg g^{-1} DW, and to produce β-carotene as a major carotenoid (Lu et al. 2006). The *crtZW* transgene alone increased astaxanthin content from 0 to 44.1 μg g^{-1} DW (total carotenoid content 57.4), whereas *Or* alone increased zeaxanthin content to 66.7 μg g^{-1} DW (total carotenoid content 88.3), but did not produce any astaxanthin. Astaxanthin levels were higher in summer than winter grown plants and could be associated with transcriptome and metabolome restructuring. Campbell et al. (2014) also made a genome-wide QTL and bulked transcriptomic analysis of population 01H15 and found a novel major QTL on chromosome 9 for increased carotenoid content.

References

Akilen R, Deljoomanesh N, Hunschede S, Smith CE, Arshad MU, Kubant R, Anderson GH (2016) The effects of potatoes and other carbohydrate side dishes consumed with meat on food intake, glycemia and satiety response in children. Nutr Diabetes 6:e195. https://doi.org/10.1038/nutd.2016.1

Akyol H, Riciputi Y, Capanoglu E, Caboni MF, Verardo V (2016) Phenolic compounds in the potato and its byproducts: an overview. Int J Mol Sci 17:835. https://doi.org/10.3390/ijms17060835

Ames M, Spooner DA (2008) DNA from herbarium specimens settles a controversy about origins of the European potato. Am J Bot 95:252–257

Amrein TM, Bachmann S, Noti A, Biedermann M, Barbosa MF, Biedermann-Brem S, Grob K, Keiser A, Realini P, Escher F, Amadò R (2003) Potential of acrylamide formation, sugars, and free asparagine in potatoes: a comparison of cultivars and farming systems. J Agric Food Chem 51:5556–5560

Andre CM, Legay S, Iammarino C, Ziebel J, Guignard C, Larondelle Y, Hausman J-F, Evers D, Miranda LM (2014) The potato in the human diet: a complex matrix with potential health benefits. Potato Res 57:201–214

Andre CM, Evers D, Ziebel J, Guignard C, Hausman J-F, Bonierbale M, zum Felde T, Burgos G (2015) In vitro bioaccessibility and bioavailability of iron from potatoes with varying vitamin C, carotenoid, and phenolic concentrations. J Agric Food Chem 63:9012–9021. https://doi.org/10.1021/acs.jafc.5b02904

Bártová V, Bárta J, Brabcová A, Zdráhal Z, Horáčková V (2015) Amino acid composition and nutritional value of four cultivated South American potato species. J Food Compost Anal 40:78–85. https://doi.org/10.1016/j.jfca.2014.12.006

Birt DF, Boylston T, Hendrich S, Jane JL, Hollis J, Li L, McClelland J, Moore S, Phillips GJ, Rowling M, Schalinske K, Whitley EM (2013) Resistant starch: promise for improving human health. Adv Nutr 4:587–601

Bonierbale MW, Plaisted RL, Tanksley SD (1988) RFLP maps based on a common set of clones reveal modes of chromosomal evolution in potato and tomato. Genetics 120:1095–1103

Bonierbale M, Grüneberg W, Amoros W, Burgos G, Salas E, Porras E, zum Felde T (2009) Total and individual carotenoid profiles in *Solanum phureja* cultivated potatoes: II. Development and application of near-infrared reflectance spectroscopy (NIRS) calibrations for germplasm characterization. J Food Compost Anal 22:509–516

Borch D, Juul-Hindsgaul N, Veller M, Astrup A, Jaskolowski J, Raben A (2016) Potatoes and risk of obesity, type 2 diabetes, and cardiovascular disease in apparently healthy adults: a systematic review of clinical intervention and observational studies. Am J Clin Nutr 104:489–498. https://doi.org/10.3945/ajcn.116.132332

Bradshaw JE (2009) A genetic perspective on yield plateau in potato. Potato J 36:79–94

Bradshaw JE (2016) Plant breeding: past, present and future. Springer, Cham

Bradshaw JE (2017) Review and analysis of limitations in ways to improve conventional potato breeding. Potato Res 60:171–193

Bradshaw JE, Bonierbale M (2010) Potatoes. In: Bradshaw JE (ed) Root and tuber crops, handbook of plant breeding 7. Springer, New York, NY, pp 1–52

Bradshaw JE, Hackett CA, Pande B, Waugh R, Bryan GJ (2008) QTL mapping of yield, agronomic and quality traits in tetraploid potato (*Solanum tuberosum* subsp. *tuberosum*). Theor Appl Genet 116:193–211

Breithaupt DE, Bamedi A (2002) Carotenoids and carotenoid esters in potatoes (*Solanum tuberosum* L.): new insights into an ancient vegetable. J Agric Food Chem 50:7175–7181

Brown CR (1993) Outcrossing rate in cultivated autotetraploid potato. Am Potato J 70:725–734

Brown CR (2005) Antioxidants in potato. Am J Potato Res 82:163–172

Brown CR, Edwards CG, Yang CP, Dean BB (1993) Orange flesh trait in potato: inheritance and carotenoid content. J Am Soc Hortic Sci 118:145–150

Brown CR, Wrolstad R, Durst R, Yang CP, Clevidence B (2003) Breeding studies in potatoes containing high concentrations of anthocyanins. Am J Potato Res 80:241–250

Brown CR, Kim TS, Ganga Z, Haynes K, De Jong D, Jahn M, Paran I, De Jong W (2006) Segregation of total carotenoid in high level potato germplasm and its relationship to beta-carotene hydroxylase polymorphism. Am J Potato Res 83:365–372

Brown CR, Haynes KG, Moore M, Pavek MJ, Hane DC, Love SL, Novy RG, Miller JC (2010) Stability and broad-sense heritability of mineral content in potato. Iron. Am J Potato Res 87:390–396. https://doi.org/10.1007/s12230-010-9145-4

Brown CR, Haynes KG, Moore M, Pavek MJ, Hane DC, Love SL, Novy RG, Miller JC (2011) Stability and broad-sense heritability of mineral content in potato. Zinc. Am J Potato Res 88:238–244. https://doi.org/10.1007/s12230-011-9188-1

Brown CR, Vales I, Yilma S, James S, Charlton B, Culp D, Hane D, Shock C, Feibert E, Pavek M, Knowles R, Novy R, Whitworth J, Stark J, Miller JC, Holm D, Quick R, Navarre R (2012a) "AmaRosa," a red skinned, red fleshed fingerling with high phytonutrient value. Am J Pot Res 89:249–254. https://doi.org/10.1007/s12230-012-9248-1

Brown CR, Haynes KG, Moore M, Pavek MJ, Hane DC, Love SL, Novy RG, Miller JC (2012b) Stability and broad-sense heritability of mineral content in potato: calcium and magnesium. Am J Potato Res 89:255–261. https://doi.org/10.1007/s12230-012-9240-9

Brown CR, Haynes KG, Moore M, Pavek MJ, Hane DC, Love SL, Novy RG (2013) Stability and broad-sense heritability of mineral content in potato: potassium and phosphorus. Am J Potato Res 90:516–523

Brown CR, Haynes KG, Moore M, Pavek MJ, Hane DC, Love SL, Novy RG, Miller JC (2014) Stability and broad-sense heritability of mineral content in potato: copper and sulfur. Am J Potato Res 91:618–624

Burgos G, Amoros W, Morote M, Stangoulis J, Bonierbale M (2007) Iron and zinc concentration of native Andean potato cultivars from a human nutrition perspective. J Sci Food Agric 87:668–675. https://doi.org/10.1002/jsfa.2765

Burgos G, Salas E, Amoros W, Auqui M, Muñoa L, Kimura M, Bonierbale M (2009a) Total and individual carotenoid profiles in *Solanum phureja* of cultivated potatoes: I. Concentrations and relationships as determined by spectrophotometry and HPLC. J Food Compost Anal 22:503–508

Burgos G, Auqui S, Amoros W, Salas E, Bonierbale M (2009b) Ascorbic acid concentration of native Andean potato varieties as affected by environment, cooking and storage. J Food Compost Anal 22:533–538

Burgos G, Amoros W, Salas E, Muñoa L, Sosa P, Díaz C, Bonierbale M (2012) Carotenoid concentrations of native Andean potatoes as affected by cooking. Food Chem 133:1131–1137

Burgos G, Muñoa L, Sosa P, Bonierbale M, zum Felde T, Díaz C (2013) In vitro bioaccessibility of lutein and zeaxanthin of yellow fleshed boiled potatoes. Plant Foods Hum Nutr 68:385–390

Burton WG (1989) The potato, 3rd edn. Longman Scientific and Technical, Harlow

Campbell R, Pont SD, Morris JA, McKenzie G, Sharma SK, Hedley PE, Ramsay G, Bryan GJ, Taylor MA (2014) Genome-wide QTL and bulked transcriptomic analysis reveals new candidate genes for the control of tuber carotenoid content in potato (Solanum tuberosum L.). Theor Appl Genet 127:1917–1933

Campbell R, Morris WL, Mortimer CL, Misawa N, Ducreux LJ, Morris JA, Hedley PE, Fraser PD, Taylor MA (2015) Optimising ketocarotenoid production in potato tubers: effect of genetic background, transgene combinations and environment. Plant Sci 234:27–37

Carroll CP (1982) A mass-selection method for the acclimatization and improvement of edible diploid potatoes in the United Kingdom. J Agric Sci Camb 99:631–640

Carroll CP, De Maine MJ (1989) The agronomic value of tetraploid F1 hybrids between potatoes of group Tuberosum and group Phureja/Stenotomum. Potato Res 32:447–456

Chakraborty S, Chakraborty N, Agrawal L, Ghosh S, Narula K, Shekhar S, Naik PS, Pande PC, Chakrborti SK, Datta A (2010) Next-generation protein-rich potato expressing the seed protein gene *AmA1* is a result of proteome rebalancing in transgenic tuber. Proc Natl Acad Sci U S A 107:17533–17538. https://doi.org/10.1073/pnas.1006265107

Chawla R, Shakya R, Rommens CM (2012) Tuber-specific silencing of *asparagine synthetase-1* reduces the acrylamide-forming potential of potatoes grown in the field without affecting tuber shape and yield. Plant Biotechnol J 10:913–924

Chitchumroonchokchai C, Diretto G, Parisi B, Giuliano G, Failla ML (2017) Potential of golden potatoes to improve vitamin A and vitamin E status in developing countries. PLoS One 12(11):e0187102. https://doi.org/10.1371/journal.pone.0187102

Cho K-S, Jeong H-J, Cho J-H, Park Y-E, Hong S-Y, Won H-S, Kim H-J (2013) Vitamin C content of potato clones from Korean breeding lines and compositional changes during growth and after storage. Hortic Environ Biotechnol 54:70–75

Chung YS, Palta P, Bamberg J, Jansky S (2016) Potential molecular markers associated with tuber calcium content in wild potato germplasm. Crop Sci 56:576–584

Colditz GA, Branch LG, Lipnick RJ, Willett WC, Rosner B, Posner BM, Hennekens CH (1985) Increased green and yellow vegetable intake and lowered cancer deaths in an elderly population. Am J Clin Nutr 41:32–36

Dale MFB, Mackay GR (1994) Inheritance of table and processing quality. In: Bradshaw JE, Mackay GR (eds) Potato genetics. CAB International, Wallingford, pp 285–315

Davies HV (2002) Commercial developments with transgenic potato. In: Valpuesta V (ed) Fruit and vegetable biotechnology. Woodhead Publishing Limited, Cambridge, pp 222–249

De Jong WS, De Jong DM, Bodis M (2003) A fluorogenic 5′ nuclease (TaqMan) assay to assess dosage of a marker tightly linked to red skin color in autotetraploid potato. Theor Appl Genet 107:1384–1390. https://doi.org/10.1007/s00122-003-1420-z

De Jong WS, Eannetta NT, De Jong DM, Bodis M (2004) Candidate gene analysis of anthocyanin pigmentation loci in the *Solanaceae*. Theor Appl Genet 108:423–432. https://doi.org/10.1007/s00122-003-1455-1

Diretto G, Tavazza R, Welsch R, Pizzichini D, Mourgues F, Papacchioli V, Beyer P, Giuliano G (2006) Metabolic engineering of potato tuber carotenoids through tuber-specific silencing of lycopene epsilon cyclase. BMC Plant Biol 6:13. https://doi.org/10.1186/1471-2229-6-13

Diretto G, Al-Babili S, Tavazza R, Papacchioli V, Beyer P, Giuliano G (2007a) Metabolic engineering of potato carotenoid content through tuber-specific overexpression of a bacterial mini-pathway. PLoS One 2(4):e350

Diretto G, Welsch R, Tavazza R, Mourgues F, Pizzichini D, Beyer P, Giuliano G (2007b) Silencing of beta-carotene hydroxylase increases total carotenoid and beta-carotene levels in potato tubers. BMC Plant Biol 7:11. https://doi.org/10.1186/1471-2229-7-11

Dodds KS (1962) Classification of cultivated potatoes. In: Correll DS (ed) The potato and its wild relatives. Texas Research Foundation, Renner, TX, pp 517–539

Ducreux LJ, Morris WL, Hedley PE, Shepherd T, Davies HV, Millam S, Taylor MA (2005) Metabolic engineering of high carotenoid potato tubers containing enhanced levels of β-carotene and lutein. J Exp Bot 56:81–89

Ek KL, Wang S, Copeland L, Brand-Miller JC (2014a) Discovery of a low-glycaemic index potato and relationship with starch digestion *in vitro*. Br J Nutr 111:699–705

Ek KL, Wang S, Brand-Miller JC, Copeland L (2014b) Properties of starch from potatoes differing in glycemic index. Food Funct 10:2509–2515

Englyst HN, Kingman SM, Cummings JH (1992) Classification and measurement of nutritionally important starch fractions. Eur J Clin Nutr 46:S33–S50

Ezekiel R, Singh N, Sharma S, Kaur A (2013) Beneficial phytochemicals in potato – a review. Food Res Int 50:487–496

Fairweather-Tait S (1983) Studies on the availability of iron in potatoes. Br J Nutr 50:15–23

Foster-Powell K, Holt SHA, Brand-Miller JC (2002) International table of glycemic index and glycemic load values: 2002. Am J Clin Nutr 76:5–56

Friedman M (1996) Nutritional value of proteins from different food sources. A review. J Agric Food Chem 44:6–29

Friedman M, Levin CE (2016) Glycoalkaloids and calystegine alkaloids in potatoes. In: Singh J, Kaur L (eds) Advances in potato chemistry and technology, 2nd edn. Elsevier, Amsterdam, pp 167–194

Friedman M, McDonald GM (1997) Potato glycoalkaloids: chemistry, analysis, safety, and plant physiology. Crit Rev Plant Sci 16:55–132

Galdón BR, Mesa DR, Rodríguez EMR, Romero CD (2010) Amino acid content in traditional potato cultivars from the Canary Islands. J Food Compost Anal 23:148–153. https://doi.org/10.1016/j.jfca.2009.08.009

Gaziano GM, Manson JE, Branch LG, Colditz GA, Willett WC, Buring JE (1995) A prospective study of consumption of carotenoids in fruits and vegetables and decreased cardiovascular mortality in the elderly. Ann Epidemiol 5:255–260

Goodrich CE (1863) The origination and test culture of seedling potatoes. Trans NY State Agric Soc 23:89–134

Goyer A, Navarre DA (2007) Determination of folate concentrations in diverse potato germplasm using a trienzyme extraction and a microbiological assay. J Agric Food Chem 55:3523–3528

Goyer A, Navarre DA (2009) Folate is higher in developmentally younger potato tubers. J Sci Food Agric 89:579–583

Goyer A, Sweek K (2011) Genetic diversity of thiamin and folate in primitive cultivated and wild potato (*Solanum*) species. J Agric Food Chem 59:13072–13080. https://doi.org/10.1021/jf203736e

Hammond BR, Johnson EJ, Russell RM, Krinsky NI, Yeum KJ, Edwards RB, Snodderly DM (1997) Dietary modification of human macular pigment density. Invest Ophthalmol Vis Sci 38:1795–1801

Hawkes JG (1990) The potato: evolution, biodiversity & genetic resources. Belhaven Press, London

Haynes KG (2000) Inheritance of yellow-flesh intensity in diploid potatoes. J Am Soc Hortic Sci 125:63–65

Haynes KG, Potts WE, Chittams JL, Fleck DL (1994) Determining yellow-flesh intensity in potatoes. J Am Soc Hortic Sci 119:1057–1059

Haynes KG, Sieczka JB, Henninger MR, Fleck DL (1996) Clone × environment interactions for yellow-flesh intensity in tetraploid potatoes. J Am Soc Hortic Sci 121:175–177

Haynes KG, Clevidence BA, Rao D, Vinyard BT (2011) Inheritance of carotenoid content in tetraploid × diploid potato crosses. J Am Soc Hortic Sci 136:265–272

Haynes K, Yencho G, Clough M, Henninger M, Sterrett S (2012) Genetic variation for potato tuber micronutrient content and implications for biofortification of potatoes to reduce micronutrient malnutrition. Am J Potato Res 89:192–198

Hellwege EM, Czapla S, Jahnke A, Willmitzer L, Heyer AG (2000) Transgenic potato (*Solanum tuberosum*) tubers synthesise the full spectrum of inulin molecules naturally occurring in globe artichoke (Cynara scolymus). Proc Natl Acad Sci U S A 97:8699–8704

Henry CJK, Lightowler H, Strik C, Storey M (2005) Glycaemic index values for commercially available potatoes in Great Britain. Br J Nutr 94:917–921

Holton TA, Cornish EC (1995) Genetics and biochemistry of anthocyanin biosynthesis. Plant Cell 7:1071–1083

Hosaka K (2004) Evolutionary pathway of T-type chloroplast DNA in potato. Am J Potato Res 81:153–158

Hougas RW, Peloquin SJ, Ross RW (1958) Haploids of the common potato. J Hered 49:103–107

Huaman Z, Golmirzaie A, Amoros W (1997.) The potato) In: Fuccillo D, Sears L, Stapleton P (eds) Biodiversity in trust: conservation and use of plant genetic resources in CGIAR centres. Cambridge University Press, Cambridge, UK, pp 21–28

Huaman Z, Hoekstra R, Bamberg JB (2000a) The inter-genebank potato database and the dimensions of available wild potato germplasm. Am J Potato Res 77:353–362

Huaman Z, Ortiz R, Gomez R (2000b) Selecting a *Solanum tuberosum* subsp. *andigena* core collection using morphological, geographical, disease and pest descriptors. Am J Potato Res 77:183–190

Huaman Z, Ortiz R, Zhang D, Rodriguez F (2000c) Isozyme analysis of entire and core collections of *Solanum tuberosum* subsp. *andigena* potato cultivars. Crop Sci 40:273–276

Itkin M, Heinig U, Tzfadia O, Bhide AJ, Shinde B, Cardenas PD, Bocobza SE, Unger T, Malitsky S, Finkers R, Tikunov Y (2013) Biosynthesis of antinutritional alkaloids in solanaceous crops is mediated by clustered genes. Science 341:175–179

Jansky S (2009) Breeding, genetics and cultivar development. In: Singh J, Kaur L (eds) Advances in potato chemistry and technology. Academic, Burlington, pp 27–62

Johns T, Alonso JG (1990) Glycoalkaloid change during the domestication of the potato, *Solanum Section Petota*. Euphytica 50:203–210

Juhaszi Z, Dancs G, Marincs F, Vossen M, Allefs S, Banfalvi Z (2014) Vitamin C, B5, and B6 contents of segregating potato populations detected by GC-MS: a method facilitating breeding potatoes with improved vitamin content. Plant Breed 133:515–520. https://doi.org/10.1111/pbr.12169

Jung CS, Griffiths HM, De Jong DM, Cheng S, Bodis M, De Jong WS (2005) The potato *P* locus codes for flavonoid 3′,5′–hydroxylase. Theor Appl Genet 110:269–275

Jung CS, Griffiths HM, De Jong DM, Cheng S, Bodis M, Kim TS, De Jong WS (2009) The potato *developer* (*D*) locus encodes an R2R3 MYB transcription factor that regulates expression of multiple anthocyanin structural genes in tuber skin. Theor Appl Genet 120:45–57

Kaminski KP, Kørup K, Andersen MN, Sønderkær M, Andersen MS, Kirk HG, Nielsen KL (2016) Next generation sequencing bulk segregant analysis of potato support that differential flux into the cholesterol and stigmasterol metabolite pools is important for steroidal glycoalkaloid content. Potato Res 59:81–97. https://doi.org/10.1007/s11540-015-9314-4

Kaspar KL, Park JS, Brown CR, Mathison BD, Navarre DA, Chew BP (2011) Pigmented potato consumption alters oxidative stress and inflammatory damage in men. J Nutr 141:108–111

Kirkman MA (2007) Global markets for processed potato products. In: Vreugdenhil D (ed) Potato biology and biotechnology advances and perspectives. Elsevier, Oxford, pp 27–44

Knight TA (1807) On raising of new and early varieties of the potato (*Solanum tuberosum*). Trans Hort Soc Lond 1:57–59

Kobayashi A, Ohara-Takada A, Tsuda S, Matsuura-Endo C, Takada N, Umemura Y, Nakao T, Yoshida T, Hayashi K, Mori M (2008) Breeding of potato variety “Inca-no-hitomi” with a very high carotenoid content. Breed Sci 58:77–82

Kromann P, Valverde F, Alvarado S, Vélez R, Pisuña J, Potosí B, Taipe A, Caballero D, Cabezas A, Devaux A (2017) Can Andean potatoes be agronomically biofortified with iron and zinc fertilizers? Plant Soil 411:121–138. https://doi.org/10.1007/s11104-016-3065-0

Lachman J, Hamouz K, Orsák M, Kotíková Z (2016) Carotenoids in potatoes—a short overview. Plant Soil Environ 62:474–481

Lewis CE, Walker JRL, Lancaster JE, Sutton KH (1998a) Determination of anthocyanins, flavonoids and phenolic acids in potatoes. I: coloured cultivars of *Solanum tuberosum* L. J Sci Food Agric 77:45–57

Lewis CE, Walker JRL, Lancaster JE, Sutton KH (1998b) Determination of anthocyanins, flavonoids and phenolic acids in potatoes. II: wild, tuberous *Solanum* species. J Sci Food Agric 77:58–63

Li X-Q, Scanlon MG, Liu Q, Coleman WK (2006.) Processing and value addition) In: Gopal J, Khurana SMP (eds) Handbook of potato production, improvement, and postharvest management. Food Products Press, New York, pp 523–555

Lindhout P, Meijer D, Schotte T, Hutten RCB, Visser RGF, van Eck HJ (2011) Towards F_1 hybrid seed potato breeding. Potato Res 54:301–312

Lockyer S, Nugent AP (2017) Health effects of resistant starch. Nutr Bull 42:10–41

Louderback LA, Pavlik BM (2017) Starch granule evidence for the earliest potato use in North America. Proc Natl Acad Sci U S A 114:7606–7610. https://doi.org/10.1073/pnas.1705540114

Love SL, Pavek JJ (2008) Positioning the potato as a primary food source of vitamin C. Am J Potato Res 85:277–285

Lu W, Haynes K, Wiley E, Clevidence B (2001) Carotenoid content and color in diploid potatoes. J Am Soc Hortic Sci 126:722–726

Lu S, Van Eck J, Zhou X, Lopez AB, O'Halloran DM, Cosman KM, Li L (2006) The cauliflower *Or* gene encodes a DnaJ cysteine-rich domain-containing protein that mediates high levels of β-carotene accumulation. Plant Cell 18:3594–3605

Macdonald-Clarke CJ, Martin BR, McCabe LD, McCabe GP, Lachcik PJ, Wastney M, Weaver CM (2016) Bioavailability of potassium from potatoes and potassium gluconate: a randomized dose response trial. Am J Clin Nutr 104:346–353. https://doi.org/10.3945/ajcn.115.127225

Mahfouz MM, Cardi T, Stewart CN (2016) Next-generation precision genome engineering and plant biotechnology. Plant Cell Rep 35:1397–1399. https://doi.org/10.1007/s00299-016-2009-8

Monro J, Mishra S (2009) Nutritional value of potatoes: digestibility, glycemic index, and glycemic impact. In: Singh J, Kaur L (eds) Advances in potato chemistry and technology. Academic, Burlington, pp 371–394

Monro J, Mishra S, Blandford E, Anderson J, Genet R (2009) Potato genotype differences in nutritionally distinct starch fractions after cooking, and cooking plus storing cool. J Food Compost Anal 22:539–545

Morris WL, Ducreux L, Griffiths DW, Stewart D, Davies HV, Taylor MA (2004) Carotenogenesis during tuber development and storage in potato. J Exp Bot 55:975–982. https://doi.org/10.1093/jxb/erh121

Morris WL, Ducreux LJM, Fraser PD, Millam S, Taylor MA (2006) Engineering ketocarotenoid biosynthesis in potato tubers. Metab Eng 8:253–263

Moseley ME (2001) The incas and their ancestors: the archaeology of Peru, 2nd edn. Thames and Hudson, London

Muttucumaru N, Powers SJ, Elmore JS, Mottram DS, Halford NG (2013) Effects of nitrogen and sulfur fertilization on free amino acids, sugars, and acrylamide-forming potential in potato. J Agric Food Chem 61:6734–6742. https://doi.org/10.1021/jf401570x

Muttucumaru N, Powers SJ, Elmore JS, Dodson A, Briddon A, Mottram DS, Halford NG (2017) Acrylamide-forming potential of potatoes grown at different locations, and the ratio of free asparagine to reducing sugars at which free asparagine becomes a limiting factor for acrylamide formation. Food Chem 220:76–86

Nassar AMK, Sabally K, Kubow S, Leclerc YN, Donnelly DJ (2012) Some Canadian-grown potato cultivars contribute to a substantial content of essential dietary minerals. J Agric Food Chem 60:4688–4696

Navarre DA, Shakya R, Hellman H (2016) Vitamins, phytonutrients, and minerals in potato. In: Singh J, Kaur L (eds) Advances in potato chemistry and technology, 2nd edn. Elsevier, Amsterdam, pp 117–166

Nayak B, Berrios JDJ, Tang J (2014) Impact of food processing on the glycemic index (GI) of potato products. Food Res Int 56:35–46

Novy RG, Whitworth JL, Stark JC, Schneider BL, Knowles NR, Pavek MJ, Knowles LO, Charlton BA, Sathuvalli V, Yilma S, Brown CR, Thornton M, Brandt TL, Olsen N (2017) Payette Russet: a dual-purpose potato cultivar with cold-sweetening resistance, low acrylamide formation, and resistance to late blight and potato virus Y. Am J Potato Res 94:38–53

Orozco RF, Gallardo-Guerrero L, Hornero-Mendez D (2013) Carotenoid profiling in tubers of different potato (*Solanum* sp) cultivars: accumulation of carotenoids mediated by xanthophyll esterification. Food Chem 141:2864–2872. https://doi.org/10.1016/j.foodchem.2013.05.016

Ortiz R (2001) The state of the use of potato genetic diversity. In: Cooper HD, Spillane C, Hodgkin T (eds) Broadening the genetic base of crop production. CAB International, Wallingford, pp 181–200

Paget M, Amoros W, Salas E, Eyzaguirre R, Alspach P, Apiolaza L, Noble A, Bonierbale M (2014) Genetic evaluation of micronutrient traits in diploid potato from a base population of Andean Landrace Cultivars. Crop Sci 54:1949–1959

Pedreschi F (2009) Fried and dehydrated potato products. In: Singh J, Kaur L (eds) Adavances in potato chemistry and technology. Academic, Burlington, pp 319–337

Pęksa A, Kita A, Kułakowska K, Aniołowska M, Hamouz K, Nemś A (2013) The quality of protein of coloured fleshed potatoes. Food Chem 141:2960–2966. https://doi.org/10.1016/j.foodchem.2013.05.125

Pieterse L, Hils U (2009) World catalogue of potato varieties 2009/10. Agrimedia GmbH, Clenze

Pinhero RG, Coffin R, Yada RY (2009) Post-harvest storage of potatoes. In: Singh J, Kaur L (eds) Adavances in potato chemistry and technology. Academic, Burlington, pp 339–370

Pinhero RG, Waduge RN, Liu Q, Sullivan JA, Tsao R, Bizimungu B, Yada RY (2016) Evaluation of nutritional profiles of starch and dry matter from early potato varieties and its estimated glycemic impact. Food Chem 203:356–366

Potato Genome Sequencing Consortium (2011) Genome sequence and analysis of the tuber crop potato. Nature 475:189–195

Raatz SK, Idso L, Johnson LK, Jackson MI, Combs GF Jr (2016) Resistant starch analysis of commonly consumed potatoes: content varies by cooking method and service temperature but not by variety. Food Chem 208:297–300

Raker CM, Spooner DM (2002) Chilean tetraploid cultivated potato, *Solanum tuberosum*, is distinct from the Andean populations: microsatellite data. Crop Sci 42:1451–1458

Ramdath DD, Padhi E, Hawke A, Sivaramalingam T, Tsao R (2014) The glycemic index of pigmented potatoes is related to their polyphenol content. Food Funct 5:909–915

Reader J (2008) Propitious esculent. William Heinemann, London

Reyes LF, Miller JC, Cisneros-Zevallos L (2005) Antioxidant capacity, anthocyanins and total phenolics in purple-and red-fleshed potato (Solanum tuberosum L.) genotypes. Am J Potato Res 82:271–277

Robinson BR, Sathuvalli V, Bamberg J, Goyer A (2015) Exploring folate diversity in wild and primitive potatoes for modern crop improvement. Genes 6:1300–1314. https://doi.org/10.3390/genes6041300

Römer S, Lübeck J, Kauder F, Steiger S, Adomat C, Sandmann G (2002) Genetic engineering of a zeaxanthin-rich potato by antisense inactivation and co-suppression of carotenoid epoxidation. Metab Eng 4:263–272. https://doi.org/10.1006/mben.2002.0234

Rommens CM, Ye J, Richael C, Swords K (2006) Improving potato storage and processing characteristics through all-native DNA transformation. J Agric Food Chem 54:9882–9887

Rommens CM, Yan H, Swords K, Richael C, Ye J (2008) Low-acrylamide French fries and potato chips. Plant Biotechnol J 6:843–853

Sakamoto Y, Mori K, Matsuo Y, Mukojima N, Watanabe W, Sobaru N, Tamiya S, Nakao T, Hayashi K, Watanuki H, Nara K, Yamazaki K, Chaya M (2017) Breeding of a new potato variety 'Nagasaki Kogane' with high eating quality, high carotenoid content, and resistance to diseases and pests. Breed Sci. https://doi.org/10.1270/jsbbs.16168

Salaman RN (1910) The inheritance of colour and other characters in the potato. J Genet 1:7–46

Sandmann G (2001) Genetic manipulation of carotenoid biosynthesis: strategies, problems and achievements. Trends Plant Sci 6:14–17

Sawai S, Ohyama K, Yasumoto S, Seki H, Sakuma T, Yamamoto T, Takebayashi Y, Kojima M, Sakakibara H, Aoki T, Muranaka T, Saito K, Umemoto N (2014) Sterol side chain reductase 2 is a key enzyme in the biosynthesis of cholesterol, the common precursor of toxic steroidal glycoalkaloids in potato. Plant Cell 26:3763–3774

Shepherd LVT, Bradshaw JE, Dale MFB, McNicol JW, Pont SDA, Mottram DS, Davies HV (2010) Variation in acrylamide producing potential in potato: segregation of the trait in a breeding population. Food Chem 123:568–573

Shepherd LVT, Hackett CA, Alexander CJ, McNicol JW, Sungurtas JA, Stewart D, McCue KF, Belknap WR, Davies HV (2015) Modifying glycoalkaloid content in transgenic potato—metabolome impacts. Food Chem 187:437–443

Simmonds NW (1995) Potatoes. In: Smartt J, Simmonds NW (eds) Evolution of crop plants, 2nd edn. Longman Scientific & Technical, Harlow, pp 466–471

Sinden SL, Sanford LL, Webb RE (1984) Genetic and environmental control of potato glycoalkaloids. Am Potato J 61:141–156
Singh J, Kaur L (2016) Advances in potato chemistry and technology, 2nd edn. Elsevier, Amsterdam
Snodderly DM (1995) Evidence for protection against age-related macular degeneration by carotenoids and antioxidant vitamins. Am J Clin Nutr 62(Suppl):1448–1461
Spooner DM (2016) Species delimitations in plants: lessons learned from potato taxonomy by a practicing taxonomist. J Syst Evol 54:191–203
Spooner DM, Hijmans RJ (2001) Potato systematics and germplasm collecting, 1989–2000. Am J Potato Res 78:237–268
Spooner DM, McLean K, Ramsay G, Waugh R, Bryan GJ (2005) A single domestication for potato based on multilocus amplified fragment length polymorphism genotyping. Proc Natl Acad Sci U S A 102:14694–14699
Spooner DM, Nunez J, Trujillo G, Herrera M, Guzman F, Ghislain M (2007) Extensive simple sequence repeat genotyping of potato landraces supports a major reevaluation of their gene pool structure and classification. Proc Natl Acad Sci U S A 104:19398–19404
Spooner DM, Ghislain M, Simon R, Jansky SH, Gavrilenko T (2014) Systematics, diversity, genetics, and evolution of wild and cultivated potatoes. Bot Rev 80:283–383
Storey M, Anderson P (2014) Income and race/ethnicity influence dietary fiber intake and vegetable consumption. Nutr Res 34:844–850
Subramanian NK, White PJ, Broadly MR, Ramsay G (2017) Variation in tuber mineral concentrations among accessions of *Solanum* species held in the Commonwealth potato collection. Genet Resour Crop Evol 64:1927–1935
Sulli M, Mandolino G, Sturaro M, Onofri C, Diretto G, Parisi B, Giuliano G (2017) Molecular and biochemical characterization of a potato collection with contrasting tuber carotenoid content. PLoS One 12(9):e0184143. https://doi.org/10.1371/journal.pone.0184143
Tian J, Chen J, Ye X, Chen S (2016) Health benefits of the potato affected by domestic cooking: a review. Food Chem 202:165–175
Tierno R, Hornero-Méndez D, Gallardo-Guerrero L, López-Pardo R, Ruiz de Gallareta JI (2015) Effect of boiling on the total phenolic, anthocyanin and carotenoid concentrations of potato tubers from selected cultivars and introgressed breeding lines from native potato species. J Food Compost Anal 41:58–65
Ugent D, Dillehay T, Ramirez C (1987) Potato remains from a late Pleistocene settlement in south central Chile. Econ Bot 4:17–27
Umemoto N, Nakayasu M, Ohyama K, Yotsu-Yamashita M, Mizutani M, Seki H, Saito K, Muranaka T (2016) Two cytochrome P450 monooxygenases catalyze early hydroxylation steps in the potato steroid glycoalkaloid biosynthetic pathway. Plant Physiol 171:2458–2467
van Eck HJ (2007) Genetics of morphological and tuber traits. In: Vreugdenhil D (ed) Potato biology and biotechnology advances and perspectives. Elsevier, Oxford, pp 91–115
Van Gelder WMJ, Vinket JH, Scheffer JJC (1988) Steroidal glycoalkaloids in tubers and leaves of *Solanum* species used in potato breeding. Euphytica 39:147–158
Vales MI, Brown CR, Yilma S, Hane DC, James SR, Shock CC, Charlton BA, Karaagac E, Mosley AR, Culp D, Feibert E, Stark JC, Pavek MJ, Knowles NR, Novy RG, Whitworth JL (2012) Purple Pelisse: a specialty 'fingerling' potato with purple skin and flesh and medium specific gravity. Am J Potato Res 89:306–314
Wheeler RM (2009) Potatoes for human life support in space. In: Singh J, Kaur L (eds) Adavances in potato chemistry and technology. Academic, Burlington, pp 465–495
White PJ, Bradshaw JE, Dale MFB, Ramsay G, Hammond JP, Broadley MR (2009) Relationships between yield and mineral concentrations in potato tubers. Hort Sci 44:6–11
White PJ, Broadley MR, Hammond JP, Ramsay G, Subramanian NK, Thompson J, Wright G (2012) Bio-fortifcation of potato tubers using foliar zinc-fertiliser. J Hortic Sci Biotechnol 87:123–129
White PJ, Thompson JA, Wright G, Rasmussen SK (2017) Biofortifying Scottish potatoes with zinc. Plant Soil 411:151–165. https://doi.org/10.1007/s11104-016-2903-4

Wolters AMA, Uitdewilligen JGAML, Kloosterman BA, Hutten RCB, Visser RGF, van Eck HJ (2010) Identification of alleles of carotenoid pathway genes important for zeaxanthin accumulation in potato tubers. Plant Mol Biol 73:659–671. https://doi.org/10.1007/s11103-010-9647-y
Zhang Y, Cheng S, De Jong D, Griffiths H, Halitschke R, De Jong W (2009a) The potato *R* locus codes for dihydroflavonol 4-reductase. Theor Appl Genet 119:931–937
Zhang Y, Jung CS, De Jong WS (2009b) Genetic analysis of pigmented tuber flesh in potato. Theor Appl Genet 119:143–150
Zitnak A, Johnston GR (1970) Glycoalkaloid content of B5141-6 potatoes. Am Potato J 47:256–260
Zorrilla C, Navarro F, Vega S, Bamberg J, Palta J (2014) Identification and selection for tuber calcium, internal quality and pitted scab in segregating 'Atlantic' × 'Superior' reciprocal tetraploid populations. Am J Potato Res 91:673–687

Chapter 4
Conventional and Molecular Breeding Approaches for Biofortification of Pearl Millet

M. Govindaraj, O. P. Yadav, R. K. Srivastava, and S. K. Gupta

4.1 Introduction

Pearl millet [*Pennisetum glaucum* (L.) R. Br.] is an essential diet of more than 90 million people in the semi-arid tropics of the world where droughts and low fertility of soils cause frequent failures of other crops. It is an important nutri-rich grain cereal in the drier regions of the world grown on 26 mha by millions of farmers (IFAD 1999; Yadav and Rai 2013). This makes pearl millet the sixth most important crop in the world and fourth most important food crop of the India, next to rice, wheat, and maize with annual cultivation over an area of ~8 mha. Pearl millet is also primary food crop in sub-Saharan Africa and is grown on 15 mha (Yadav and Rai 2013). The significant increase in productivity of pearl millet in India is attributed to development and adoption of hybrids of early to medium duration maturity. More than 120 diverse hybrids/varieties have been released till date for various production environments. The heterosis breeding and improved crop management technologies increased productivity substantially achieving higher increased production of 9.80 mt in 2016–2017 from 2.60 mt in 1950–1951 in spite of declined of area under the crop by 20–30% over last two decades (Yadav et al. 2012).

Over 50% pearl millet grain production in Asia is utilized for food purpose and the 20% is used for feed, while 100% grain is used as food in west and central Africa. The per capita consumption of pearl millet in India is highest among rural population in the western Rajasthan and Gujarat, contributing to more than 50% of cereal consumption in these regions (Parthasarathy Rao et al. 2006). Pearl millet

M. Govindaraj · R. K. Srivastava · S. K. Gupta
International Crops Research Institute for the Semi-Arid Tropics (ICRISAT), Patancheru, Telanagana, India
e-mail: m.govindaraj@cgiar.org; r.k.srivastava@cgiar.org; s.gupta@cgiar.org

O. P. Yadav (✉)
Central Arid Zone Research Institute, Jodhpur, Rajasthan, India
e-mail: director.cazri@icar.gov.in

A. M. I. Qureshi et al. (eds.), *Quality Breeding in Field Crops*,
https://doi.org/10.1007/978-3-030-04609-5_4

grain is also consumed in other parts of Maharashtra and Haryana (Basavaraj et al. 2010). Likewise, the domestic consumption of pearl millet is rising steadily in Africa (Ajetomobi 2008).

Pearl millet is a rich source of energy (361 Kcal 100 g^{-1}) comparable with other cereals such as wheat (346 Kcal 100 g^{-1}), rice (345 Kcal 100 g^{-1}), maize (342 Kcal 100 g^{-1}), and sorghum (349 Kcal 100 g^{-1}) (NIN 2003). The carbohydrates in pearl millet (67.5 g 100 g^{-1}) are lower than in wheat, rice, and sorghum, but higher than in maize, while germ portion of pearl millet is larger than sorghum (NIN 2003). These differences explain pearl millet having lower starch and higher protein content. Pearl millet has high fiber content (1.2 g 100 g^{-1}), lowest glycemic index (55) among cereals (Mani et al. 1993), and has relatively higher methionine and phytochemicals and micronutrients (Mal et al. 2010; Singh et al. 2012). Pearl millet is also rich in calcium, potassium, magnesium, iron, zinc, manganese, riboflavin, thiamine, niacin, lysine, and tryptophan.

Shrinking of food basket to a few fine cereals like wheat and rice largely due to subsidized government price and distribution policies for these two cereals has contributed to inadequate intake of essential micronutrients such as iron (Fe) and Zinc (Zn). Fe deficiency affects more than 30% of the population globally, and highest prevalence of anemia is reported among preschool-age children (47%) and pregnant women (42%) (WHO 2008). An estimated 20% of the global population is at risk of inadequate Zn intake (Wessells and Brown 2012). Thus, deficiency of Fe and Zn is most prevalent worldwide. Although government-supported program in India showed marginal reduction in malnutrition over the decades, the progress is very slow as National Family Health Survey revealed unacceptably high prevalence of anemia (>55%), under-weight (35%), and stunting (38%) among children under 5 years (NFHS 2016). The intake of Fe and Zn appears to be below the recommended dietary allowance for an average Indian adult particularly in the low-income rural households including the pearl millet-consuming regions (ICMR 2002; Parthasarathy Rao et al. 2006). Interestingly, pearl millet serves as a significant source of dietary energy and contributes to 19–63% of the Fe and 16–56% of the Zn intake from all food sources to a vast population in parts of the major pearl millet growing states of India (Parthasarathy Rao et al. 2006).

Addressing the micronutrient malnutrition through supplementations and food fortification has been initiated but has not been found as a sustainable approach in developing countries due to poor purchasing power of the consumers and unsatisfactory delivery infrastructure, especially in the rural areas. Therefore, diversified food uses and biofortified crops provide cost-effective and sustainable options to reduce micronutrient malnutrition in such areas. Dietary diversity is a qualitative measure of food consumption that reflects household access to a variety of foods. However, getting people to eat more nutrient-rich fewer staples is very challenging, and affordability is constrained. Biofortification is the process of increasing the content and bioavailability of essential vitamins and minerals in staple crops, through plant breeding to improve nutritional status. This approach contributes to improving the diet quality of populations, and can be viewed as integral part of dietary diversity. Biofortification program has initiated the development and dissemination of improved crop cultivars with elevated levels of many micronutrients in several crops including pearl millet (Yadava et al. 2017). Genetic enhancement in pearl millet for increased micronutrients has focused grain Fe and Zn using conventional and molecular breeding approaches and the progress achieved is reviewed here.

4.2 Genetic Enhancement of Grain Quality Traits

The aim of core breeding has long been to increase yield potential of cultivars and has largely been accomplished by increasing grain yield through heterosis and building of resistance genes for various diseases and pests in cross-pollinated crops. Recent addition of improving grain nutritional traits is assumed to be newer area for breeders. Conventional breeding methods in combination with advanced phenotyping and biotechnological approaches enable desirable changes to improve the micronutrient content of new cultivars. The available natural genetic variation for essential nutrient content should permit breeding programs to improve the levels of minerals and vitamins in crops (Cakmak 2008; Monasterio and Graham 2000). In pearl millet, a major initiative toward the development of high-iron pearl millet cultivars has been taken involving the International Crops Research Institute for the Semi-Arid Tropics (ICRISAT) and National Agricultural Research System (NARS) partners from the public and private sector. Such efforts can only be successful and sustainable when biofortified cultivars maintain high yield productivity along with higher nutrient contents offering a benefit to both grower and consumer. Interestingly, the micronutrient traits in pearl millet (like in other cereals) are relatively more stable than grain yield and its components (Satyavathi et al. 2015; Kanatti et al. 2014a).

The extent of genetic variation for grain Fe and Zn contents in germplasm collection, identification of seed-mineral dense germplasm, nature of genotype × environment interaction, relationships between grain minerals and agronomic traits and genetic control of micronutrients would determine breeding efficiency for developing grain mineral dense cultivars. Therefore, a detailed insight is provided here to assess the progress made so far in these areas.

4.2.1 Targeted Micronutrients and Extent of Genetic Variation

Although pearl millet grain is rich in minerals and proteins, the severity of Fe and Zn deficiency and its associated health consequences show greater importance of these two micronutrients than others. The natural variation for grain Fe and Zn contents has been extensively studied in pearl millet (Table 4.1). In general, the spread of variation for grain Fe content was larger than the grain Zn content in various types of genetic materials that includes breeding progenies, populations, and cultivars. For example, too ambitious variation was observed in some studies: the grain

Table 4.1 Iron (Fe) and Zinc (Zn) content in high-Fe seed parents progenies (data are mean of 6 seasons at Patancheru)

Advanced breeding line	XRF Fe content (mg kg^{-1})		XRF Zn content (mg kg^{-1})	
	Mean	Range	Mean	Range
(EEBC S1-407-1-B-B-B-B-B-1-B-1-B-10-1 × B-bulk (3981-3989 G))-2-4-1	82	57–101	51	35–67
(EEBC S1-407-1-B-B-B-B-B-1-B-1-B-13-1 × B-bulk (3981-3989/S06 G1))-1-2-3	81	68–104	50	29–65
(EEBC S1-407-1-B-B-B-B-B-1-B-1-B-5-1 × B-bulk (3981-3989/S06 G1))-2-1-3	88	61–113	55	31–81
(ICMB 04888 × ICMB 02333)-1-1-3-2	77	68–88	49	33–64
(ICMB 95111 × EEBC S1-407-1-B-B)-17-3-1-B-B-B-B-4-B × 3981-4011 G2}-1-4-2	90	76–104	57	40–72
(ICMB 99555 × ICMB 99111)-2-1-1-B-B-B-1	81	55–101	46	33–57
(NC D2 BC7F4-34-3-1-2-B-2-B × EEBC 407)-12-1-2	86	54–97	56	36–75
(NC D2 BC7F4-34-3-1-2-B-2-B × EEBC 407)-4-2-2-2	86	62–102	52	35–66
{[(843B × ICTP 8202-161-5)-20-3-B-B-3 × B-bulk]-2-B-9 × [(ICMB 96555 × LaGrap C2 S1-32-1)-10 × IP 14758-2-1]-8-2}-1-1-1-2	86	80–92	46	34–53
{[(BESCBPT/91-40 × SPF3/S91-3)-1-2-2-3 × B-bulk]-8-1-1-3-B-B-B-B-3-1 × B-bulk (3981-4011/S06G1)}-1-3-2	104	79–123	57	38–82
AIMP 92901 S1-15-1-2-3-B-3-B-9-2-1	92	71–119	53	30–78
AIMP 92901 S1-296-2-1-1-4-2-B-7-3-1	102	90–118	57	42–72
HHVBC Tall S1-51-1-P1-3-B	100	77–129	53	38–62
ICMR 312 S1-59-1	97	83–118	60	46–77
ICMV 221 S1-366	86	80–92	58	42–73
ICMV 96490-S1-15-1-2-2-1-2	115	95–138	62	41–77
ICTP 8203 S1-386	98	84–124	61	41–83

Fe and Zn contents ranged from 40 to 580 mg kg^{-1} and from 10 to 66 mg kg^{-1}, respectively (Jambunathan and Subramanian 1988). But looking at the size of experimental materials (test entries) that were studied in these studies, varied and only few studies had adequate number of test entries to investigate the variation for grain Fe and Zn contents.

Velu et al. (2007) reported large variability for grain Fe content (30.1–75.7 mg kg^{-1}) and Zn content (24.5–64.8 mg kg^{-1}) in pearl millet breeding lines. Rai et al. (2012) also reported large variation for both Fe and Zn contents: Fe ranging from 18 to 97 mg kg^{-1} and Zn varying from 22 to 69 mg kg^{-1} in the advance breeding lines; and Fe ranging from 52 to 135 mg kg^{-1} and Zn ranging from 40 to 92 mg kg^{-1} in the population progenies. Similarly, two to threefold variation among germplasm collections was reported for both Fe (51–121 mg kg^{-1}) and Zn (46–87 mg kg^{-1}) contents (Rai et al. 2014). Interestingly, most of the high Fe and Zn accessions were from Togo and Ghana that had Fe content of 95–121 mg $kg^{-1}$and Zn content of 59–87 mg kg^{-1} indicating *iniadi* germplasm as a valuable germplasm resource for genetic improvement of Fe and Zn contents in pearl millet.

The magnitude of variation for these micronutrients among the released and commercial cultivars (18 OPVs and 122 hybrids) in India was studied (Rai et al. 2016). OPVs had a Fe range of 42–67 mg kg^{-1}, and Zn range of 37–52 mg kg^{-1} with ICTP 8203 having the highest Fe content (67 mg kg^{-1}) followed by ICMV 221 (61 mg kg^{-1}) and AIMP 92901 (56 mg kg^{-1}). ICTP 8203 had highest level of Zn content (52 mg kg^{-1}), followed by ICMV 221 and AIMP 92901 (45–46 mg kg^{-1}), whereas Fe content in hybrids varied from 46 to 56 mg kg^{-1} and Zn content from 37 to 44 mg kg^{-1}. Four high Fe and Zn hybrids were identified as Ajeet 38, Proagro XL 51, PAC 903, and 86M86 with 55–56 mg kg^{-1} Fe content and with 39–41 mg kg^{-1} Zn content. These high Fe and Zn cultivars can be readily utilized for expanded cultivation and can also be proposed to be included in development programs.

4.2.2 Micronutrient Phenotyping Protocols

Since pearl millet is a highly cross-pollinated crop, three types of seed samples (selfed, sibbed, and open-pollinated seeds) can be used for the micronutrient analysis. However, open-pollinated seed sampling is the best choice in terms of cost-effectiveness and reliable estimation, provided Al contents of samples are monitored for possible dust contamination (Rai et al. 2015a). The availability of low cost and quick throughput analytical methods for micronutrient screening is a prerequisite for successful biofortification breeding. Although variety of instrumental techniques have been used for plant mineral determination so far, breeders presently rely heavily on Atomic Absorption Spectrometry (AAS) and Inductively Coupled Plasma Optical Emission Spectrometry (ICP). Almost all the analytical laboratories of the National Agricultural Research System (NARS) in India are using AAS and very few use ICP. When a large number of samples are to be screened for a given micronutrient, a simple, rapid, and cost-efficient method can surely save the time and resources, and can increase the breeding efficiency to enhance genetic gain for that trait. Recently, Energy-Dispersive X-ray Fluorescence (XRF) has been used for plant sample analysis. XRF is a relatively non-destructive method for grain Fe and Zn contents estimation and has now been validated for pearl millet (Paltridge et al. 2012). Setting up this table-top machine requires little recurring expenditure and provides non-destructive analysis of 300 samples per day at the cost of <USD 2.0 per sample compared to the ICP method (>18 USD/sample) which takes a month time (Rai et al. 2012). High-throughput XRF facility was established in 2010 at ICRISAT, Patancheru, which enables handling a large number of breeding lines at ICRISAT and its partners' center (15,000–20,000 grain samples per year). Efficiency of XRF over ICP for high-throughput Fe and Zn estimation in pearl millet grain was demonstrated with large samples from several trials (Govindaraj et al. 2016a, b). This study showed that highly significant and positive correlations between ICP and XRF (r = >0.80**; $p < 0.01$) for both micronutrients provide the reliable screening technique and breeders can rapidly discard low Fe/Zn genotypes while generation advancement.

4.2.3 Genetics of Grain Iron and Zinc Contents

Understanding the nature of gene action and inheritance patterns of grain micronutrient is crucial for breeders to develop effective biofortification breeding strategies. Several studies in pearl millet using different mating designs showed that the inheritance of grain Fe and Zn contents is largely attributed to additive genetic variance with higher magnitude of heritability, explaining the simple inheritance pattern and simple selection for these micronutrients to be effective (Velu 2006; Arulselvi et al. 2007; Gupta et al. 2009; Govindaraj et al. 2016a, b). In general, variability among the hybrids attributable to general combining ability (σ^2GCA) was 3–4 times greater than the variability attributable to specific combining ability (σ^2SCA) for Fe and Zn contents. This proposition of σ^2GCA over σ^2SCA, in turn, contributes to greater predictability ratio which was always closer to unity for both micronutrients. This indicated that the GCA effect for both Fe and Zn contents were predominantly under additive genetic control in pearl millet (Velu et al. 2011; Rai et al. 2012; Govindaraj et al. 2013; Kanatti et al. 2014a). Highly significant and positive correlations between hybrid performance per se and mid-parental values provided further support for these micronutrients being largely under additive genetic control. In contrary, another study reported predominance of non-additive genetic variance for these micronutrients (Arulselvi et al. 2006).

Differences in reciprocal crosses is a widely used method for estimating maternal effects in trait inheritance. The differences between the direct crosses and reciprocal crosses were non-significant both for the Fe and Zn contents both in genotypes with high- and low-content genetic backgrounds (Kanatti et al. 2018). This indicated that genetics of both the micronutrients are controlled by nuclear determinants of male and female parents which showed the relatively greater importance of both nuclear than cytoplasmic contribution. Further, genetic studies revealed the high grain Fe and Zn parents had positive and significant GCA effects, while parents with low grain Fe and Zn had significant negative GCA effects (Govindaraj et al. 2013; Kanatti et al. 2014a). This pattern of genetic control suggested that the selection for higher grain micronutrients should be commenced in earlier generation while agronomic superiority can be selected in later generations. Interestingly, unlike yield traits, inbreeding has no adverse effect on micronutrient content in pearl millet (Rai et al. 2017).

4.2.4 Conventional Breeding

4.2.4.1 Source of Higher Fe and Zn Contents

Evaluations have been undertaken to identify germplasm sources for high Fe and Zn grain contents. Seed parent and restorer lines of hybrids and advanced breeding lines developed from a diverse range of germplasm have been screened to identify existing sources for high Fe/Zn in elite agronomic backgrounds. Consequently, large variability was observed for both micronutrients among seed parents progenies and ten lines have been identified having very high Fe and Zn contents (Table 4.1). Except for two progenies that involved a NCD2 (Nigerian Composite Dwarf) progeny as one of the parents in the cross, all other progenies were derived from crosses that had both parents developed from *iniadi* germplasm, with a progeny from Extra-early B-composite (EEBC) involved in most of the crosses. These identified progenies with such high levels of Fe and Zn contents would serve as ready-to-use donor source for B × B crosses to develop counterpart A-line. Similarly, selected sources from restorer progenies for high-Fe and Zn are given in Table 4.2. Previous evaluation studies in pearl millet have also shown that breeding lines, hybrid parents, and improved populations having high Fe and Zn contents were often based largely on *iniadi* germplasm (Velu et al. 2011; Govindaraj et al. 2013; Kanatti et al. 2014a; Rai et al. 2015b). *Iniadi* refers to an early-maturing and large-seeded landrace found in the adjoining parts of Togo, Ghana, Benin, and Burkina Faso and such source also known by various local names such as *nara*, *nata*, *ignati*, *ignate*, *ignie*, *misse*, and *likoun* (Andrews and Kumar 1996).

The positive and significant correlation between per se performance of the parents for micronutrients and their GCA effects indicates that per se performance of the parents, in general, is a good indicator of hybrid performance (Rai et al. 2012).

Table 4.2 Iron (Fe) and Zinc (Zn) content in high-Fe restorer parents progenies (data are mean of 4–8 seasons at Patancheru)

	XRF Fe content (mg kg^{-1})		XRF Zn content (mg kg^{-1})	
Advanced breeding line	Mean	Range	Mean	Range
AIMP 92901 S1-15-1-2-3-B-3-B-9-2-1	92	71–119	53	30–78
AIMP 92901 S1-296-2-1-1-4-2-B-7-3-1	102	90–118	57	42–72
HHVBC Tall S1-51-1-P1-3-B	100	77–129	53	38–62
ICMR 312 S1-59-1	97	83–118	60	46–77
ICMV 221 S1-366	86	80–92	58	42–73
ICMV 96490-S1-15-1-2-2-1-2	115	95–138	62	41–77
ICTP 8203 S1-386	98	84–124	61	41–83
LaGrap C2-S1-14-4-1-3-4-4	89	72–112	57	43–70
MRC HS-130-2-2-1-B-B-3-B-B-B-1-3-1	110	93–128	64	42–76
SDMV 90031-S1-11-1-1-3-3-B-4-B-2-1-B	87	64–107	57	38–79
(EERC-HS-8)-B-2-1-2-1	100	85–119	48	28–78
(MC 94 C2-S1-3-1-3-3-1-2-1 × ICMR 312 S1-3-2-3-2-1-1-B-B)-B-46-P1-1	92	87–96	51	36–69
(MC 94 C2-S1-3-1-3-3-1-2-1 × SDMV 90031 S1-3-3-2-2-2-2-2)-B-8-2-1	76	65–87	62	44–74
(MC 94 C2-S1-3-2-2-2-1-3-B-B × AIMP 92901 S1-488-2-1-1-4-B-B)-B-30-1-3	79	66–105	54	41–67
[(IPC 1617 × SDMV 90031-S1-84-1-1-1-1) × AIMP 92901 S1-296-2-1-1-3-B-1]-4-4-2-1	90	69–121	59	36–79

Using both of these parameters, 863B, ICMB 98222, ICMB 99222, ICMB 02333, ICMB 04999, IPC 1650, IPC 843, IPC 774, IPC 1178, IPC 689, and IPC 735 were identified as moderate-to-high Fe lines and the best general combiners for use in hybrid breeding. The above results suggest that population progenies with higher levels (>70 mg kg^{-1}) are available as donor sources for further genetic enhancement of these micronutrients, and lines with high Fe and Zn can be identified in breeding material with elite genetic backgrounds for their direct use in hybrid parents development.

4.2.4.2 Genotype by Environment Interaction

Interaction between genetic and environmental factors (G × E interactions) affects expression of any quantitative trait. Early breeding efforts in biofortification were hindered by gaps regarding appropriate methods for micronutrient traits assessment and the effects of variable environmental factors on biofortified traits. This was primarily due to perception that biofortified traits are qualitative rather than quantitative in nature. There are now evidences that there are significant G × E interactions in expression of biofortified traits (Reynolds et al. 2005; Govindaraj et al. 2013; Kanatti et al. 2014a). Recent studies have shown significant role of environment and genotype × environment (G × E) interaction in determining the levels of grain Fe and Zn contents in pearl millet (Rai et al. 2016). While G × E interactions for Fe and Zn appear to play an important role, the genetic variance contribution was twice than that due to G × E for these micronutrients (Govindaraj et al. 2016b). Most studies in pearl millet showed that G × E interactions accounted for 10–30% of the variation for Fe and Zn contents and it is possible to identify the genotypes with high and stable mineral content across environments.

Complexity of soil micronutrient status may partly contribute to environmental interaction for expression of these traits. Analyzing soil and grain samples from the target environments explains the underlying factors of G × E interactions and the magnitude of micronutrient trait expression. A large number of multi-location evaluations under biofortification program at ICRISAT indicated that all the locations had sufficient levels of Fe and Zn and other important minerals and did not establish relationship of micronutrient content in the grain with the soil available micronutrient status (Govindaraj et al. 2013; Kanatti et al. 2014a). Differences in soil Fe and Zn contents between contrast environment (rainy and summer) were also not reflected in the grain Fe and Zn contents (Gupta et al. 2009) indicating that soil micronutrient status above critical limits has no influence on grain mineral contents. In spite of G × E challenges, there is a growing evidence that breeding for increased levels of micronutrient across environment is feasible with high yield in pearl millet because of positive correlation between Fe and Zn contents and reported higher heritability of these micronutrients than grain yield (Govindaraj 2011; Govindaraj et al. 2016b).

4.2.4.3 Trait Association

Biofortification aims to address nutrient deficiencies as an integral part of core breeding program, but there is need to understand the potential impact of higher micronutrient contents on other important traits. For instance, selection for increased micronutrient content may be expected to negatively affect yield or other important agronomic and end-use characters. This happens if genes that increase micronutrient content are linked with genes that have a deleterious effect on other desired traits, or it could occur as a consequence of negative trait associations. Association between grain Fe and Zn has largely been studied in pearl millet and other crops and highly positive and significant correlations between Fe and Zn have been revealed (Gregorio et al. 2000; Ozkan et al. 2007; Velu et al. 2011). The correlations between Fe and Zn contents in pearl millet varied from 0.43 to 0.97 (Velu et al. 2011; Rai et al. 2012; Govindaraj et al. 2013 and Kanatti et al. 2014b). Such high positive correlations among micronutrients indicate that improvement in Fe content may simultaneously improve the Zn content owing to similar transport and chelation process affecting the accumulation of both Fe and Zn contents in pearl millet seeds. Studies also reported significant positive association of Fe and Zn contents with grain weight in pearl millet (Velu et al. 2007, 2008a, b; Kanatti et al. 2014b), while other studies in pearl millet observed non-significant association grain Fe and Zn contents with 1000-grain weight (Gupta et al. 2009 and Rai et al. 2012; Yadav et al. 2016). Thus, the genetic enhancement of these micronutrients is possible without compromising on grain size.

The Fe and Zn contents had negative and mostly non-significant correlation with grain yield in pearl millet (Rai et al. 2016; Kanatti et al. 2014b; Yadav et al. 2016). In those studies where correlation was negative, it was weak enough in the magnitude, indicating that if these were the results of adverse genetic associations, high-yielding hybrids with high Fe and Zn contents can be bred by making selection for these traits in larger segregating populations and progenies as compared to those used for breeding for grain yield alone. These weak negative relationships resulted from dilution effects when dealing with selfed seeds where grain yield was reduced and micronutrients were overestimated (Govindaraj et al. 2012). On the other side, this trend could be of unidirectional selections as most correlations so far reported are in those lines/cultivars that were bred exclusively for yield (as target trait). Hence, further research involving random sets of lines derived from random-mated populations constituted from crosses between high-Fe/Zn and low-Fe/Zn lines but high yielding is required to examine the magnitude and direction of association of these micronutrients with grain yield. Commercial hybrids (86M86, XL51, and Ajeet 38) bred for higher yield and widely cultivated in India have higher yield and higher Fe content (>50 mg kg^{-1} Fe) among released cultivars shows the possibility of combining grain yield and micronutrient in cultivars (Rai et al. 2016).

4.2.4.4 Population Improvement

Pearl millet is a highly cross-pollinated crop with 70–80% outcrossing and development of open-pollinated varieties (OPVs) as commercial cultivar is an option. Both Fe and Zn contents being largely under additive genetic control, inter and intra-population improvement is highly effective. Although individual plant selection is not very effective for grain yield, S_1 progenies selection is an effective population improvement method for grain yield in pearl millet. A study confirmed the efficiency of single plant selection for Fe and Zn contents in four diverse OPVs (ICTP 8203, IBV 3, AIMP 92901, and ICMR 312) (Govindaraj et al. 2012). Selfed grains produced from S_0 plants and S_1 progenies were assessed for Fe and Zn content and correlation between the S_0 plants and the mean of S_1 progenies across environments was positive and highly significant in all four populations, both for Fe (r = 0.58–0.75) and Zn content (r = 0.61–0.73). Therefore, individual plant progeny selection is effective for both Fe and Zn contents for intra-population improvement as followed for grain yield improvement. For inter- and intra-population improvement, a study revealed that one cycle of selective random matting had improved grain Fe and Zn in C_1 over C_0 bulks with an increase of 8% (Fe and Zn)) in AIMP 92901 and ICMR 312 (Govindaraj 2011). Interestingly, such selection for high Fe and Zn significantly increased 1000-grain mass by 5–14% in these two populations and had no adverse effect on grain yield. Similarly, ICTP8203-10-2, a population developed by recombining 11S3 progenies, had 71 mg kg^{-1} Fe content (9% higher than original) and 2.2 t ha^{-1} grain yield (11% higher than original). Based on national testing, this population was released as 'Dhanashakti' and is the first biofortified crop cultivar for Fe in public domain in India and few other high-Fe OPVs are under development at ICRISAT with much higher Fe and Zn contents (Rai et al. 2014).

4.2.4.5 Hybrid Breeding

The higher level of outcrossing and heterosis supported with availability of commercially viable cytoplasmic-nuclear male sterility system, hybrids breeding has been very effective in increasing pearl millet productivity in India. Heterosis, defined as the superiority in performance of hybrids over its parents (mostly higher parent), is largely explained either due to dominance or over-dominance effects. There is no better-parent heterosis for Fe and Zn reported so far in pearl millet since predominance of additive gene action in the genetic control of these traits, which indicates that there would be little opportunity to exploit better-parent heterosis for improving these micronutrients. However, development of hybrids with high Fe and Zn contents highly require incorporation of high Fe and Zn genes into both parental lines of hybrids where the mid-parent heterosis of a hybrid is gradually increased. Therefore, to breed high iron/zinc hybrids, all potential parental lines should be characterized for these micronutrients and only selected lines should be hybridized.

Unlike grain yield, performance per se of lines is significantly and positively correlated with general combining ability for Fe and Zn in pearl millet, implying the lines selected for high Fe and Zn will also be high general combiners for these micronutrients (Velu et al. 2011; Govindaraj et al. 2013; Kanatti et al. 2014a, 2016). Development of inbred lines with high Fe and Zn depends on the level of and variability for these micronutrients in the base population (whether F_2s or OPVs or composites), and on the magnitude, direction, and pattern of inbreeding effects. It has been observed that inbreeding had either no significant effect or had marginally increased both micronutrients (Rai et al. 2017). In contrast to the low heritability and inbreeding depression of grain yield, micronutrient contents are highly heritable and hybrids can be readily improved through hybrid parents breeding. So far, the best source of high Fe and Zn contents in pearl millet is found to be *iniadi* germplasm (Velu et al. 2011; Rai et al. 2012; Govindaraj et al. 2013). Considering the additive gene action and one source of germplasm genes introgressed in both parental lines, it is expected to reduce genetic diversity between male and female groups for other important traits. This will also lead to reduced heterosis for yield traits, which are predominantly under non-additive gene control. Thus, genomics approaches for selective introgression of genes for Fe and Zn contents in the parental lines without disrupting the diversity for other traits can play a major role in future biofortification breeding. New sources, other than *iniadi*, of Fe and Zn contents in the germplasm collections are also being explored at ICRISAT for genetic diversification for high Fe and Zn. ICRISAT and NARS have developed several biofortified seed and restorer parents with elite agronomic backgrounds in diverse cytoplasmic systems. The Fe content in these seed parents is 69–110 and 42–55 mg kg^{-1} Zn content while restorer had 74–110 mg kg^{-1} Fe and 41–62 mg kg^{-1} Zn (Table 4.3).

So for biofortification breeding at ICRISAT and NARS is intensively supported by HarvestPlus Challenge Program of the CGIAR which set Fe targets for pearl millet as 77 mg kg^{-1} with increment of 30 mg kg^{-1} over first pearl millet variety (WC-C75). However, a recent study reported baseline for Fe content is 42 mg kg^{-1} among hybrids and thus target would be 72 mg kg^{-1} (Rai et al. 2016). Besides using parental lines with high Fe, hybrids being developed with these targets are being tested at national level. A special hybrid trial at national level is being conducted to encourage mainstreaming of this trait in public and private sector breeding programs. In addition, many more hybrids have been identified (Rai et al. 2016) and are in pipeline for testing. It is important to note that higher adoption of biofortified pearl millet hybrids/varieties in a long run largely depend on higher Fe/Zn contents coupled with high yield, downy mildew resistance and drought tolerance.

4.2.4.6 Improved Cultivars

Identification of appropriate germplasm and populations with highest Fe and Zn contents is very important for demonstrating the biofortification breeding. Use of such materials would continue until parental lines with higher Fe and Zn contents

Table 4.3 High-iron seed and restorer parents developed at ICRISAT (data are mean of four seasons)

Line	50% flowering (days)	XRF Fe content (mg kg^{-1})	XRF Zn content (mg kg^{-1})	1000-grain weight (g)	CMS
Seed parents					
ICMA/B 1501	39	76	42	13.2	A4
ICMA/B 1502	43	92	50	13.6	A1
ICMA/B 1503	43	69	43	15.0	A4
ICMA/B 1504	47	97	55	15.5	A1
ICMA/B 1505 (15222)	41	110	55	15.5	A1
ICMA/B 1506 (15444)	45	96	53	9.9	A4
ICMA/B 1507	43	92	50	10.1	A4
ICMA/B 1508	53	73	44	15.0	A1
Restorer parents					
ICMR 1201	48	79	41	10.5	A1
ICMR 1202	50	89	47	14.2	A1
ICMR 1203	52	101	58	7.9	A4
ICMR 1301	55	91	52	12.7	A1
ICMR 1501	55	86	42	9.9	A1
ICMR 1502	51	110	62	12.4	A1
ICMR 1503	51	99	47	13.7	A4
ICMR 1504	57	96	51	8.8	A1
ICMR 1505	55	74	41	6.9	A4

are developed through targeted breeding for these micronutrients in high-yielding backgrounds. With partnership with NARS, improved version of ICTP8203 was released as Dhanashakti for all India level in 2014 (Rai et al. 2014), and has now been adopted by 60,000 ha. Dhanashakti has been accepted by farmers not only for Fe and Zn content, but also for higher grain yield, earliness and bold seed (Table 4.4). ICMV 221, another popular cultivar, has now been further improved for Fe content and is under testing. The improved version of ICMV 221 has 70 mg kg^{-1} Fe content (11% higher than ICMV 221), 58 mg kg^{-1} Zn content (9% higher than ICMV 221), and grain yield 4.2 t ha^{-1} (5% higher than ICMV 221) (Govindaraj and Rai 2016).

Several A/B pairs with high Fe content (range 65–77 mg kg^{-1}) have been identified with ICMA 98222 and ICMA 99222 as the best general combiner for high-Fe hybrid breeding. By exploiting these lines with advanced high-Fe breeding lines as potential restorers, several high-Fe hybrids have been developed with good yield potential. Based on multi-location and multi-year testing, two hybrids, viz., ICMH 1201 and ICMH 1301 have been identified for commercialization. ICMH 1201 had 75 mg kg^{-1} Fe content and 3.6 t ha^{-1} grain yields. ICMH 1201 flowered only 3 days later than ICTP 8203, so it fits in the early-maturity group and production system. Performance of ICMH 1301 showed 72 mg kg^{-1} Fe content and 3.6 t ha^{-1} grain yield.

Table 4.4 Performance of biofortified pearl millet cultivars for Fe and Zn content, grain yield, and time to 50% flower across environments

	XRF Fe content (mg kg^{-1})		XRF Zn content (mg kg^{-1})		Grain yield (t ha^{-1})		50% flowering (days)	
Cultivar	Mean	Range	Mean	Range	Mean	Range	Mean	Range
Dhanashakti	71	40–106	40	18–71	2.20	0.65–5.65	45	40–55
ICTP 8203[a]	65	31–94	40	19-71	1.97	0.63–4.30	45	38–56
ICMV 221 Fe 11-2	70	62–85	58	53–63	4.24	3.38–5.40	43	40–47
ICMV 221[a]	63	59–69	53	51-58	3.98	3.12–4.81	39	38–41
ICMH 1201	75	47–102	39	18–69	3.58	1.49–6.20	48	41–56
86M86[a]	56	39–71	37	18–61	4.37	2.60–6.47	54	47–64
ICTP 8203[a]	71	50–102	43	19-77	2.58	1.33–5.06	45	38–55
ICMH 1301	77	31–107	42	24–64	3.26	1.32–7.03	52	42–62
86M86[a]	58	39-85	37	18-56	4.10	2.60–6.95	54	44–64
ICTP 8203[a]	75	40–117	44	19-77	2.46	1.28–6.36	45	36–53

[a]Check entry

These hybrids are cultivated by >30,000 farmers using truthfully labeled seed (TLS), mostly in Maharashtra and Rajasthan. Much greater progress in breeding high-Fe hybrids with high-grain yield is expected in the near future by utilizing A- and R-lines that are being developed through targeted breeding for high-Fe content.

4.2.5 Molecular Breeding

4.2.5.1 Availability of Molecular Markers

Pearl millet had been a crop of limited genomic resources in the past. However, since last decade, substantial progress has been made in generation of molecular markers. Like in other crops, PCR-based simple sequence repeats (SSR) markers are the extensively used markers for various genetic and mapping studies of pearl millet due to their abundance in the genome, highly polymorphic nature and easy assay. Transcriptome projects have brought enormous information on expressed sequence tags (ESTs), which were used as targets for the identification of SSR markers known as EST-SSR markers through the computational approach thereby making the SSR marker development rapid, easy, and inexpensive. These markers were subsequently used for various genetic studies in pearl millet. PCR-based screening of a bacterial artificial chromosome (BAC) library constructed using nuclear DNA from pearl millet using five sequence-tagged site (STS), led to the identification of 6 microsatellite markers (Allouis et al. 2001). Twenty-five SSRs were isolated from bacterial artificial clones (BA Clone) of pearl millet without any sub-cloning using 3′ anchored SSR primers and isolation of flanking sequences by suppression PCR (Qi et al. 2001), while screening of a small-insert partial genomic library with a (CT)15 oligonucleotide probe resulted in the development of 18 SSR markers (Budak et al. 2003). Similarly, 44 markers were developed from a (CA) n-enriched small-insert genomic library (Qi et al. 2004). Yadav et al. (2007) developed 32 SSR markers for pearl millet and assessed them for polymorphism in parental lines of 4–6 mapping populations (Yadav et al. 2008). A set of 58 SSR markers was developed in pearl millet and ESTs made available (Senthilvel et al. 2008). SNPs accounted for about two-thirds of the variation while InDels accounted for one-third of the variation. This demonstrated the use of syntenic information to develop SSCP-SNP markers (Bertin et al. 2005). Subsequently, Rajaram et al. (2013) developed 99 EST-SSR markers from transcriptomics work.

4.2.5.2 Linkage Maps

A genetic/linkage map depicts the arrangement of molecular markers. Linkage map is the basic framework in different linkage groups based on the recombination frequency among the markers, which are essential for the mapping of QTL. The first RFLP linkage map in pearl millet was published by Liu et al. (1994). A linkage map consisting of seven linkage groups and a total of 181 loci was made using an inter-varietal F2 population, with an average inter-marker distance of about 2 cM. Qi et al. (2004) developed an integrated consensus linkage map from the genetic maps constructed for four different crosses using 353 RFLP and 65 SSR markers. The mapping of 21 polymorphic SSR markers mapped using existing mapping populations (ICMB 841-P3 × 863B-P2 and 81B-P8 × IPC 804) revealed that most EST-SSR markers map to distal regions of linkage groups to cover the previous gaps (Senthilvel et al. 2008).

A linkage map was constructed using 258 DArT and 63 SSR marker data using a RIL population from cross H 77/833-2 and PRLT 2/89-33 (Supriya et al. 2011). With an objective of developing a linkage map with more evenly distributed markers and greater marker coverage of the gaps in earlier maps, Pedraza-Garcia et al. (2010) constructed a map using 196 PCR-based DNA markers (66 SRAPs, 63 RAPDs, 27 ISSRs, 31 pearl millet, six sorghum, and three maize SSRs) of nine linkage groups with an average genetic distance of 9.2 cM between markers. Rajaram et al. (2013) also constructed Linkage maps using 99 newly developed EST-SSR markers and previously mapped 17 EST-SSR, 53 genomic SSR, and two STS markers.

A consensus map of 174 loci (899 cM) was developed by integrating the individual linkage maps using MergeMap, which showed a well-conserved locus order for nearly all linkage groups. The linkage maps constructed using codominant SSRs and dominant DArTs span 1748.7 cM (ICMB 841 × 863B map consisting of 305 markers) (Kumar et al. 2016). Longest SSR-based skeleton linkage map for F_2-derived F_3 progenies with a length of 1018.7 cM accommodating only 44 well-distributed markers has been reported. Large map using SSRs with map length of 748 cM for an F_2 population was reported by Gulia (2004). Liu et al. (1994) reported by the shortest F_2-based map (287.7 cM) reported so far had 181 RFLP markers.

Yadav et al. (2004) incorporated 91 marker loci in a population of F_2 individuals, with a map length of 617.4 cM using (ICMB 841 × 863B) population. Senthilvel et al. (2008) have used the same F_2 population to obtain a map length of 677 cM with 112 loci. Vengadessan et al. (2013) constructed a linkage map, primarily based on SSCP-SNP markers, using 188 $F_{2:3}$ mapping population progenies produced from a cross between pearl millet inbred lines having diverse parentage.

The skeleton linkage map covered 1019 cM and it comprised of 44 markers distributed across the seven linkage groups. Average adjacent-marker intervals ranged from 14 cM on LG1 to 38 cM on LG6, with an overall mean of 23 cM. Rajaram et al. (2013) developed 99 new EST-SSR markers (IPES series) and constructed linkage maps of four F7 recombinant inbred populations (RIP) based on four crosses along with previously mapped EST-SSR (17), genomic SSR (53), and STS (2) markers. A total of 176 loci detected by 171 primer pairs were mapped among the four crosses. A consensus map of 174 loci (899 cM) detected by 169 primer pairs was constructed using Merge Map to integrate the individual linkage maps.

4.2.5.3 Quantitative Trait Loci (QTL)

Development of a mapping population is the most critical factor in the construction of a molecular map. Decisions on selection of parents and mating design for the development of mapping population and the type of markers used depend upon the objectives of experiments, availability of markers, and the molecular map. An appropriate mapping population, with suitable marker system and the data analyzing software are the key rations for molecular mapping and breeding. Genetic map construction requires: appropriate mapping population; pairwise recombination frequencies calculation on population; establishment of linkage groups; estimation of map distances; and determine map order. Size of mapping population is also essential factor in mapping as limited population sizes used in many QTL detection experiments may have led to underestimation of QTL number, overestimation of QTL effects, and failure to quantify QTL interactions (Beavis 1998; Melchinger et al. 1998; Utz et al. 2000). The size of the population may be determined by the gene effect to be detected as well as the type of population. While the analysis of large population would enable the detection of small-effect QTLs, the basic purpose of mapping would be served if one can detect the major QTL with large effect and this would require, in general, a mapping population of a size 200–300 individuals.

Mapping software packages, such as Mapmaker (Lander and Botstein 1989; Lander et al. 1987), Mapmanager (Manly and Elliott 1991), and Joinmap (Stam 1993) have been developed to analyze the genetic data for map construction. These software packages use genetic data of segregating mapping populations to estimate recombination frequency followed by determination of linear arrangement of genetic markers. Different types of mapping populations that are often used in linkage mapping are: F2 population; F2-derived F3 (F2:F3) populations; Backcrosses; Doubled haploids (DHs); Recombinant Inbred Lines (RILs); and

Near-isogenic Lines (NILs). The F2 mapping population can be developed with possible combinations of parental alleles (Lander et al. 1987). A recombinant inbred line (RIL) can be obtained from an F2 generation by successive self-pollinations using the single seed descent method (SSD) (Burr et al. 1988). The resulting inbred lines are highly homozygous and the segregation ratio for each locus tends to be 1:1 (AA:aa) representing an 'immortal' or permanent mapping family and can thus be used in experiments with replications in several environments allowing for more accurate estimates of genetic components and identification of QTL vs. environment interactions. Disadvantages of recombinant inbred lines are that at least six generations are required to obtain the line and the inability to estimate dominance effects of mapped quantitative trait loci (QTL) due to the absence of heterozygous genotypes.

In pearl millet, several F2:3 and F2:4 mapping population have been developed from diverse inbred lines of Asian, American, and African origin (Hash et al. 2002). Liu et al. (1994) developed the first F2 population of 133 individuals for the study of downy mildew. It was used by Devos et al. (2000) for comparative mapping of pearl millet with foxtail millet and rice. Further to understand the genetic control of domestication trait, Poncet et al. (2000) developed a population of 250 F2 individuals from cultivated and wild F1 hybrid (*P. glaucum* spp. *monodii*). Poncet et al. (2002) developed another F2 population having 168 individuals. Yadav et al. (2002, 2004) developed two mapping population by crosses of H77/833-2 × PRLT 2/89-33 (early maturing inbred line, 150 F2 individuals) and ICMB 841 × 863B agronomically elite inbred seed parent, 106 F6 individuals (Kumar et al. 2016).

The QTL analysis is based on association between trait value and marker allele. Number of studies have been reported for detecting QTLs with traits like downy mildew (Jones et al. 1995, 2002; Gulia 2004), rust and blast (Morgan et al. 1998), drought tolerance (Yadav et al. 2002, 2003, 2004; Bhattacharjee et al. 2002; Bidinger et al. 2007; Kholová et al. 2012), flowering time (Kumar et al. 2017; Saïdou et al. 2009), panicle length (Poncet et al. 2000; Kumar et al. 2017), and 1000-grain mass (Yadav et al. 2002; Bidinger et al. 2007; Kumar et al. 2017). Research on identifying QTLs and candidate genes for elevated levels of Fe and Zn in pearl millet is limited at this time (Manwaring et al. 2016). A recent study has identified the QTLs for higher Fe and Zn contents in ICMB 841 × 863B cross on LG3 (Kumar et al. 2016).

4.2.5.4 Marker-Assisted Selection (MAS)

Despite the fact conventional breeding approaches will continue to make valuable aids to the genetic improvement of important traits in pearl millet, the efficiency of such concerted efforts can be increased extensively through the supplementation of MAS approaches. A number of QTL and associated markers have been identified for downy mildew resistance (Hash and Witcombe 2001; Jones et al. 2002; Hash and Witcombe 2002). NILs of H 77/833-2 introgressed with various putative QTL regions from PRLT 2/89-33 were used for validation of major drought tolerance QTL in the LG2 target region (Serraj et al. 2005).

The validated QTL on LG3 for higher grain Fe and Zn contents (Kumar et al. 2016) has been the target for marker-assisted breeding (MAB). Using the linked flanking markers, this QTL along with downy mildew resistance QTLs has been moved into genetic background of pollen parent of hybrid HHB 67 Improved. These double QTL introgression lines were crossed with the seed parent of HHB 67 Improved to generate HHB 67 Improved like hybrids. These QTL introgression lines along with the improved test-cross hybrids are being tested in the national testing system in India.

4.3 Future Prospects

Considering pearl millet adaptation traits and productivity gains in the drylands over the decades, it would continue to be an important food crop for India and sub-Saharan Africa. It is an ideal native food crop to expand the Indian and global food basket to meet healthy food and nutritional demand of the growing population. National policy measures such as inclusion of pearl millet under public distribution system are essential, besides promotion of pearl millet in poultry/animal feed and breweries to increase incentives to growers for higher production. Creating public awareness about the nutritional values of pearl millet is urgently needed otherwise consumers are likely to prefer non-native crops for daily energy and nutritional requirements. Major area of future focus in biofortification of pearl millet should include:

- The enhanced nutrient contents of new cultivars have to be achieved without any trade off with higher productivity. This would translate into adding one additional trait in breeding program. Micronutrient traits screening is highly expensive while dealing with larger germplasm. Thus breeding for such additional traits may delay progress for productivity traits when resources are limited to breeders. Micronutrient traits are apparently not affected by genetic erosion and involve little maintenance breeding after the genes are incorporated into elite backgrounds. With availability of XRF tool, cost of biofortification breeding will decrease over time, and micronutrient content built into the gene pool will not affect future breeding for productivity traits. In order to achieve it, a higher investment in breeding would be required on a long-term basis.
- Genomic approaches should now be an integral part of breeding program particularly for nutritional traits to use diagnostic markers given that pearl millet genome has been sequenced now (Varshney et al. 2017). Genomic markers can be used to make the biofortification breeding more efficient through marker-assisted selection in near future. This would also help in improving high-yielding cultivars with low iron and zinc under wider cultivation through introgression of micronutrient genes and become essentially derived variety.

- Seed companies have well established network and dominate the pearl millet hybrid seed market in India. Since, hybrids occupy approximately 90% of the area under improved cultivars, first-biofortified variety (OPV) has limited potential to make a mega impact. To address this, public–private partnership (PPP) model needs to be strengthened by institutional policy of nutrition commitments and special price allocation for mineral-dense seeds with subsidized rates in the markets to promote biofortified cultivars.
- ICRISAT has played a key role in diversifying the hybrid parents and its contribution to achieving higher yield gain at farm level through the PPP model. Seed companies, those have research and development division, capture more than 80% of the pearl millet hybrid seed market in India. Thus, the sustainability of biofortified pearl millet will mainly depend on mainstreaming of biofortification with seed companies, state seed corporations, ICAR institutes, and state agricultural universities. Hence, cultivar product concept of partners requires considering micronutrient as a generic trait in their breeding program and this joint effort will address development of high yielding and micronutrient rich pearl millet cultivars.
- There are good prospects for large scale on-farm field and food product demonstrations through state agricultural universities, agricultural departments and Krishi Vigyan Kendras (KVKs); large-scale production and procurements of biofortified cultivar grains for Anghanwadi (childcare center); and integration of biofortified grains in mid-day meal scheme. Several governments sponsored programs such as National Food Security Mission and Integrated Child Development Program would provide window for PDS system model to address the iron and zinc deficiency.

References

Ajetomobi JO (2008) Total factor productivity of agricultural commodities in economic community of West African States (ECOWAS): 1961–2005. In: CSAE conference 2008. Economic development in Africa held 16th–18th Mar 2008 at St. Catherine's College, Oxford

Allouis S, Qi X, Lindup S, Gale MD, Devos KM (2001) Construction of a BAC library of pearl millet, Pennisetum glaucum. Theor Appl Genet 102:1200–1205

Andrews DJ, Kumar AK (1996) Use of the West African pearl millet landrace Iniadi in cultivar development. Plant Genet Resources Newsl 105:15–22

Arulselvi S, Mohanasundram K, Selvi B, Malarvizhi P (2006) Heterosis for grain yield components and grain quality characters in pearl millet. ISMN 47:36–38

Arulselvi S, Mohanasundaram K, Selvi B, Malarvizhi P (2007) Genetic variability studies and interrelationships among nutritional quality characters, phytate phosphorus and grain yield in the seeds of pearl millet (*Pennisetum glaucum* (L.) R. Br.). Indian J Genet Plant Breed 67:37–40

Basavaraj G, Parthasarathy Rao P, Bhagavatula S, Ahmed W (2010) Availability and utilization of pearl millet in India. J SAT Agric Res 8:1–6

Beavis WD (1998) QTL analyses: power, precision and accuracy. In: Paterson AH (ed) Molecular analysis of complex traits. CRC Press, Boca Raton, FL, pp 145–161

Bertin I, Zhu JH, Gale MD (2005) SSCP-SNP in pearl millet—a new marker system for comparative genetics. Theor Appl Genet 110:1467–1472

Bhattacharjee R, Bramel PJ, Hash CT, Kolesnikova-Allen M, Khaid IS (2002) Assessment of genetic diversity within and between pearl millet landraces. Theor Appl Genet 105:666–673

Bidinger FR, Nepolean T, Hash CT, Yadav RS, Howarth CJ (2007) Identification of QTLs grain yield of pearl millet [*Pennisetum glaucum* (L.) R. Br.] in environments with variable moisture during grain filling. Crop Sci 47:969–980

Budak H, Pedraza F, Cregan PB, Baenziger PS, Dweikat I (2003) Development and utilization of SSRs to estimate the degree of genetic relationships in a collection of pearl millet germplasm. Crop Sci 43:2284–2290

Burr B, Burr FA, Thompson KH, Albertson MC, Stuber CW (1988) Gene mapping with recombinant inbreds in maize. Genetics 118:519–526

Cakmak I (2008) Enrichment of cereal grains with zinc: agronomic or genetic biofortification? Plant Soil 302:1–17

Devos KM, Pittaway TS, Reynolds A, Gale MD (2000) Comparative mapping reveals a complex relationship between the pearl millet genome and those of foxtail millet and rice. Theor Appl Genet 100:190–198

Govindaraj M (2011) Genetics of grain iron and zinc concentration in pearl millet (*Pennisetum glaucum* (L.) R. Br.). Ph.D. thesis, Tamil Nadu Agricultural University

Govindaraj M, Rai KN (2016) Breeding biofortified pearl millet cultivars with high iron density. Indian Farm 65:53–55

Govindaraj M, Rai KN, Shanmugasundaram P, Rao AS (2012) Efficiency of single plant selection for grain iron and zinc density in pearl millet. Eur J Plant Sci Biotechnol 6:114–117

Govindaraj M, Rai KN, Shanmugasundaram P, Dwivedi SL, Sahrawat KL, Muthaiah AR, Rao AS (2013) Combining ability and heterosis for grain iron and zinc densities in pearl millet. Crop Sci 53:507–517

Govindaraj M, Rai KN, Pfeiffer WH, Kanatti A, Shivade H (2016a) Energy-dispersive X-ray fluorescence spectrometry for cost-effective and rapid screening of pearl millet germplasm and breeding lines for grain iron and zinc density. Commun Soil Sci Plant Anal 47:2126–2134

Govindaraj M, Rai KN, Shanmugasundaram P (2016b) Intra-population genetic variance for grain iron and zinc contents and agronomic traits in pearl millet. Crop J 4:48–54

Gregorio GB, Senadhira D, Htut H, Graham RD (2000) Breeding for trace mineral density in rice. Food Nutr Bull 21:383–386

Gulia SK (2004) QTL mapping for improvement of downy mildew [*Sclerospora graminicola* (Sacc.) J. Schroet.] resistance (DMR) in pearl millet [*Pennisetum glaucum* (L.) R. Br.] ICMB 89111. Ph.D thesis, Chaudhary Charan Singh Haryana Agricultural University, Hisar

Gupta SK, Velu G, Rai KN, Sumalini K (2009) Association of grain iron and zinc content with grain yield and other traits in pearl millet (*Pennisetum glaucum* (L.) R. Br.). Crop Improv 36:4–7

Hash CT, Witcombe JR (2001) Pearl millet molecular marker research. Int Sorghum Millets Newsl 42:8–15

Hash CT, Witcombe JR (2002) Gene management and breeding for downy mildew resistance. In: Leslie JF (ed) Sorghum and millets diseases. Iowa State Press, Ames, IO, pp 27–36

Hash CT, Schaffert RE, Peacock JM (2002) Prospects for using conventional techniques and molecular biological tools to enhance performance of 'orphan' crop plants on soils low in available phosphorus. Plant Soil 245:135–146

ICMR (2002) Nutrient requirements and recommended dietary allowances for Indians. Indian Council of Medical Research, New Delhi, p 83

IFAD (1999) Farmer participatory testing of technologies to increase sorghum and pearl millet production in the Sahel. www.ifad.org/grants

Jambunathan R, Subramanian V (1988) Grain quality and utilization in sorghum and pearl millet. In: Proceedings of the workshop on biotechnology for tropical crop improvement. ICRISAT, Patancheru, pp 1330–1339

Jones ES, Liu CJ, Gale MD, Hash CT, Witcombe JR (1995) Mapping quantitative trait loci for downy mildew resistance in pearl millet. Theor Appl Genet 91:448–456

Jones ES, Breese WA, Liu CJ, Singh SD, Shaw DS, Witcombe JR (2002) Mapping quantitative trait loci for resistance to downy mildew in pearl millet: Field and glasshouse screens detect the same QTL. Crop Sci 42:1316–1323

Kanatti A, Rai KN, Radhika K, Govindaraj M, Sahrawat KL, Rao AS (2014a) Grain iron and zinc density in pearl millet: combining ability, heterosis and association with grain yield and grain size. Springerplus 3:763

Kanatti A, Rai KN, Radhika K, Govindaraj M, Sahrawat KL, Srinivasu K, Shivade H (2014b) Relationship of grain iron and zinc content with grain yield in pearl millet hybrids. Crop Improv 41:91–96

Kanatti A, Rai KN, Radhika K, Govindaraj M (2016) Tester effect on combining ability and its relationship with line performance per se for grain iron and zinc densities in pearl millet. Crop Sci 56:1–8

Kanatti A, Govindaraj M, Rai KN, Rao AS (2018) Maternal effect on grain iron and zinc densities in pearl millet. In: 1st National Genetic Congress on Genetics for Sustainable Food, Health and Nutrition Security December 14–16, IARI, New Delhi (Accepted)

Kholová J, Nepolean T, Hash CT, Supriya A, Rajaram V, Senthilvel S, Kakkera A, Yadav RS, Vadez V (2012) Water saving traits co-map with a major terminal drought tolerance quantitative trait locus in pearl millet (*Pennisetum glaucum* (L.) R. Br.). Mol Breed 30:1337–1353

Kumar S, Hash CT, Thirunavukkarasu N, Singh G, Rajaram V, Rathore A, Senapathy S, Mahendrakar MD, Yadav RS, Srivastava RK (2016) Mapping quantitative trait loci controlling high iron and zinc content in self and open pollinated grains of pearl millet [*Pennisetum glaucum* (L.) R. Br.]. Front Plant Sci 7:1636

Kumar S, Hash CT, Nepolean T, Satyavathi TS, Singh G, Mahendrakar MD, Yadav RS, Srivastava RK (2017) Mapping QTLs controlling flowering time and important agronomic traits in pearl millet [*Pennisetum glaucum* (L.) R. Br.]. Front Plant Sci 8:1731

Lander ES, Botstein D (1989) Mapping Mendelian factors underlying quantitative traits by using RFLP linkage maps. Genetics 121:185–199

Lander E, Green P, Abrahamson J, Barlow A, Daly MJ, Lincoln SE, Newburg L (1987) MAPMAKER: an interactive computer package for constructing primary genetic linkage maps of experimental and natural populations. Genomics 1:174–181

Liu CJ, Witcombe JR, Pittaway TS, Nash M, Hash CT, Busso CS, Gale MD (1994) An RFLP-based genetic map of pearl millet (*Pennisetum glaucum*). Theor Appl Genet 89:481–487

Mal B., Padulosi S. and Ravi S.B. 2010. Minor millets in South Asia: learnings from IFAD-NUS Project in India and Nepal, Rome: Bioversity International

Mani UV, Prabhu BM, Damle SS, Mani I (1993) Glycemic index of some commonly consumed foods in Western India. Asia Pac J Clin Nutr 2:111–114

Manly KF, Elliott RW (1991) RI manager, a microcomputer program for analysis of data from recombinant inbred strains. Mamm Genome 1:123–127

Manwaring HR, Bligh HFJ, Yadav R (2016) The challenges and opportunities associated with biofortification of pearl millet (*Pennisetum glaucum*) with elevated levels of grain iron and zinc. Front Plant Sci 7:1944

Melchinger AE, Utz HF, Schön CC (1998) Quantitative trait locus (QTL) mapping using different testers and independent population samples in maize reveals low power of QTL detection and large bias in estimates of QTL effects. Genetics 149:383–403

Monasterio I, Graham RD (2000) Breeding for trace minerals in wheat. Food Nutr Bull 21:392–396

Morgan RN, Wilson JP, Hanna WW, Ozias-Akin P (1998) Molecular markers for rust and Pyricularia leaf spot disease resistance in pearl millet. Theor Appl Genet 96:413–420

NFHS (National Family Health Survey) (2016) NFHS-4-fact sheets for key indicators. International Institute for Population Sciences, Mumbai

NIN (2003) Nutritive value of Indian foods. In: Gopalan C, Deosthale YG (eds) Nutritive value of Indian foods. National Institute of Nutrition, Hyderabad

Ozkan H, Brandolini A, Torun A, AltIntas S, Eker S, Kilian B, Braun HJ, Salamini F, Cakmak I (2007) Natural variation and identification of microelements content in seeds of Einkorn wheat (*Triticum monococcum*). Dev Plant Breed 12:455–462

Paltridge NG, Palmer LJ, Milham PJ, Guild GE, Stangoulis JCR (2012) Energy-dispersive X-ray fluorescence analysis of zinc and iron concentration in rice and pearl millet grain. Plant Soil 361:251–260

Parthasarathy Rao P, Birthal PS, Reddy BVS, Rai KN, Ramesh S (2006) Diagnostics of sorghum and pearl millet grains-based nutrition in India. ISMN 46:93–96

Pedraza-Garcia F, Specht JE, Dweikat I (2010) A new PCR-based linkage map in pearl millet. Crop Sci 50:1754–1760

Poncet V, Lamy F, Devos KM, Gale MD, Sarr A, Robert T (2000) Genetic control of domestication traits in pearl millet (*Pennisetum glaucum* L., Poaceae). Theor Appl Genet 100:147–159

Poncet V, Martel E, Allouis S, Devos M, Lamy F, Sarr A (2002) Comparative analysis of QTLs affecting domestication traits between two domesticated × wild pearl millet (*Pennisetum glaucum* L., Poaceae) crosses. Theor Appl Genet 104:965–975

Qi X, Lindup S, Pittaway TS, Allouis S, Gale MD, Devos KM (2001) Development of simple sequence repeat markers from bacterial artificial chromosomes without sub cloning. BioTechniques 31(2):355–358

Qi X, Pittaway TS, Lindup S, Liu H, Waterman E, Padi FK, Devos KM (2004) An integrated genetic map and a new set of simple sequence repeat markers for pearl millet, *Pennisetum glaucum*. Theor Appl Genet 109:1485–1493

Rai KN, Govindaraj M, Rao AS (2012) Genetic enhancement of grain iron and zinc content in pearl millet. Quality assurance and safety of crops and food 4:119–125

Rai KN, Patil HT, Yadav OP, Govindaraj M, Khairwal IS, Cherian B, Rajpurohit BS, Rao AS, Kulkarni MP (2014) Dhanashakti: a high-iron pearl millet variety. Indian Farm 64:32–34

Rai KN, Govindaraj M, Pfeiffer WH, Rao AS (2015a) Seed set and xenia effects on grain iron and zinc density in pearl millet. Crop Sci 55:821–827

Rai KN, Velu G, Govindaraj M, Upadhyaya HD, Rao AS, Shivade H, Reddy KN (2015b) *Iniadi* pearl millet germplasm as a valuable genetic resource for high grain iron and zinc densities. Plant Genet Resour 13:75–82

Rai KN, Yadav OP, Govindaraj M, Pfeiffer WH, Yadav HP, Rajpurohit BS, Patil HT, Kanatti A, Rathore A, Rao AS, Shivade H (2016) Grain iron and zinc densities in released and commercial cultivars of pearl millet (*Pennisetum glaucum*). Indian J Agric Sci 86:11–16

Rai KN, Govindaraj M, Kanatti A, Rao AS, Shivade H (2017) Inbreeding effects on grain iron and zinc concentrations in pearl millet. Crop Sci 57:1–8

Rajaram V, Nepolean T, Senthilvel S, Varshney RK, Vadez V, Srivastava RK, Shah TM, Supriya A, Kumar S, Kumari BR, Bhanuprakash A, Narasu ML, Riera-Lizarazu O, Hash CT (2013) Pearl millet [*Pennisetum glaucum* (L.) R. Br.] consensus linkage map constructed using four RIL mapping populations and newly developed EST-SSRs. BMC Genomics 14(159):1–15.

Reynolds MP, Pellegrineschi A, Skovmand B (2005) Sink-limitation to yield and biomass: a summary of some investigations in spring wheat. Ann Appl Biol 146:39–49

Saïdou AA, Mariac C, Luong V, Pham JL, Bezançon G, Vigouroux Y (2009) Association studies identify natural variation at PHYC linked to flowering time and morphological variation in pearl millet. Genetics 182:899–910

Satyavathi CT, Sankar SM, Singh SP, Bhowmick P, Bhat J, Singh O, Anuradha N (2015) Stability analysis of grain iron and zinc content in pearl millet (*Pennisetum glaucum* (L.) R. Br). Int J Trop Agric 33:1387–1394

Senthilvel S, Jayashree B, Mahalakshmi V, Kumar PS, Nakka S, Nepolean T, Hash CT (2008) Development and mapping of simple sequence repeat markers for pearl millet from data mining of expressed sequence tags. BMC Plant Biol 8:119

Serraj R, Hash CT, Rizvi SMH, Sharma A, Yadav RS, Bidinger FR (2005) Recent advances in marker-assisted selection for drought tolerance in pearl millet. Plant Prod Sci 8:334–337

Singh KP, Mishra A, Mishra HN (2012) Fuzzy analysis of sensory attributes of bread prepared from millet-based composite flours. LWT Food Sci Technol 48:276–282

Stam P (1993) Construction of integrated genetic linkage maps by means of a new computer package: JoinMap. Plant J 3:739–744

Supriya A, Senthilvel S, Nepolean T, Eshwar K, Rajaram V, Shaw R, Narasu ML (2011) Development of a molecular linkage map of pearl millet integrating DArT and SSR markers. Theor Appl Genet 123:239–250

Utz HF, Melchinger AE, Schön CC (2000) Bias and sampling error of the estimated proportion of genotypic variance explained by quantitative trait loci determined from experimental data in maize using cross validation and validation with independent samples. Genetics 154:1839–1849

Varshney RK, Shi C, Thudi M, Mariac C, Wallace J, Qi P, Zhang H, Zhao Y, Wang X, Rathore A, Srivastava RK, Chitikineni A, Fan G, Bajaj P, Punnuri S, Gupta SK, Wang H, Jiang Y, Couderc M, Katta MAVSK, Paudel DR, Mungra KD, Chen W, Harris-Shultz KR, Garg V, Desai N, Doddamani D, Kane N, Conner JA, Ghatak A, Chaturvedi P, Subramaniam S, Yadav OP, Berthouly-Salazar C, Hamiduo F, Wang J, Liang X, Clotault J, Upadhyaya HD, Cubry P, Rhoné B, Gueye MC, Sunkar R, Dupuy C, Sparvoli F, Cheng S, Mahala RS, Singh B, Yadav RS, Lyons E, Datta SK, Hash CT, Devos KM, Buckler E, Bennetzen JL, Paterson AH, Ozias-Akins P, Grando S, Wang J, Mohapatra T, Weckwerth W, Reif JC, Xi L, Vigouroux Y, Xu X (2017. Pearl millet genome sequence provides a resource to improve agronomic traits in arid environments) Nat Biotechnol 35:969–974. https://doi.org/10.1038/nbt.3943

Velu G (2006) Genetic variability, stability and inheritance of grain iron and zinc content in pearl millet (*Pennisetum glaucum* (L.) R. Br.). Ph.D., thesis, Tamil Nadu Agricultural University, Coimbatore, India

Velu G, Rai KN, Muralidharan V, Kulkarni VN, Longvah T, Raveendran TS (2007) Prospects of breeding biofortified pearl millet with high grain iron and zinc content. Plant Breed 126:182–185

Velu G, Rai KN, Sahrawat KL (2008a) Variability for grain iron and zinc content in a diverse range of pearl millet populations. Crop Improv 35:186–191

Velu G, Rai KN, Sahrawat KL, Sumalini K (2008b) Variability for grain iron and zinc contents in pearl millet hybrids. J SAT Agric Res 6:1–5

Velu G, Rai KN, Muralidharan V, Longvah T, Crossa J (2011) Gene effects and heterosis for grain iron and zinc density in pearl millet (*Pennisetum glaucum* (L.) R. Br). Euphytica 180:251–259

Vengadessan V, Rai KN, Kannan Bapu JR, Hash CT, Bhattacharjee R, Senthilvel S, Vinayan MT, Nepolean T (2013) Construction of genetic linkage map and QTL analysis of sink-size traits in pearl millet (*Pennisetum glaucum*). ISRN Genetics 52:724–728

Wessells KR, Brown KH (2012) Estimating the global prevalence of zinc deficiency: results based on zinc availability in national food supplies and the prevalence of stunting. PLoS One 7:e50568

WHO (2008) Worldwide prevalence of anaemia 1993–2005: WHO global database. WHO, Geneva

Yadav OP, Rai KN (2013) Genetic improvement of pearl millet in India. Agric Res 2:75–92

Yadav OP, Mitchell SE, Zamora A, Fulton TM, Kresovich S (2007) Development of new simple sequence repeat markers for pearl millet. J Semi-Arid Trop Agric Res 3:34–37

Yadav RS, Hash CT, Bidinger FR, Cavan GP, Howarth CJ (2002) Quantitative trait loci associated with traits determining grain and stover yield in pearl millet under terminal drought-stress conditions. Theor Appl Genet 104:67–83

Yadav RS, Bidinger FR, Hash CT, Yadav YP, Yadav OP, Bhatnagar SK, Howarth CJ (2003) Mapping and characterization of QTL3E interactions for traits determining grain and stover yield in pearl millet. Theor Appl Genet 106:512–520

Yadav RS, Hash CT, Bidinger FR, Devos KM, Howarth CJ (2004) Genomic regions associated with grain yield and aspects of post-flowering drought tolerance in pearl millet across stress environments and tester background. Euphytica 136:265–277

Yadav OP, Mitchell SE, Fulton TM, Kresovich S (2008) Transferring molecular markers from sorghum, rice and other cereals to pearl millet and identifying polymorphic markers. J Semi-Arid Trop Agric Res 6:1–4

Yadav OP, Rai KN, Bidinger FR, Gupta SK, Rajpurohit BS, Bhatnagar SK (2012) Pearl millet (*Pennisetum glaucum*) restorer lines for breeding dual-purpose hybrids adapted to arid environments. Indian J Agric Sci 82:922–927

Yadav OP, Rai KN, Yadav HP, Rajpurohit BS, Gupta SK, Rathore A, Karjagi CG (2016) Assessment of diversity in commercial hybrids of pearl millet in India. Indian J Plant Genet Resour 29:130–136

Yadava DK, Choudhury PR, Hossain F, Kumar D (2017) Biofortified varieties: sustainable way to alleviate malnutrition. Indian Council of Agricultural Research, New Delhi

Chapter 5
Genetic Approaches to Improve Common Bean Nutritional Quality: Current Knowledge and Future Perspectives

Eleonora Cominelli, A. Paula Rodiño, Antonio M. De Ron, and Francesca Sparvoli

5.1 Introduction

The common bean (*Phaseolus vulgaris* L.) is the third most important food legume species worldwide, surpassed only by soybean (*Glycine max* (L.) Merr.) and peanut (*Arachis hypogea* L.), and it is the first one for direct human consumption, both as a dry food legume, due to the high protein content of the grain, and also as vegetable as snap bean for the fresh pod (De Ron et al. 2015). Among the main food crops, the common bean shows the greatest variation in growth habit, seed characteristics (size, shape, and color) and maturation time. This variability enables its production in a wide range of cropping systems and environments as diverse as the Americas, Africa, the Middle East, China, and Europe (Blair et al. 2010a), where this crop is highly consumed often as the most important source of dietary protein, as well as in traditional recipes of the Middle East and the European Mediterranean region (Broughton et al. 2003; Casquero et al. 2006). This legume is a component of healthy diets of the Mediterranean basin and is also gaining importance in the USA where dry bean consumption has been increasing due to greater interest in "ethnic" and healthy foods (Blair and Izquierdo 2012). Recently, the role of bean in human diet is being focused not only in its protein content but in the functional properties of some components also and different authors have reported that its consumption could contribute to reduce risk of diseases such as obesity, diabetes, cardiovascular problems and colon, prostate and breast cancer (Hangen and Bennink 2003;

E. Cominelli · F. Sparvoli (✉)
Institute of Agricultural Biology and Biotechnology (IBBA), Italian National Research Council (CNR), Milan, Italy
e-mail: cominelli@ibba.cnr.it; sparvoli@ibba.cnr.it

A. P. Rodiño · A. M. De Ron
Biology of Agrosystems, Misión Biológica de Galicia (MBG), Spanish National Research Council (CSIC), Pontevedra, Spain
e-mail: aprodino@mbg.csic.es; amderon@mbg.csic.es

A. M. I. Qureshi et al. (eds.), *Quality Breeding in Field Crops*,
https://doi.org/10.1007/978-3-030-04609-5_5

Thompson et al. 2009). These health benefits could be due to the fiber content in the grain but also to antioxidant compounds as the phenolic ones.

Dry bean provides an economical source of calories and proteins, with an average protein content of 22% despite being deficient in the sulfur amino acids, as other legumes (Kelly and Bliss 1975; Escribano et al. 1990). Dry bean is one of the major constituents in human diet, normally complementing cereals, especially in developing countries where there is a lack of animal protein in the diet and beans are the most economic source of protein; moreover, dry and fresh beans are a good source of vitamins, minerals, and dietary fiber. Additionally, this pulse crop has high market value, which mainly depends on the quality of the pod and the grain, thus being an interesting crop from the point of view of the consumer, farmer, and processor. For the consumer, bean is important for its adequate nutritive composition and its variable uses in different culinary forms. For the farmer, the crop contributes nitrogen to the soil, dry seed and fresh pod of specific landraces attract a high market price while the fresh pod crop can also be produced during the coolest season in glasshouses. For the processor, common bean has many possibilities, such as canned or frozen grain and pod.

Some temperate countries are in a position to produce more dry and fresh beans, which can easily be absorbed by local markets or be exported to other areas. Therefore, breeding on quality in bean germplasm, including different varieties and cropping systems, is required.

5.2 Culinary Quality and Market Classes

One of the factors influencing bean culinary quality is seed texture, which can be influenced by many factors such as variation in germplasm (Hosfield and Uebersax 1980), growing and storage conditions (Burr et al. 1968), and finally soaking and processing conditions. Dry seeds are first soaked and then cooked to render them palatable, inactivate heat labile antinutrients, and permit digestion and assimilation of protein and starch (Burr et al. 1968; Quenzer et al. 1978). Another important factor relating to quality is the nutritional composition of dry seed bean, which includes both protein content and quality. It has been determined that protein content ranges between 18% and 31% of total dry matter, depending on the accession and growing conditions (Mutschler and Bliss 1981). Variability in culinary and nutritional traits of bean has been previously reported (Hosfield and Uebersax 1980; Hosfield et al. 1984), in addition to significant interactions between genotype and site and environmental effects for these traits.

Water absorption is an important attribute because it is related to culinary quality. Seed has to uptake water before cooking to become tender and allow for uniform expansion. Water absorption is related to cooking time and palatability (Quenzer et al. 1978) and to a term defined by Bourne (1967) as hard seed. Hence, water absorption may be a useful and rapid indirect selection method to screen germplasm for cooking time, avoiding the 'hard-to-cook' problem present in some bean varieties.

Consumers have progressively shown specific preferences for various combinations of size and shape of bean seeds and pods, and the market reflects this trend by giving preference to types of good quality rather than high yield. The conservation by farmers of this germplasm is strongly influenced by the taste, shape, and color of the pods and grains, which make them easily recognizable; although a large majority of old varieties have been replaced by improved varieties which are economically more viable. The introduction of an edible podded bean with long, wide, and tender pods could have a great deal of interest among farmers and may offer some possibilities for expanded commercial production and consumption.

Bean markets and consumers in different countries show particular preferences for grain size, shape, color, and cooking time of beans. The polymorphism of common bean is so great that, in each region, and even in each locality, different varieties with similar characteristics correspond to different names. There are many ethnic varieties or "heirloom" varieties, which are characteristic of an area or region, and they can be designated with different names. This germplasm has derived from ancient types by conscious or unconscious selection by farmers and are currently adapted to the agroecological conditions under which they have been grown for centuries. Therefore, a classification often used for common bean is based upon the commercial types, which is based predominantly on characteristics of grain color and size, and is related to market preferences. The wide range of seed characteristics has been formalized in the bean world into distinct commercial or 'international market classes' (Table 5.1) that are recognized according to consumer preferences, production, and market price. Some authors have described the major worldwide market classes (Voysest 2000; Santalla et al. 2001) that often include improved germplasm and thus tend to show a low level of variability.

Breeding for commercial varieties in common bean usually occurs within each market class in order to retain their preferred seed size, shape, color, and pattern. However, the range of commercially available bean cultivars and varieties in different market classes is constantly changing and new cultivars are being released for their increased yield potential, disease resistance, and improved grain quality. In some countries, such as Spain, dry bean production is concentrated in market classes with added value 'Alubia de riñón', 'Faba Asturiana' and 'Faba Galaica', 'Pinta de León', 'Tolosana', 'Ganxet', etc. that have acquired considerable importance due to their good seed quality and these varieties are protected by specific legal regulations or 'Geographic Protected Indication' (De Ron et al. 2016). Additionally, another type of dry bean should be considered, the "nuña", probably ancestral, originated in the Andes. The characteristic that makes it different from other varieties of dry bean is that by heating its grain for 2–3 min, it is roasted, enlarged, and acquires an edible texture, so it can be consumed without cooking. Unlike the current products in the market of snacks, the nuñas have lower fat content and a higher protein content; hence, they have a higher added value due to their condition of healthy and innovative food, having multiple possibilities of processing, with different aromas and condiments, with salt, candied, in salads, desserts, etc. (Van Beem et al. 1992).

Table 5.1 Main international common bean market classes (adapted from Santalla et al. 2001)

White seed	**Yellow seed**
Small white	Small yellow
Navy	Garbancillo
Great northern	Canario bola
Marrow	Azufrado
Large great northern	**Brown seed**
Hook	Chumbinho
Canellini	Brown marrow
White kidney	Brown garbanzo
Favada	Brown mottled
White (bi-colored) seed	Manteca
Hen eye	**Pink seed**
Rounded caparron	Rosada
Red Caparron	Light red kidney
Kindney caparron	**Red seed**
Favada pinto	Small red
Cream seed	Sangretoro
Carioca	Guernikesa
Mulatihno	Dark red kidney
Dark garbanzo	Red Pinto
Sargaço	Large red mottled
Mottled canellini	**Purple seed**
Viscado	Morado
Pinto	Purple caparron
Ojo de cabra	**Black seed**
Bayo gordo	Black turtle
Cranberry	Negro brillante
Canela	Black canellini
Large cranberry	Black mottled

5.3 Improvement of Protein Quality

Although beans possess several health benefits and a valuable nutritional composition, they also contain other substances (such as lectins, phytate, digestive enzyme inhibitors, phenolic compounds) considered as antinutritional, that may cause adverse negative effect to those who consume them as staple food and/or improperly processed/cooked. Seeds accumulate most of these compounds as a defense mechanism against the attack of parasites, insects, fungi, and herbivorous animals mainly, although some of them, i.e. phytate and lectins, serve as reserve to provide phosphorous and amino acids at seed germination From a nutritional point of view, these compounds contribute to lowering bean nutritional value reducing protein and amino acid digestibility (Sarwar Gilani et al. 2012) as well as decreasing mineral bioavailability (Doria et al. 2012; Petry et al. 2013).

Bean seeds are mainly consumed as a dietary source of protein and fiber, hence the type and amount of seed proteins determine a consistent part of their nutritional value. Storage proteins are the most abundant proteins in bean seeds. They are represented by the globulins phaseolin (7S fraction, vicilin family) and legumin (11S fraction), and by albumins belonging to the family of lectin and lectin-related proteins [phytohemagglutinin (*P*HA), α-amylase inhibitor (α*AI*), and arcelin (*A*rl), APA proteins] coded by the single Mendelian APA locus (Sparvoli et al. 2015).

Generally, in legume seeds globulins and albumins constitute the major protein fractions accounting for about 35–90% and 10–37% of the seed dry weight, respectively. Differently from all other legumes, in common bean seeds the globulins to albumins ratio is unbalanced in favor of albumins (35–39% and 27.6–36.6% of seed dry weight, respectively); furthermore, legumin is less abundant (10% of total seed proteins) than phaseolin which accounts for up 50% of total seed proteins (Vitale and Bollini 1995).

The nutritive value of both phaseolin and APA proteins is limited by their low content in sulfur amino acids (phaseolin is also poor in tryptophan) as well as their high resistance to enzymatic proteolysis (Marquez and Lajolo 1981; Jivotovskaya et al. 1996). Heat treatment may significantly improve the hydrolysis of both classes of proteins, however this is not always sufficient to ensure a good nutritional quality (Genovese and Lajolo 1996, 1998). Moreover, in the case of APA proteins it should be taken into account that it is necessary to abolish their biological activity, especially for the PHA lectins, which, if not properly inactivated by cooking, can exert toxic effects on the digestive tract causing vomiting, diarrhea, bloating, and nausea that eventually interfere with nutrient absorption, thus leading to a decrease in the nutritional power of beans (Vasconcelos and Oliveira 2004).

5.3.1 Modulation of Seed Storage Proteins Content: Sulfur Amino Acids Increase and Phaseolin Digestibility

Different studies, based on breeding programs or genetic manipulation, have been undertaken with the aim to improve bean seed protein quality in relation to the enhancement of limiting sulfur amino acids. Most of them have been focused on phaseolin, as this is the most abundant protein in bean seeds (Gepts and Bliss 1984; Aragao et al. 1999; Montoya et al. 2010). However, interesting results have also been obtained from studies based on the manipulation of protein seed composition through the elimination of specific seed proteins by breeding approaches (Osborn and Bliss 1985; Confalonieri et al. 1992; Taylor et al. 2008; Campion et al. 2009a).

Gepts and Bliss (1984) studied the relationship between available methionine concentration and the levels of phaseolin using genetic materials differing for their phaseolin content and showed that higher phaseolin levels lead to increased available methionine concentration. Later, the biotechnological approach of overexpressing the methionine-rich 2S albumin of Brazil nut allowed an increase in methionine of 14% and 23% in two bean transgenic lines (Aragao et al. 1999), however since this protein is a strong allergen, this material could not be exploited. Furthermore, results from studies using a similar approach showed that the expression of foreign proteins is limited by sulfur supply and often causes a shift of sulfur from endogenous sulfur-rich proteins (Streit et al. 2001; Tabe and Droux 2002).

Another strategy to increase the content of sulfur amino acids in seeds consists in modifying protein fractions by decreasing those with low contents of limiting amino acids. Taking advantage of genetically related common bean lines integrating a progressive deficiency in major seed storage proteins (Osborn and Bliss 1985; Delaney and Bliss 1991; Burrow et al. 1993; Osborn et al. 2003), Taylor et al. (2008) evaluated the impact of seed storage protein (phaseolin, phytohemagglutinin, and/or arcelin) deficiency on protein accumulation and amino acid composition in mature seeds of common bean, with a particular emphasis on sulfur amino acids (Fig. 5.1). The lines that were used in this study were in the genetic background of cv. Sanilac. The SARC1 line contains the S-type phaseolin derived from cv. 'Sanilac' and the phytohemagglutinin (PHA) and arcelin (Arl) derived from the Arcelin1 wild parent (G12882). The SARC1 line accumulates high amount of Arl, has lower content of PHA and phaseolin (less than 50%) than the parent 'Sanilac', and does not accumulate αAI, a trait derived from the APA locus of the wild parent (Osborn and Bliss 1985; Romero-Andreas et al. 1986; Marsolais et al. 2010). The phaseolin null (PN) trait from *P. coccineus* cv. 'Mexican Red Runner' was then introgressed in the SARC1 line and the resulting phaseolin null line was named

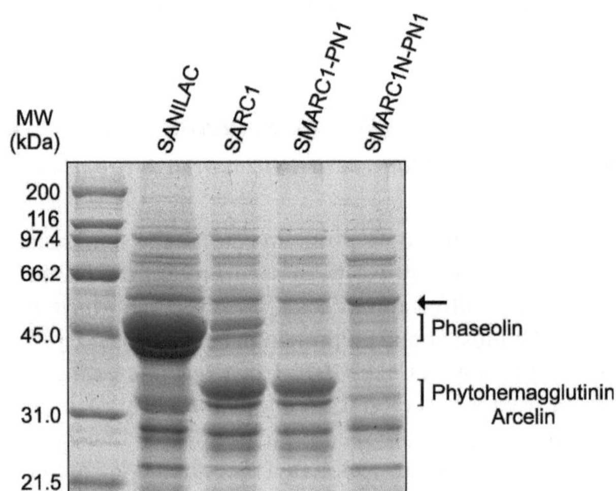

Fig. 5.1 Protein profiles of seed storage protein-deficient lines and commercial parent Sanilac. The size of the markers is indicated on the left. Bracket indicates the position of the major seed storage proteins. Arrow indicates the position of a 54 kDa band (corresponding to legumin α subunit) up-regulated in plants deficient in phaseolin, phytohemagglutinin, and arcelin (SMARC1N-PN) (from Taylor et al. 2008)

SMARC1-PN. A third line, named SMARC1N-PN, was obtained introgressing the PHA-null trait of cv. 'Great Northern 1140' in the SMARC1-PN line and therefore was devoid of phaseolin, PHA and Arl, but not αAI, as in this case the APA locus was that of 'Great Northern 1140' which expresses αAI (Taylor et al. 2008; Marsolais et al. 2010).

The absence of major seed storage proteins (SMARC1N-PN line) had little impact on seed protein content, as this was compensated by increases in other proteins, while it was observed a progressive increase in soluble protein content, indicating a change in protein composition. In addition, several changes in free amino acids content, including a reduction of the non-protein sulfur amino acids S-methyl-Cys and γ-glutamyl-S-methyl-Cys, were observed. Deficiency of phaseolin and major lectins (PHA and Arl) resulted in a nearly twofold increase in sulfur amino acids, particularly cysteine and methione increased by 70% and 10%, respectively. These results indicate that a preferential interconversion of sulfur stored as γ-glutamyl-S-methyl-Cys to the protein Cys pool occurs in low storage proteins bean lines and suggest that S-methyl-Cys may arise from S-methylation of Cys during seed maturation (Taylor et al. 2008). Integration of proteomic and transcriptomic data obtained on developing seeds of SMARC1N-PN lines showed that the absence of major storage proteins was compensated by an increase of sulfur-rich proteins, such as legumin, albumins, and defensin D1, together with the up-regulation of genes involved in multiple sulfur metabolic pathways, including enzymes responsible for sulfate assimilation (Marsolais et al. 2010; Yin et al. 2011).

Resistance to proteolysis of phaseolin was the main focus of the work of Montoya et al. (2008a). Raw phaseolin is poorly digestible both in vitro (10–27%) and in vivo (28–36%). Of course thermal treatment improves its digestibility, however an extensive analysis on 43 different phaseolin types showed that the degree of hydrolysis (DH) ranges between 57% and 96%, depending on the phaseolin type. Further, the same authors showed that the estimated nutritional value of heated phaseolin is influenced more by its DH value than by its amino acid composition. It was suggested that an improvement of beans nutritional value could be achieved by breeding and selecting for those phaseolin types with high DH after heat treatment (Montoya et al. 2008a). In fact, almost 90% of cultivated beans contain S (Mesoamerican germplasm) or T (Andean germplasm) phaseolins (Singh et al. 1991), which are among the ten phaseolins with the lowest DH, having therefore the lowest nutritional value. It was hypothesized that variations in DH values could be ascribed to differences in phaseolin subunit composition, subunit precursor origin (α or β), and trypsin susceptibility between subunits (Montoya et al. 2008b, 2009). In particular, it appeared that higher content of α phaseolin subunits correlated with higher thermal stability, as the lowest DH value (50%) was observed for S type phaseolin, which has a higher content of α subunits (αα), compared to T (αββ) and I (ββ) type of phaseolins (both with DH values of 70%). A possible explanation might be due to the fact that the α subunit is more glycosylated than the β one and this could make α phaseolin less accessible to the action of hydrolytic enzymes.

5.3.2 Modulation of Seed Storage Proteins Content: Lectin Removal and Exploitation of α-Amylase Inhibitor

The strategy based on modulation of seed storage proteins was also undertaken to improve seed nutritional quality with the aim to reduce or eliminate antinutritional proteins. Besides phaseolin the second most abundant class of storage proteins is that comprising the APA proteins (PHA, αAI, and Arl), which may represent up to 10% of seed proteins. The APA proteins are bioactive proteins that have evolved from a common ancestor undergoing to functional specialization (Lioi et al. 2003). PHA, considered the true bean lectin, is a tetramer with the ability to agglutinate cells and is highly toxic to monogastric animals (Vitale and Bollini 1995). It consists of two types of polypeptide chains called E and L subunits (PHA-E and PHA-L, respectively), which possess erythroagglutinating and leukoagglutinating activity, respectively. α-Amylase inhibitor is a dimeric glycoprotein, which inhibits the activity of certain mammalians, including humans, and insect α-amylases by completely blocking access to the active site of the enzyme (Pueyo et al. 1993; Le Berre-Anton et al. 1997; Barrett and Udani 2011). Arcelins are present only in some wild Mesoamerican accessions, where they are very abundant, and have been proposed to play a role in the Mexican bean weevil *Zabrotes subfasctiatus* and the bean weevil *Acanthoscelides obtectus* resistance (Osborn et al. 1988; Lioi et al. 2003).

From a nutritional point of view APA proteins are antinutritional factors. Raw lectins (but also lectins from cooked beans) are resistant to the digestive process (Bollini et al. 1999; Morari et al. 2008; Galimberti et al. unpublished). Lectins can reduce the bioavailability of dietary important micronutrients such as Fe and Zn (Welch and House 1984; Welch and Graham 2004) with negative effects on human health (The World Health Report 2002). Phytohemagglutinin damages the gut wall, causes coliform overgrowth in the lumen (Pusztai et al. 1993), and reduces the fractional rate of protein synthesis in skeletal muscle (Palmer et al. 1987; Bardocz et al. 1992). Lectin binding to the gut wall is also associated with a disruption of the brush border, reduced epithelial cell viability, hyperplasia in the crypts, and increase in the weight of the tissue (Pusztai 1991). Among the three lectins, PHA is the most dangerous for mammalians health. Heat treatment applied to seeds or flours (i.e. cooking beans for human consumption) denatures large part of seed lectins but not all of them completely: in beans, a residual activity causing reduced protein digestibility and toxicity can be detected also after cooking (Bender and Reaidi 1982). Finally, although the risks of acute toxicity are low, prolonged exposure to lectins, also at low levels, may be harmful to human health and monogastric animals. Arcelins are very stable proteins and their resistance to proteolysis has been suggested to be one of the mechanisms by which they are toxic to the larvae of seed feeding insects (Zaugg et al. 2013). No data are available on the possible toxicity of arcelins to humans, and anyway their presence has been reported only in some wild Mesoamerican accessions (Lioi et al. 2003; Zaugg et al. 2013). αAI, also known as phaseolamin, is able to inhibit mammalian α-amylases, thus inhibiting starch digestion by blocking access to a basic active site of the α-amylase enzyme (Santimone et al. 2004). As αAI prevents the digestion of complex carbohydrates, it is widely used as basic active ingredient of commercial starch-blocker preparations for the control of body weight (Barrett and Udani 2011). However, αAI has also been shown to be effective in reducing post-prandial plasma levels of glucose, insulin, C-peptide, and gastric inhibitory polypeptide in healthy subjects as well as in individuals affected by diabetes mellitus (Layer et al. 1985, 1986).

Genetic elimination of lectins has been proposed to improve grain nutritional quality and positive effects are expected especially in those cases where they cannot be properly cooked or, less frequently, when bean seeds are provided as raw flour to feed animals (Bollini et al. 1999). Presence of αAI is undesirable in those cases in which the raw seeds or flour are used for animal feeding, while in the case beans are for human consumption, the presence of active αAI might turn to be useful to control starch assimilation and post-prandial plasma levels of glucose.

Bean lines (BC_4F_5) devoid of active PHA were obtained by crossing the cv. 'Pinto UI111', carrying an inactive PHA, the so-called 'pinto lectin', with the recurrent wt parent cv. 'Taylor's Horticultural' (Confalonieri et al. 1992). Most likely, this PHA-null trait is coded by the same APA locus carried by the 'Great Northern 1140', in fact in both cases the locus codes only for a pinto lectin and an active αAI (Voelker et al. 1986; Pandurangan et al. 2016). The protein quality of this genetic material was analyzed using an animal system, in which raw and cooked beans were used for feeding rats (Bollini et al. 1999). Results showed that true protein digestibility and protein digestibility corrected amino acid score were both higher for raw and cooked lectin-null beans compared to the parent 'Taylor's Horticultural' (Table 5.2).

Although most of the bean genotypes carry a canonical APA locus coding for PHA-E, PHA-L, and αAI (Hoffman and Donaldson 1985), variability in gene composition and organization at the APA locus is quite frequent and genotypes carrying different combinations might be identified. For example, the APA locus of the BAT 93 genotype carries genes coding for the pinto lectin, αAI, and PHA-E, but not

Table 5.2 Protein quality of raw and cooked lectin-null bean lines (adapted from Bollini et al. 1999)

	Raw beans		Cooked beans	
Protein quality	Taylor's H.	Lectin-null	Taylor's H.	Lectin-null
Chemical score[a]	0.92	0.92	0.91	0.91
Limiting aa	Sulfur aa	Sulfur aa	Sulfur aa	Sulfur aa
True protein digestibility[b]	17.55 ± 2.41[c]	40.13 ± 2.29[d]	66.77 ± 1.4[d]	72.23 ± 1.51* (4)
Protein digestibility corrected aa score[a]	0.16	0.37	0.61	0.66

*Lectin-null bean versus recurrent parent: $p < 0.02$
[a]According to FAO/WHO (1991)
[b]Mean ± SE
[c]Adult rat weight: 100 g
[d]Weaning rat weight: 60 g

PHA-L (Pandurangan et al. 2016; Joshi et al. 2017), while a wild accession (G6388) almost devoid of PHA-E, αAI and accumulating low levels of PHA-L was also identified (Sparvoli et al. 1994). This wild accession was used to develop breeding lines combining the 'no lectin' trait with 'reduced polyphenols and tannins', with the aim to produce starting materials for the development of biofortified beans to improve human nutrition and animal feeding (Campion et al. 2009a). Indeed, one of these breeding lines was used to develop an EMS mutagenized population from which a *low phytic acid* mutant (*lpa1*) was isolated (Campion et al. 2009b).

Some of these materials, the cv. 'Lady Joy' derived from breeding lines developed by Confalonieri et al. (1992) and the *lpa1* mutant line were recently used to produce unprocessed flours for the preparation of biscuits with improved nutritional properties and to verify the advantage of their use compared to normal beans (Sparvoli et al. 2016). Results showed that use of unprocessed flour from normal beans must be avoided, since lectin activity is still present after baking. Furthermore, it was demonstrated the advantage of using 'Lady Joy' since biscuits containing flour of this cultivar were nutritionally better than the control, having a better amino acid score, higher fiber amount, lower starch content and predicted glycemic index (pGI). In fact, baking did not fully inactivate α-AI, further contributing to lowering the pGI of the biscuits. Replacement of 'Lady Joy' bean flour with that of *lpa1*, having a 90% reduction of phytic acid and devoid of α-AI, contributed to about a 50% reduction of phytic acid content (Sparvoli et al. 2016).

5.4 Bean Biofortification: Breeding for High Iron Beans (HIB) and Development of *low phytic acid* (*lpa*) Mutants

In common bean seed the majority of iron (Fe), from 75% to 95%, is localized in cotyledons; the seed coat can contain from 4% to 22% of total Fe; the embryo axis, although containing a high concentration of Fe, contributes only for 2–3% to total seed weight (Ariza-Nieto et al. 2007). The average Fe concentration in common bean seeds is high (55 μg/g; Beebe et al. 2000; Islam et al. 2002). However, different studies reported low Fe absorption from beans, in the range of 1–3% (Donangelo et al. 2003; Beiseigel et al. 2007). In many countries, such as for example Burundi, Kenia and Rwanda, common bean is highly consumed as a relevant source of plant protein, reaching 180 g per capita and day, and constitutes an important component of the diet together with cereals. This kind of diet provide non-heme Fe that is poorly bioavailable, differently from diets where meat, poultry, and fish are present, that provide heme Fe that is more bioavailable. Consequently, Fe deficiency (ID), defined as a hemoglobin concentration below the optimum value in an individual, is the most common and widespread nutritional disorder in the world. Insufficient Fe intake may cause major pathophysiological complications, such as: stunted growth, impaired physical and cognitive development, and increased risk of morbidity and mortality in children (FAO/WHO 2004). Fe biofortification is considered a sustainable and long-term cost-effective strategy to combat ID. The term biofortification indicates the different strategies aimed to increase the level and/or bioavailability of minerals and vitamins in the edible parts of the plants. Biofortification is a multidisciplinary strategy that takes advantage of the expertise in different fields: plant genetics, biochemistry, genomics, breeding, and medical, economic and social sciences. It can be achieved with different approaches: the application of fertilizer to soil or leaves; conventional plant breeding; or genetic engineering, which includes genetic modification and transgenesis (Murgia et al. 2012; Bouis and Saltzman 2017). For Fe the first approach has not yet been applied successfully (Cakmak et al. 2010). Concerning the other two approaches, three different goals may be pursued to Fe biofortify crops: increasing Fe concentration; reducing the concentration of the so-called "antinutrients", such as phytic acid (PA) and PP (PP), known to decrease absorption of dietary Fe; increasing the concentration of compounds favoring Fe absorption (Murgia et al. 2012, Bouis and Saltzman 2017). HarvestPlus, a program of the Consultative Group for International Agricultural Research (CGIAR), chose common bean as one of the target species to be Fe biofortified (Asare-Marfo et al. 2013). According to HarvestPlus definition, the Biofortification Priority Index (BPI) is a "composite, crop-specific index accounting for the intensity and level of production and consumption of a specific crop in any given country and the deficiency levels for the micronutrient(s) with which the specific crop can be enriched". Hence, for each crop, countries with high BPI should be considered for prioritization for biofortification interventions. As shown in Fig. 5.2, the majority of the countries having "top" BPI for common bean are the main consumers in Africa, followed by countries in the region of Latin America and the Caribbean (LAC) and Asia (Prasai and Asare-Marfo 2015)

In the last years, many efforts have been spent in order to achieve common bean Fe biofortification mainly with two purposes: increase common bean seed Fe concentration, inside HarvestPlus Program, and decrease PA content. Here, we summarize the most promising results that contributed to common bean Fe biofortification. Moreover, we discuss the possible new strategies that may be considered in the future to improve this goal, also taking into account growing environmental challenges that may constrain biofortification efforts.

5.4.1 Increasing Fe Concentration

The steps necessary to increase Fe concentration in common bean were the same of any crop improvement program, as summarized in Fig. 5.3, as reviewed by Blair (2013). The main activities were performed under the frame of the HarvestPlus Program that started in 2003, with an explorative phase of screening of Fe concentration variability and pre-breeding activities between 1994 and 2002, under the Pan Africa Research Alliance (PABRA), led by the International Center of Tropical Agriculture (CIAT, Cali, Colombia) (Mulambu et al. 2017).

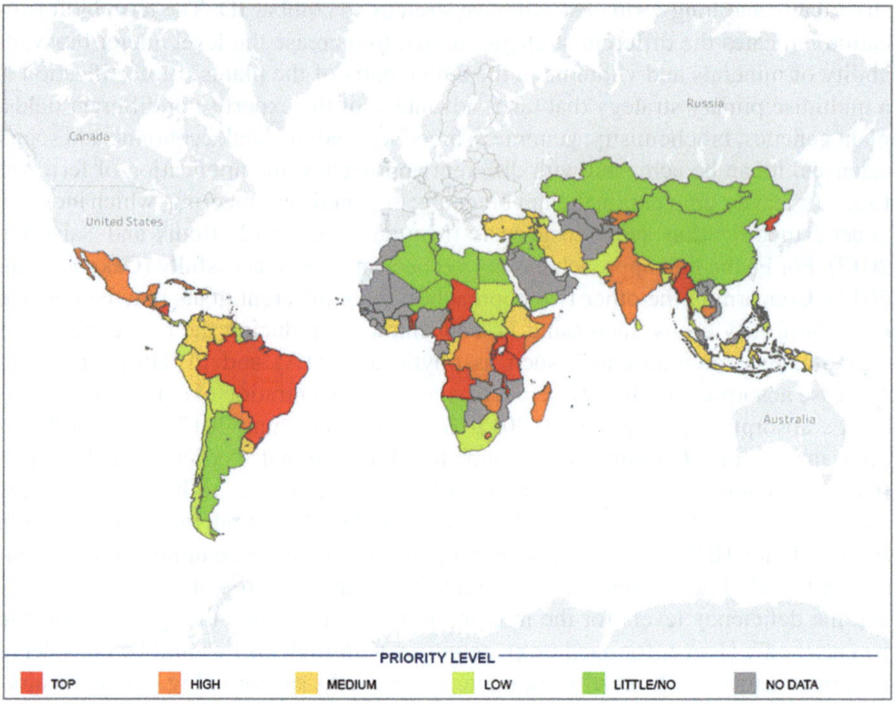

Fig. 5.2 Common bean Biofortification Priority Index. Picture courtesy of HarvestPlus (Prasai and Asare-Marfo 2015; Asare-Marfo et al. 2013)

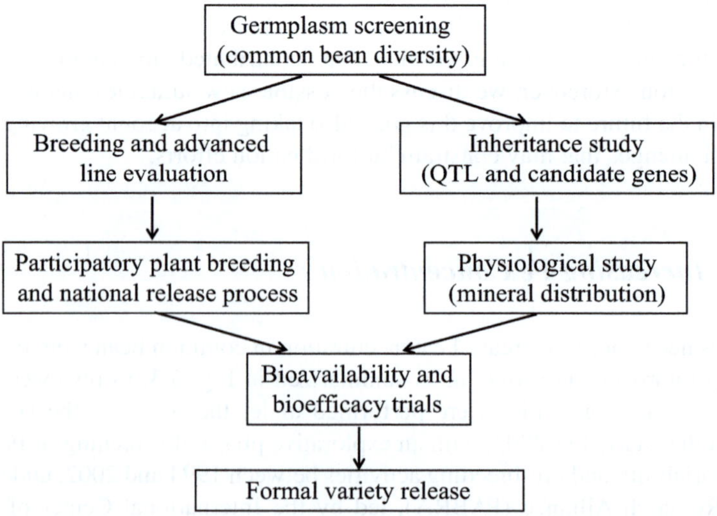

Fig. 5.3 Steps followed to develop new varieties with increased Fe concentration. Picture reproduced from Blair (2013)

5.4.1.1 Germplasm Screening

The most complete overview and reliable information on the Fe content in common bean was provided by two similar studies in which the common bean core collection of CIAT, containing about 1100 genotypes, was screened. The collection includes wild and cultivated accessions belonging to both the Andean and the Mesoamerican gene pools (Beebe et al. 2000; Islam et al. 2002). These studies revealed a high degree of variability for this trait, ranging from 35 to 92 μg/g of Fe seed concentration, with an average of 55 μg/g. These studies did not highlight a clear relationship between Fe content and geographic distribution, although accessions from the Andean gene pool or inter-gene-pool hybrids tended to have higher Fe content than those from the Mesoamerican one. Moreover, wild bean accessions had only a narrow advantage than cultivated ones in Fe content (Beebe et al. 2000).

In other studies locally available germplasm was screened for Fe concentration variability: a range of Andean varieties (Blair et al. 2005); a large collection of more than 350 Rwandan genotypes (Blair et al. 2010a); a collection from ten countries belonging to the East and Central AFRICAN Bean Research Network (ECABREN) (Mulambu et al. 2017).

Moreover, screening of related species *Phaseolus coccineus* and *Phaseolus dumosus* was performed in order to identify high grain Fe genotypes to cross them with common bean-adapted materials. A *P. dumosus* accession registered a very high Fe concentration of 127 μg/g (Beebe et al. 2005).

Interestingly, a statistically significant correlation (60–80%) was identified between the Fe and Zn levels, suggesting that across different genotypes some genetic factors for different minerals are co-segregating and that selection for Fe will in fact result in an increase in Zn (Beebe et al. 2000), as later confirmed by QTL studies, as discussed below (Blair et al. 2009a, 2010b; Guzman-Maldonado et al. 2003).

An important aspect that was evaluated in different studies was the stability of the high Fe trait, in relationship with the environment. Some studies highlighted a relevant genotype by environment interaction for this trait (Cichy et al. 2009; de Araújo et al. 2003), while other analyses reported it is not so strong (Ribeiro et al. 2008; Blair et al. 2011). However, some highly adaptable and stable genotypes were identified (Beebe et al. 2000; Martins et al. 2016).

Many of the identified high-Fe lines were of non-commercial seed types, therefore, the identification of potential commercial parents, through the screening of advanced lines for each gene pool, is very important (Blair 2013).

5.4.1.2 Breeding Progress and Variety Release

The main goal of biofortification of common beans has been to produce varieties with 80% more Fe content, reaching 94 μg/g, starting from the average concentration in the germplasm of 55 μg/g. A second goal has been to maintain the properties that farmers and consumers require in a variety, such as adaptation to stresses and seed shape or color. Both the Andean and the Mesoamerican gene pools of common bean, and both bush and climbing beans have been used in different breeding programs. Different breeding techniques or strategies have been used, such as backcrossing, recurrent selection, and various permutations of gamete and pedigree selection (Blair 2013).

During the HarvestPlus Phase I (2003–2008) pre-breeding feasibility and trait hereditability studies were performed at CIAT. During this phase the first strategy was applied to the Andean gene pool and the resulting *Nu*tritional Improved *A*ndean (NUA) lines represent the first result of a breeding program, aimed to improve nutritional quality in common bean (Blair et al. 2010c). The NUA35 (determinate, type I growth habit, with 81 μg/g Fe content) and NUA56 (indeterminate, type II growth habit, with 76 μg/g) lines resulted from the cross of the G14519 high seed Fe accession from the CIAT (Islam et al. 2002) with a CAL96 commercial cultivar in Colombia and Uganda. These lines are also improved for Zn concentration, compared to CAL96, and show good adaptation to a range of environments in Latin America and Carribean (LAC) and African countries (Blair et al. 2010c; Blair 2013).

During the HarvestPlus Phase II (2009–2013) different HIB lines were developed, including intermediate development stage products and advanced materials (Mulambu et al. 2017). Then, breeding programs were started in Rwanda and in Democratic Republic of Congo (DRC) to test the adaptation of HIB lines (locally developed and from CIAT) under local growing conditions, both at experiment stations and in farmers' fields. Between 2010 and 2016 different varieties were released in Rwanda, DRC, and Uganda. Different trials to evaluate adaptability and resistance/tolerance to pests and diseases were performed. Different bush and climbing beans, well-adapted to local conditions and that suit farmer and consumer preferences were officially released in 2016. A list of the HIB varieties released under HarvestPlus in Rwanda (Rwanda Agriculture Board, RAB), DRC (Institut National pour l'Etude et la Recherche Agronomique, INERA) and Uganda (National Agricultural Research Organization, NARO) and their main characteristics are reported in Table 5.3. In Rwanda, about 100 climbing and bush bean lines are currently in advanced line validation trials to identify agronomically competitive varieties (Mulambu et al. 2017).

In parallel similar studies and trials were performed in other countries, such as for example in Guatemala (Pérez et al. 2015) and different varieties were released also in LAC countries, as shown in Table 5.4.

Moreover, interspecific crosses between *P. vulgaris* and *P. domosus, P. coccineus* or *P. acutifolius* were also conducted to introgress high Fe trait found in these related species into common bean-adapted materials (Beebe 2012; (Blair 2013).

Table 5.3 HIB varieties released under HarvestPlus in Rwanda (RAB), DRC (INERA), and Uganda (NARO)

Name	Year of release	Origin	Growth type	Grain color	Fe content ppm (%)	Grain yeld (kg/ha)	Adaptation; agronomic properties
HM21-7	2008 (DR Congo)	CIAT	Bush	Red mottled	62 (27)	1000–1500	Low to mid altitude; R: AB, AC, BCMV; T: ALS, drought
MORE 88002	2016 (Uganda)	CIAT	Bush	Yellow	70 (45)	1000–1200	Low to mid altitude; under assessment
RWR 2154	2010 (Rwanda) 2016 (Uganda)	RAB	Bush	Sugar	71 (47)	1200–1700	Low to mid altitude; R: AB, AC, BCMV; T: ALS
RWR 2245	2010 (Rwanda) 2011 (DR Congo) 2016 (Uganda)	RAB	Bush	Red mottled	76 (59)	1000–1500	Low to mid altitude; R: AB, AC, BCMV; T: ALS, RR
PVA 1438	2013 (DR Congo)	INERA	Bush	Red kidney	79 (66)	1000–1500	Mid to high altitude; R:CBB, RR; T: BCMV
MAC 44	2010 (Rwanda) 2016 (Uganda)	CIAT	Climber	Red mottled	78 (64)	2500–3000	Mid to high altitude; R: AC; T: AB; ALS, BCMV, RR
RWV 1129	2010 (Rwanda)	RAB	Climber	Salmon	77 (61)	3000	Mid to high altitude; R:AC, BCMV, RR; t: AB, ALS
RWV 3006	2012 (Rwanda)	RAB	Climber	White	78 (64)	3800	Mid to high altitude; R:AB, AC, ALS, BCMV
RWV 3316	2012 (Rwanda)	RAB	Climber	Red	87 (84)	4000	High altitude; R: AC, BCMV; T: AB, ALS
RWV 3317	2012 (Rwanda)	RAB	Climber	Sugar	74 (54)	4000	High altitude; R: AC, BCMV; T: AB, ALS
MAC 42	2012 (Rwanda)	CIAT	Climber	Sugar	91 (94)	3500–4000	High altitude; R: AC, BCMV; T: AB, ALS
RWV 2887	2012 (Rwanda)	INERA	Climber	Dark red	85 (80)	3800	High altitude; R: AC, BCMV; T: AB, ALS

(continued)

Table 5.3 (continued)

Name	Year of release	Origin	Growth type	Grain color	Fe content ppm (%)	Grain yeld (kg/ha)	Adaptation; agronomic properties
CAB 2	2010 (Rwanda)	CIAT	Climber	White	95 (98)	4000–4500	High altitude; R: AC, BCMV; T: AB, ALS
Namulenga	2013 (DR Congo)	CIAT	Climber	Zebra	76 (60)	2500–3000	Mid to high altitude; R: AC, BCMV, RR; T: AB, ALS
Cod MLV 059	2012 (DR Congo)	INERA	Climber	Red mottled	84 (77)	2000–3000	Mid to high altitude; R: AC, CBB, RR; T: LAS, BCMV
Cuarentino	2013 (DR Congo)	INERA	Climber	White	100 (114)	2000–3000	Mid to high altitude; R: AC, BCMV, CBB; T: RR
Nyiramuhundo	2016 (Uganda)	Rwanda	Climber	Yellow-orange	67 (39)	1200-1700	Mid to high altitude; under assessment

Data from Mulambu et al. (2017). The different abbreviations stand for: *R* resistance, *T* tolerance, *AB* asochyta blight, *AC* anthracnose, *ALS* angular leaf spot, *BCMV* bean common mosaic virus, *CBB* common bacterial blight, *RR* root rot

Table 5.4 Testing and release status of common bean Fe biofortified varieties developed under the HarvestPlus program (December 2016)

Status of varieties	Region	
Tested in *n* countries	Africa	8
	LAC	9
	Total	17
Released in n countries	Africa	4
	LAC	8
	Total	12
Number of varieties released	Africa	26
	LAC	18
	Total	44

5.4.1.3 Inheritance of Fe Seed Accumulation

Another important aspect in a biofortification program is the study of the inheritance and physiology of the trait of interest. Different quantitative trait loci (QTL) studies have been performed, using intergene pool populations (derived from crosses between Andean and Mesomerican genotypes) (Blair et al. 2009a), intragene pool populations (crosses among Andean or Mesoamerican genotypes) (Blair et al. 2010b, 2011; Cichy et al. 2009), and a population deriving from a cross between a wild and a cultivated genotype (Blair and Izquierdo 2012; Guzman-Maldonado et al. 2003). Different QTLs for Fe were found, some of them are common to the different studies, some others are specific of the studied population. Interestingly, several QTL for Fe and Zn content colocalized or overlapped, suggesting common mechanisms for mineral uptake and/or loading (Blair 2013). More recently, an association mapping approach was used to identify molecular markers associated with Fe, Zn, and protein content in an Indian core collection of common bean genotype (Mahajan et al. 2017). All the mentioned studies were performed on the whole seed, including cotyledonary tissues, embryo axis, and seed coat. Interestingly, it was shown that both the distribution and the inheritance of micronutrients are different in the maternally derived seed coat tissue versus the cotyledonary tissues. On the same analyzed materials, the range of Fe concentration in seed coat was very large, ranging from 20 to 263 μg/g, compared to values between 54.5 and 93 μg/g found in cotyledonary tissue (Blair et al. 2013).

Possible candidate genes included in the identified QTLs are: phaseolin, the major common bean storage protein; ferritin, a Fe-binding protein in different species, although, as discussed below it does not have a major role in common bean; Fe reductase, a root protein that reduce Fe^{3+} to Fe^{2+}, the first step for Fe uptake in strategy I plants (Blair 2013).

5.4.2 Decreasing Antinutritional Compounds Concentration

The major compounds that decrease Fe bioavailability are PA and polyphenols (PP). Here, we summarized the main efforts done in the last years in order to characterize the role of these two major classes of antinutrients and to develop common bean materials with low PA content.

5.4.2.1 Phytic Acid

Ferritin has been suggested as the major iron-storing protein in legume seeds (Marentes and Grusak 1998). However, different studies, using different approaches, scaled down this role for ferritin in common bean. On one side, a study on Fe localization in three *Phaseolus* species at microscopic level revealed that Fe mainly accumulates in the cytoplasm of the cells surrounding the provascular tissue, and does not co-localize with ferritin that is mainly accumulated in amyloplasts (Cvitanich et al. 2010). With a different approach, consisting in an isotope dilution method to quantify ferritin, Hoppler et al. (2014) found that in common bean seed only 15–30% of total Fe is present in the form of ferritin-bound, while 70–85% is in the form of non-ferritin-bound Fe. When Fe concentration increases, a correspondent increase in the non-ferritin bound Fe was observed. Moreover, a positive correlation between non-ferritin bound Fe and phytate was also shown (Hoppler et al. 2014). All these results suggest that the majority of Fe is bound to PA.

PA chemically corresponds to *myo*-inositol 1,2,3,4,5,6 hexa*kis*phosphate that, being highly negatively charged, is a strong cation chelator, stored as the main source of phosphorous (P) in the seed. After germination, P becomes available for the developing seedlings, due to the activity of phytases, enzymes that hydrolyze PA into inorganic P and lower *myo*-inositol phosphates ($InsP_s$). Phytases are present in ruminants, but absent in monogastric animals, including humans. For this reason PA decreases the nutritional value of the seeds by limiting mineral bioavailability for these animals nutrition. In plant vegetative tissues PA plays important roles in the regulation of different cell processes together with lower and higher $InsP_s$, such as hormone activity, abiotic and biotic stress response, calcium and sugar signaling, phosphorus homeostasis, photomorphogenesis, chromatin modification and remodeling, and mRNA nuclear export (Sparvoli and Cominelli 2015).

In common bean, PA is mainly accumulated in the cotyledons (95–98%), only a small portion is located in the embryo (1–3%) and the seed coat (0.5–4%), reaching 3% of total seed weight (Blair et al. 2012; Petry et al. 2015). PA concentration in common bean vary in wild or cultivated germplasm as well as in segregating populations (Guzman-Maldonado et al. 2003; Blair et al. 2009b) and ranges from 4 to 26 mg/g, with a mean content of about 10 mg/g and a PA: iron molar ratio ranging from 6:1 to 33:1 (Petry et al. 2015). The PA level depends not only on the bean variety, but also on phosphorous concentration in the soil (Blair et al. 2012). The majority of common bean genes involved in PA biosynthesis and transport were initially isolated and sequenced from 'Taylor's Horticultural' and the 905 breeding line (Mesoamerican gene pool), and mapped on the common bean reference genetic map of McClean et al. (2008), using an in silico mapping strategy against the soybean genome (Fileppi et al. 2010). *PvMRP1* and *PvMRP2* paralog genes, coding for two PA-ABC family putative transporters, were also identified (Panzeri et al. 2011). Other putative common bean genes involved in PA biosynthesis and transport were recently described (Cominelli et al. 2017). Moreover, QTL analysis for PA and P content, performed on a mapping population, derived from a cross between an Andean (G19833) and a Mesoamerican (DOR364) parent, showed association between PA concentration QTL and some of the already described genes. Interestingly, different studies reported that most of the seed PA- and P-related QTL were independent on the Fe and Zn seed mineral QTL, suggesting possibilities of improving common bean seed nutritional quality, both increasing Fe and decreasing PA concentration (Blair et al. 2012). The availability of the whole common bean genome from two different genotypes, representing the Andean and Mesoamerican gene pools (Vlasova et al. 2016; Schmutz et al. 2014), allowed the identification of a higher number of genes putatively involved in these processes. Available transcriptomic data helped to select the ones most probably responsible for PA accumulation in the seed (Cominelli et al. 2017).

Although one of the strategies proposed to increase the bioavailability of minerals from seeds is the development of *low phytic acid* (*lpa*) mutants, different *lpa* mutants were described to be affected by negative pleiotropic effects, such as low germination rates, reduced seed development and weight, and stunted vegetative growth, due to the mentioned roles of PA in the regulation of different cellular processes. Consequently, breeding programs aimed at obtaining *lpa* crops should take into account aspects regarding the agronomic potential of these mutants (Sparvoli and Cominelli 2015). *lpa* mutations belong to three different classes, depending on the affected genes: (1) mutations disturbing the first steps of the PA biosynthetic pathway; (2) mutations perturbing the last part of the pathway; and (3) mutations affecting PA transport. Mutants belonging to the first and third classes are generally characterized by decreased PA levels and a molar equivalent increase in inorganic P. A specific characteristic of only the second class of mutants is accumulation of lower $InsP_s$ (inositol phosphates with up to five phosphate residues) (Sparvoli and Cominelli 2015).

In common bean the screening of two ethyl methanesulfonate-mutagenized populations for high inorganic P phenotype, a typical feature of *lpa* mutants, allowed the isolation of two allelic mutants affecting the *PvMRP1* PA transporter: *lpa1* (originally isolated in the 905 breeding line background) and *lpa1*^{2} (isolated in the BAT 93 genotype) and of other putative mutants whose characterization is still in progress (Campion et al. 2009b; Panzeri et al. 2011; Cominelli et al. 2018). The *lpa1* and *lpa1*^{2} mutants are characterized by a 75–90% reduction in PA seed concentration; for *lpa1* a 25% reduction of raffinosaccharides content, a 30% reduction of *myo*inositol and a sevenfold increase of free Fe cations were also shown (Campion et al. 2009b, 2013; Panzeri et al. 2011). Furthermore, a negative feedback on the expression of key genes of the PA biosynthetic pathway was observed in the *lpa1* mutant (Panzeri et al. 2011). The *lpa* mutations affecting orthologous PA MRP transporters in other crops show negative pleiotropic effects, such as reduced germination, stunted growth, reduced tolerance to stresses, showing a very limited potential of these genotypes for use in breeding (Sparvoli and Cominelli 2015). Different agronomic traits, such as seedling emergence, dry seed yield, seed weight, and plant growth duration, were evaluated in the common bean *lpa1*, but no significant differences were observed, compared to the reference genotype (Campion et al. 2009b, 2013). Preliminary data suggest that also the allelic *lpa1*^{2} mutant behaves in the same way. *PvMRP1* is highly expressed in seeds at different developmental stages, while the expression of its paralog, *PvMRP2*, is hardly detectable. Conversely, in different vegetative organs both genes are expressed at similar levels, suggesting that *PvMRP2* is able to complement *PvMRP1* function in organs other than the seed in the *lpa1* and *lpa1*^{2} mutants. Interestingly, a very refined regulation of the expression of *PvMRP1* and *PvMRP2* genes at different transcriptional, post-transcriptional, and translational levels was suggested by the GUS activity pattern in transgenic *Arabidopsis thaliana* and *Medicago truncatula* lines harboring the putative promoter sequences of the two genes fused upstream of *GUS* reporter, and by *in silico* analysis of the promoter and 5′UTR sequences (Cominelli et al. 2018). It was recently shown that *lpa1* plants establish an efficient nitrogen fixing symbiosis with *Rhizobium etli* CE3. The nitrogen-fertilized *lpa1* plants show milder stress symptoms in response to water stress and increased biomass after the rehydration recovery period, compared to the wild-type plants. Moreover, the mutant genotype has higher drought resistance index, under symbiosis conditions, than the wild type. Transcriptional analysis revealed higher nodule function gene expression under control conditions and higher stress-related gene expression in nodules and bacteroids of *lpa1* plants, compared to the wild-type plants (Chiozzotto et al. 2018). These data suggest that PA has an important role also in nodule function, an aspect that was never previously reported. Moreover, these results open new perspectives to obtain new varieties nutritionally improved and able to cope with water stress.

Molecular markers for these two mutations were developed and breeding programs aimed to introgress these mutations in cultivated varieties are in progress (Sparvoli et al. unpublished results).

Other 28 putative *lpa* mutants were isolated from the EMS mutagenized population in the BAT 93 background. All are characterized by a modest increase in free inorganic P concentration (between 51% and 176%, compared to the wild type concentration). Three hypotheses were formulated concerning the genes that can be affected in these mutants: (1) genes involved in the intermediate or late steps of the biosynthetic pathway; (2) weak mutations in early genes or in *PvMRP1*; (3) redundant genes of the pathway. A candidate genes approach will be used to identify the affected genes (Cominelli et al. 2018).

5.4.2.2 Polyphenols

Phenolic compounds represent about 11% of common bean seed weight and are predominantly located in the seed coat. They are responsible for the seed color, a trait, together with color pattern, that shows a huge variability in this species. Extensive genetic analyses have identified specific genes that control these two traits. Many of these genes exhibit epistatic interactions with other genes, interactions that define the many seed coat patterns and colors observed within the species (McClean et al. 2002). PP present in common bean belong to different subclasses, including: anthocyanins, flavanols, flavanones, flavons, flavonols, isoflavonoids, hydroxybenzoic acids, hydroxycinammic acids and stilbenes, as recently reviewed (Ganesan and Xu 2017). Many of these compounds, studied also in common bean, are known for their beneficial effects on human health, such as anti-oxidant, anti-diabetic, anti-obesity, anti-inflammatory, anti-mutagenic, and anti-cancerogenic properties. However, they are also considered antinutritional factors, because they reduce digestion, binding to proteins, and Fe bioavailability (Ganesan and Xu 2017). However, recent research identified specific PP, present in common bean seed coat, that, at least in an in vitro system, are able to promote Fe uptake, such as catechin, 3,4-dihydroxybenzoic acid, kaempferol, and kaempferol 3-glucoside. Nevertheless, the inhibitory compounds have a more potent effect that outweighs the ability of promoting compounds to increase Fe uptake (Hart et al. 2015, 2017). To our knowledge no specific breeding program was undertaken to develop materials with a particular PP profile in order to reduce negative effect on Fe bioavailability. However, these results should be taken under considerations for breeding bean lines with improved Fe nutritional qualities.

5.4.3 Bioavailability and Bioefficacy Trials

In the last years, bioavailability trials and, in the case of HIB, also bioefficacy trials, showed that efforts done in the development of HIB and *lpa* mutants have been important achievements in the biofortification of common bean, as recently reviewed (Boy et al. 2017). Hence, this crop can be considered a good vehicle for Fe biofortification (Petry et al. 2015). Stable isotope absorption studies, mainly in Rwandese women with low iron status, were conducted to evaluate bioavailability from HIB developed in the frame of the HarvestPlus program. Fe bioavailability from HIB, consumed with potatoes or rice, ranges from 3.8% to 7.3% (Petry et al. 2012, 2014, 2016). However, it was calculated that the total amount of Fe absorbed from HIB (234–431 μg per day) represents up to 30% of the physiologic requirement for non-pregnant non-lactating women of reproductive age (FAO/WHO 2004).

From trial studies the role of PA as a very strong inhibitor of Fe bioavailability resulted clear. PA increases with the increase of Fe concentration in HIB, as the gained amount of Fe is stored as a non-bioavailable form of Fe bound to PA. When the PA concentration in HIB was 2 μg/g the additional Fe absorbed from HIB was 80% compared to control beans (Petry et al. 2016), when its concentration was 5.4 μg/g the increase in Fe absorbed was irrelevant (Petry et al. 2012, 2013, 2014). Trials with partial and completely dephytinized HIB and with the *lpa1* beans confirmed these results (Petry et al. 2013, 2014, 2016). Although in the study from Petry et al. (2013) a 50–60% increase of Fe bioavailability resulted from the *lpa1* mutant seeds compared to the control ones, in a more recent study, *lpa1* beans caused gastrointestinal disease in the majority of the participants. It was hypothesized that the problems were caused by the presence of residues of leucoagglutinating phytohemagglutinin (PHA-L) in these beans (Petry et al. 2016). Sparvoli et al. (unpublished results) showed that the stability of PHA-L is highly increased, due to the presence of a high concentration of free cations, because they are not complexed in the *lpa1* mutant with only 10% of PA, compared to the reference genotype. The phenomenon is strongly attenuated in the presence of EDTA and it is possible to mimic the phenomenon in the wild type, adding calcium in excess. The *lpa1* mutant is also characterized by a "hard to cook" phenotype that was already correlated to a higher concentration of cations in the cell wall-middle lamella area of the parenchyma cells that can stabilize some components of the seed coat, resulting in seeds not softening during cooking, as previously shown for cowpea (Kruger et al. 2015). This phenomenon is evident only when the *lpa1*mutation is present in a genetic background harboring only the PHA-L protein and not when only the PHA-E (erythroagglutinating phytohemagglutinin) or both PHA-L and PHA-E are present. Hence, this aspect should be taken into account in breeding programs aimed to introgress the *lpa1* mutation.

Some concerns with the use of *lpa* crops exist such there is evidence that PA is a broad-spectrum antineoplastic agent, acting in different steps of cancer development and progression. Hence, *lpa* crops may be useful in areas where micronutrient deficiencies are prevalent so as to increase Fe and Zn bioavailability, while high amounts of phytates may have a health benefit in societies where source of more bioavailable form of Fe are present in the diet and cancer and obesity are on the rise (Blair 2013).

In some trials the role of PP as inhibitors of Fe bioavailability was also investigated. Their inhibitory action was more evident in the absence of PA. When a quantity of bean seed coats, containing 50 and 200 mg of PP, was added to a bread meal (with no PA) there is a decrease in Fe absorption of 14% and 45%, respectively

(Petry et al. 2010). In another study two different common bean genotypes with similar PA concentration but differing two times in PP content, were used. Fe absorption was only 27% lower from the high PP genotype and no difference was observed in Fe absorption between the two accessions, when rice and potatoes were added to the meal (Petry et al. 2012). *lpa1* mutants and the corresponding reference genotypes in two genetic backgrounds, a colored and a white one, were used. The inhibitory effect of PP on Fe absorption was not evident in this study and the highest Fe absorption was obtained from colored *lpa1* beans (Petry et al. 2013).

The inhibitory effect of PA on Fe absorption is clear, while for PP it has not yet clearly defined, maybe depending on the high heterogeneity of polyphenol compounds that may have different effects on Fe bioavailability.

HIB were used also in two bioefficacy trials on Rwandese women Fe-depleted, aged 18–27 years, consuming HIB for 18 weeks. In the first one it was assessed that consumption of HIB increases Fe status (Haas et al. 2016). The second study showed that HIB diet positively affected cognitive performance in terms of speed of spatial selective attention, improvement in speed, efficiency, and memory (Murray-Kolb et al. 2017).

Interestingly, it was shown that HIB consumption in broiler chickens (*Gallus gallus*) modified the composition of the gut microbiota. Although no phylogenetic diversity was observed between the microbiome of chicken fed with HIB or with control beans, there were differences in the composition of microbiota between the two groups of chickens. Particularly, the HIB diet decreased the number of taxa participating in bacterial Fe uptake and increased abundance of bacteria involved in phenolic catabolism and in the production of butyrate. Importantly, no negative alteration in the gut microbiota was observed in chicken fed with HIB, a consequence that was observed with other nutritional methods of Fe supplementation (Reed et al. 2017).

5.5 Future Perspectives

Consumers are becoming better educated about nutrition and more sophisticated in choosing foods that are wholesome and nutritious. Breeding strategies for processing and consumer acceptance have been suggested (Adams and Bedford 1973) and breeding for culinary quality has been proposed and carried out (Ghaderi et al. 1984). Breeders can manipulate culinary and nutritional quality if the existing genotypic variability is significant with respect to environmental effects and if a useful screening method is available. Besides genotypic variability, also the complexity of the trait, which might be under the control of multiple genes, needs to be faced by the breeders.

Improvement of the common bean means possessing in-depth knowledge of its genetic diversity, the genome and gene functions, to enable the analysis of pathways and networks in response to fluctuating environmental conditions. Various genomic resources for common bean are available and include physical maps, bacterial artificial chromosome libraries, anchored physical and genetic maps, expressed sequence tags, and the recently published complete genome sequence (Schmutz et al. 2014; Vlasova et al. 2016). However, these approaches require precise phenotypic data. Complex interactions between the crop genotype, environmental factors in combination with plant population dynamics and crop management greatly affect plant phenotypes in field experiments. Hence, novel techniques should be kept cost-effective and robust under varying field conditions and should allow for the monitoring of various and complex traits.

However, for some nutritional traits (antinutritional factors) good results have been achieved and since these are relatively simple traits the classical breeding approaches could produce interesting results. For example, highly nutritious (non-toxic, low pGI, and biofortified) beans and flours might be obtained combining the PHA-null trait (pinto lectin and active αAI) with the *lpa1* mutation.

Concerning Fe biofortification, a promising achievement might derive from breeding simultaneously for high Fe and low PA. As previously reported, both strategies, used individually, have been shown successful to increase Fe bioavailability (Petry et al. 2015). Most PA- and P-related QTLs are independent of Fe QTLs (Blair et al. 2012). Therefore, the possible strategy to combine both traits is supposed to give also better results (Petry et al. 2014). Moreover, the knowledge of QTLs specific for Fe accumulation in the seed coat and the finding that some PP may increase Fe bioavailability, suggest also breeding strategies in order to further increase Fe accumulation and bioavailability acting also in the seed coat (Sperotto and Ricachenevsky 2017). Recently obtained data highlighted which are the good combinations between the *lpa1* mutation and the PHA-E and/or PHA-L proteins in order to avoid the "hard to cook" phenotype and the induced digestive problems in human subjects. These will be the combinations that have to be preferred for future breeding programs to introgress the *lpa1* mutation. Moreover, as already shown for the *lpa1* mutant that shows milder stress symptoms in response to drought conditions (Chiozzotto et al. 2018), some other possible positive pleiotropic effects may be associated with *lpa* mutations. The availability of the whole common bean genome sequence allows the use of reverse genetic approaches, such as Targeting Induced Local Lesions in Genome (TILLING), also in its high-throughput version of TILLING by sequencing, or genome editing, in order to isolate new mutants. Some interesting possible targets may be genes involved in the synthesis or transport of PA, as recently discussed (Cominelli et al. 2017). Moreover, genes in model species were described to have a role in seed iron localization, such as *Vacuolar Iron Transporter* (*VIT*), *NICOTIANAMINE SYNTHASE* (*NAS*) and genes coding for regulators of this process, such as *bHLH39* and *BRUTUS/HRZ-like*, as recently reviewed (Sperotto and Ricachenevsky 2017) may be target of genetic manipulation.

Acknowledgement We kindly acknowledge Dr. Roberto Bollini for critical reading of the chapter.

References

Adams MW, Bedford CL (1973) Breeding food for improved processing and consumers acceptance properties. In: Milner N (ed) Nutritional improvement of food legumes by breeding. Proceedings United Nations Protein Advisory Group, New York, NY, pp 299–304

Aragao FJL, Barros LMG, Sousa MV et al (1999) Expression of a methionine-rich storage albumin from the Brazil nut (Bertholletia excelsa H.B.K., Lecythidaceae) in transgenic bean plants (*Phaseolus vulgaris* L., Fabaceae). Gen Mol Biol 22:445–449

Ariza-Nieto M, Blair M, Welch R et al (2007) Screening of iron bloavallability patterns in eight bean (*Phaseolus vulgaris* L.) genotypes using the Caco-2 cell *in vitro* model. J Agric Food Chem 55:7950–7956

Asare-Marfo D, Birol E, Gonzalez C (2013) Prioritizing countries for biofortification interventions using country-level data. HarvestPlus working paper no. 11. International Food Policy Research Institute (IFPRI), Washington, DC

Bardocz S, Brown DS, Grant G et al (1992) Effect of the padrenoreceptor agonist clenbuterol and phytohaemagglutinin on growth, protein synthesis and polyamine metabolism of tissues of the rat. Br J Pharmcol 106:476–482

Barrett ML, Udani JK (2011) A proprietary alpha-amylase inhibitor from white bean (*Phaseolus vulgaris*): a review of clinical studies on weight loss and glycemic control. Nutr J 10:1–10

Beebe S (2012) Common bean breeding in the tropics. In: Janick J (ed) Plant breeding reviews, vol 36. Wiley-Blackwell, Hoboken, NJ, pp 357–426

Beebe S, Gonzalez AV, Rengifo J (2000) Research on trace minerals in the common bean. Food Nutr Bull 21:387–391

Beebe S, Cajiao C, Mosquera O (2005) Identification of high mineral accessions in sister species of common bean. In: Centro Internacional de Agricultura Tropical. Project IP-1. Bean improvement for the tropics. Annu Rep. CIAT, Cali, pp 59–60

Beiseigel JM, Hunt JR, Glahn RP et al (2007) Iron bioavailability from maize and beans: a comparison of human measurements with Caco-2 cell and algorithm predictions. Am J Clin Nutr 86:388–396

Bender AE, Reaidi GB (1982) Toxicity of kidney beans (*Phaseolus vulgaris*) with particular reference to lectins. J Plant Foods 4:15–22

Blair MW (2013) Mineral biofortification strategies for food staples: the example of common bean. J Agric Food Chem 61(35):8287–8294

Blair MW, Izquierdo P (2012) Use of the advanced backcross-QTL method to transfer seed mineral accumulation nutrition traits from wild to Andean cultivated common beans. Theor Appl Genet 125:1015–1031

Blair MW, Astudillo C, Restrepo J et al (2005) Analísis multi-locacional de liñeas de frījol arbustivo con alto contenido de hierro en el departamento de Nariño. Fitotec Colombiana 5:20–27

Blair MW, Astudillo C, Grusak M et al (2009a) Inheritance of seed iron and zinc concentrations in common bean (*Phaseolus vulgaris* L.). Mol Breed 23:197–207

Blair MW, Sandoval T, Caldas G (2009b) Quantitative trait locus analysis of seed phosphorus and seed phytate content in a recombinant inbred line population of common bean. Crop Sci 49:237–246

Blair MW, González LF, Kimani PM, Butare L (2010a) Genetic diversity, inter-gene pool introgression and nutritional quality of common beans (*Phaseolus vulgaris* L.) from Central Africa. Theor Appl Genet 121:237–248

Blair MW, Medina J, Astudillo C et al (2010b) QTL for seed iron and zinc concentration and content in a Mesoamerican common bean (*Phaseolus vulgaris* L.) population. Theor Appl Genet 121:1059–1070

Blair MW, Monserrate F, Beebe S et al (2010c) Registration of high mineral common bean germplasm lines NUA35 and NUA56 from the red-mottled seed class. J Plant Registr 4:55–59

Blair MW, Astudillo C, Rengifo J et al (2011) QTL analyses for seed iron and zinc concentrations in an intra-genepool population of Andean common beans (*Phaseolus vulgaris* L.). Theor Appl Genet 122:511–521

Blair MW, Herrera A, Sandoval T et al (2012) Inheritance of seed phytate and phosphorus levels in common bean (*Phaseolus vulgaris* L.) and association with newly-mapped candidate genes. Mol Breed 30:1265–1277

Blair MW, Izquierdo P, Astudillo C (2013) A legume biofortification quandary: variability and genetic control of seed coat micronutrient accumulation in common beans. Front Plant Sci 4:275

Bollini R, Carnovale E, Campion B (1999) Removal of anti nutritional factors from bean (*Phaseolus vulgaris* L.) seeds. Biotechnol Agron Soc Environ 3:217–219

Bouis HE, Saltzman A (2017) Improving nutrition through biofortification: a review of evidence from HarvestPlus, 2003 through 2016. Glob Food Sec 12:49–58

Bourne MC (1967) Size, density and hardshell in dry beans. Food Technol 21:335–338

Boy E, Haas JD, Petry N et al (2017) Efficacy of iron-biofortified crops. Afr J Food Agric Nutr Dev 17:11879–11892

Broughton WJ, Hernández G, Blair M et al (2003) Beans (*Phaseolus* spp.) – model food legumes. Plant and Soil 252:55–128

Burr HK, Kon S, Morris HJ (1968) Cooking rates of dry beans as influenced by moisture content and temperature and time of storage. Food Technol 22:336–338

Burrow MD, Ludden PW, Bliss FA (1993) Suppression of phaseolin and lectins in seeds of common bean *Phaseolus vulgaris* L.: increased accumulation of 54 kDa polypeptides is not associated with higher methionine concentrations. Mol Genet Genomics 241:431–439

Cakmak I, Pfeiffer W, McClafferty B (2010) Biofortification of durum wheat with zinc and iron. Cereal Chem 87:10–20

Campion B, Perrone D, Galasso I (2009a) Common bean (*Phaseolus vulgaris* L.) lines devoid of major lectin proteins. Plant Breed 128:199–204

Campion B, Sparvoli F, Doria E et al (2009b) Isolation and characterisation of an *lpa* (*low phytic acid*) mutant in common bean (*Phaseolus vulgaris* L.). Theor Appl Genet 118:1211–1221

Campion B, Glahn R, Tava A (2013) Genetic reduction of antinutrients in common bean (*Phaseolus vulgaris* L.) seed, increases nutrients and in vitro iron bioavailability without depressing main agronomic traits. Field Crop Res 141:27–37

Casquero PA, Lema M, Santalla M et al (2006) Performance of common bean landraces from Spain in the Atlantic and Mediterranean environments. Genet Resour Crop Evol 53:1021–1032

Chiozzotto R, Ramírez M, Talbi C et al (2018) Characterization of the symbiotic nitrogen-fixing common bean low phytic acid (*lpa1*) mutant response to water stress. Genes 9:2

Cichy K, Caldas G, Snapp S et al (2009) QTL analysis of seed iron, zinc, and phosphorus levels in an Andean bean population. Crop Sci 49:1742–1750

Cominelli E, Orozco-Arroyo G, Sparvoli F (2017) Phytic acid biosynthesis and transport in Phaseolus vulgaris: exploitation of new genomic resources. In: Marsolais F, Perez de la Vega M, Santalla M (eds) The common bean genome – compendium of plant genomes. Springer, Cham, pp 167–186

Cominelli E, Confalonieri M, Carlessi M et al (2018) Phytic acid transport in *Phaseolus vulgaris*: a new low phytic acid mutant in the *PvMRP1* gene and study of the PvMRPs promoters in two different plant systems. Plant Sci 270:1–12

Confalonieri M, Bollini R, Berardo N et al (1992) Influence of phytohemagglutinin on the agronomic performance of beans (*Phaseolus vulgaris* L). Plant Breed 109:329–334

Cvitanich C, Przybyłowicz WJ, Urbanski DF et al (2010) Iron and ferritin accumulate in separate cellular locations in Phaseolus seeds. BMC Plant Biol 10:26

de Araújo R, Miglioranza E, Montalvan R et al (2003) Genotype × environment interaction effects on the iron content of common bean grains. Crop Breed Appl Biotechnol 3:269–274

De Ron AM, Papa R, Bitocchi E et al (2015) Common bean. In: De Ron AM (ed) Grain legumes, series: handbook of plant breeding. Springer, New York, NY, pp 1–36

De Ron AM, Santalla M, Rodiño AP et al (2016) Judía. In: Ruiz de Galarreta JI, Prohens J, Tierno R (eds) Las variedades locales en la mejora genética de plantas. Servicio Central de Publicaciones del Gobierno Vasco, Vitoria, pp 155–170

Delaney DE, Bliss FA (1991) Selection for increased percentage phaseolin in common bean: 1. Comparison of selection for seed protein alleles and S1 family recurrent selection. Theor Appl Genet 81:301–305

Donangelo CM, Woodhouse LR, King SM et al (2003) Iron and zinc absorption from two bean (*Phaseolus vulgaris* L.) genotypes in young women. J Agric Food Chem 51:5137–5143

Doria E, Campion B, Sparvoli F et al (2012) Anti-nutrient components and metabolites with health implications in seeds of 10 common bean (*Phaseolus vulgaris* L. and *Phaseolus lunatus* L.) landraces cultivated in southern Italy. J Food Compos Anal 26:72–80

Escribano MR, Santalla M, De Ron AM (1990) Preliminary study of quality characters in populations of common bean from the northwestern Iberian Peninsula. An Aula Dei 20(1-2):189–198

FAO/WHO (1991) Protein quality evolution. Report of joint FAO/WHO Expert Consultation. Food and Nutrition Paper 51. Rome, FAO/WHO

FAO/WHO (2004) Vitamin and mineral requirements in human nutrition. Geneva, FAO/WHO

Fileppi M, Galasso I, Tagliabue G et al (2010) Characterisation of structural genes involved in phytic acid biosynthesis in common bean (*Phaseolus vulgaris* L.). Mol Breed 25:453–470

Ganesan K, Xu B (2017) Polyphenol-rich dry common beans (*Phaseolus vulgaris* L.) and their health benefits. Int J Mol Sci 18:11

Genovese MI, Lajolo FM (1996) Effect of bean (*Phaseolus vulgaris*) albumins on phaseolin in vitro digestibility, role of trypsin inhibitors. J Food Biochem 20:275–294

Genovese MI, Lajolo FM (1998) Influence of naturally acid-soluble proteins from bean (*Phaseolus vulgaris* L.) on *in vitro* digestibility determination. Food Chem 62:315–323

Gepts P, Bliss FA (1984) Enhanced available methionine concentration associated with higher phaseolin levels in common bean seeds. Theor Appl Gen 69:47–53

Ghaderi A, Hosfield GL, Adams MW et al (1984) Variability in culinary quality component interrelationship, and breeding implications in navy and pinto beans. J Am Hort Sci 109:85–90

Guzman-Maldonado S, Martinez O, Acosta-Gallegos J et al (2003) Putative quantitative trait loci for physical and chemical components of common bean. Crop Sci 43:1029–1035

Haas JD, Luna SV, Lung'aho MG et al (2016) Consuming iron biofortified beans increases iron status in rwandan women after 128 days in a randomized controlled feeding trial. J Nutr 146:1586–1592

Hangen L, Bennink MR (2003) Consumption of black beans and navy beans (*Phaseolus vulgaris*) resduced azoxymethane-induced colon cancer in rats. Nutr Cancer 44:60–65

Hart JJ, Tako E, Kochian LV et al (2015) Identification of black bean (*Phaseolus vulgaris* L.) polyphenols that inhibit and promote iron uptake by Caco-2 cells. J Agric Food Chem 63:5950–5956

Hart JJ, Tako E, Glahn RP (2017) Characterization of polyphenol effects on inhibition and promotion of iron uptake by Caco-2 cells. J Agric Food Chem 65:3285–3294

Hoffman LM, Donaldson DD (1985) Characterization of two *Phaseolus vulgaris* phytohemagglutinin genes closely linked on the chromosome. EMBO J 4:883–889

Hoppler M, Egli I, Petry N et al (2014) Iron speciation in beans (*Phaseolus vulgaris*) biofortified by common breeding. J Food Sci 79:C1629–C1634

Hosfield GL, Uebersax MA (1980) Variability in physicochemical properties and nutritional components of tropical and domestic dry bean germplasm. J Am Soc Hort Sci 105:246–252

Hosfield GL, Uebersax MA, Isleib TG (1984) Seasonal and genotypic effects on yield and physicchemical seed characteristics related to food quality in dry, edible beans. J Am Soc Hort Sci 109:182–189

Islam F, Basford K, Jara C et al (2002) Seed compositional and disease resistance differences among gene pools in cultivated common bean. Genet Resour Crop Evol 49:285–293

Jivotovskaya AV, Senyuk VI, Rota VI et al (1996) Proteolysis of phaseolin in relation to its structure. J Agric Food Chem 44:3768–3772

Joshi J, Pandurangan S, Diapari M et al (2017) Comparison of gene families: storage and other seed proteins. In: Perez de la Vega M, Santalla M, Marsolais F (eds) The common bean genome. Compendium of plant genome series. Springer, pp 201–218

Kelly JD, Bliss FA (1975) Quality factors affecting the nutritive value of bean seed protein. Crop Sci 15:757–760

Kruger J, Minnis-Ndimba R, Mtshali C et al (2015) Novel in situ evaluation of the role minerals play in the development of the hard-to-cook (HTC) defect of cowpeas and its effect on the in vitro mineral bioaccessibility. Food Chem 174:365–371

Layer P, Carlson G, Dimagno E (1985) Partially purified white bean amylase inhibitor reduces starch digestion *in vitro* and inactivates intraduodenal amylase in humans. Gastroenterology 88:1895–1902

Layer P, Zinsmeister A, Dimagno E (1986) Effects of decreasing intraluminal amylase activity on starch digestion and postprandial gastrointestinal function in humans. Gastroenterology 91:41–48

Le Berre-Anton V, Bompard-Gilles C, Payan F et al (1997) Characterization and functional properties of the α-amylase inhibitor (α-AI) from kidney bean (*Phaseolus vulgaris*) seeds. Biochim Biophys Acta 1343:31–40

Lioi L, Sparvoli F, Galasso I et al (2003) Lectin-related resistance factors against bruchids evolved through a number of duplication events. Theor Appl Genet 107:814–822

Mahajan R, Zargar S, Salgotra R et al (2017) Linkage disequilibrium based association mapping of micronutrients in common bean (*Phaseolus vulgaris* L.): a collection of Jammu & Kashmir, India. 3 Biotech 7:295

Marentes E, Grusak MA (1998) Iron transport and storage within the seed coat and embryo of developing seeds of pea (*Pisum sativum* L.). Seed Sci Res 8:367–375

Marquez UM, Lajolo FM (1981) Composition and digestibility of albumin, globulins and glutelins from *Pahseolus vulgaris* L. J Agric Food Chem 29:1068–1074

Marsolais F, Pajak A, Yin F et al (2010) Proteomic analysis of common bean seed with storage protein deficiency reveals up-regulation of sulfur-rich proteins and starch and raffinose metabolic enzymes, and down-regulation of the secretory pathway. J Proteomics 73:1587–1600

Martins S, Melo P, Faria L (2016) Genetic parameters and breeding strategies for high levels of iron and zinc in *Phaseolus vulgaris* L. Genet Mol Res 15:2

McClean PE, Lee RK, Otto C et al (2002) Molecular and phenotypic mapping of genes controlling seed coat pattern and color in common bean (*Phaseolus vulgaris* L.). J Hered 93:148–152

McClean P, Cannon S, Gepts P, Hudson M, Jackson S, Rokhsar D, Schmutz J, Vance C (2008) Towards a whole genome sequence of common bean (*Phaseolus vulgaris* L.): background, approaches, applications. http://arsftfbean.uprm.edu/bic/wpcontent/uploads/2018/04/Bean_Genomics_Status_2008.pdf

Montoya CA, Leterme P, Victoria NF et al (2008a) The susceptibility of phaseolin to in vitro proteolysis is highly variable across *Phaseolus vulgaris* bean varieties. J Agric Food Chem 56:2183–2191

Montoya CA, Leterme P, Beebe S et al (2008b) Phaseolin type and heat treatment influence the biochemistry of protein digestion in rat intestine. Br J Nutr 99:531–539

Montoya CA, Lallès JP, Beebe S et al (2009) Susceptibility of phaseolin (*Phaseolus vulgaris*) subunits to trypsinolysis and influence of dietary level of raw phaseolin on protein digestion in the small intestine of rats. Br J Nutr 101:1324–1332

Montoya CA, Lallès JP, Beebe S et al (2010) Phaseolin diversity as a possible strategy to improve the nutritional value of common beans (Phaseolus vulgaris). Food Res Int 43:443–449

Morari D, Stepurina T, Rotari V (2008) Calcium ions make phytohemagglutinin resistant to trypsin proteolysis. J Agric Food Chem 56:3764–3771

Mulambu J, Andersson M, Palenberg M et al (2017) Iron beans in Rwanda: crop development and delivery experience. Afr J Food Agric Nutr Dev 17:12026–12050

Murgia I, Arosio P, Tarantino D et al (2012) Biofortification for combating 'hidden hunger' for iron. Trends Plant Sci 17:47–55

Murray-Kolb LE, Wenger MJ, Scott SP et al (2017) Consumption of iron-biofortified beans positively affects cognitive performance in 18- to 27-year-old Rwandan Female College students in an 18-week randomized controlled efficacy trial. J Nutr 147:2109–2117

Mutschler MA, Bliss FA (1981) Inheritance of bean seed globulin content and its relationship to protein content on quality. Crop Sci 21:289–294

Osborn TC, Bliss FA (1985) Effects of genetically removing lectin seed protein on horticultural and seed characteristics of common bean. J Am Soc Hortic Sci 110:484–488

Osborn TC, Alexander DC, Sun SSM et al (1988) Insecticidal activity and lectin homology of arcelin seed protein. Science 240:207–210

Osborn TC, Hartweck LM, Harmsen RH et al (2003) Registration of *Phaseolus vulgaris* genetic stocks with altered seed protein compositions. Crop Sci 43:1570–1571

Palmer RM, Pusztai A, Bain P et al (1987) Changes in rates of tissue protein synthesis in rats induced in vivo by consumption of kidney bean lectins. Comp Biochem Physiol 88C:179–183

Pandurangan S, Diapari M, Yin F et al (2016) Genomic analysis of storage protein deficiency in genetically related lines of common bean (*Phaseolus vulgaris*). Front Plant Sci 7:389

Panzeri D, Cassani E, Doria E et al (2011) A defective ABC transporter of the MRP family, responsible for the bean *lpa1* mutation, affects the regulation of the phytic acid pathway, reduces seed myo-inositol and alters ABA sensitivity. New Phytol 191:70–83

Pérez S, Oparinde A, Birol E et al (2015) Consumer acceptance of an iron bean variety in Northwest Guatemala: the role of information and repeated messaging. International association of agricultural economists conference, 9–14 Aug 2015, Milan, Italy

Petry N, Egli I, Zeder C et al (2010) Polyphenols and phytic acid contribute to the low iron bioavailability from common beans in young women. J Nutr 140:1977–1982

Petry N, Egli I, Gahutu JB et al (2012) Stable iron isotope studies in Rwandese women indicate that the common bean has limited potential as a vehicle for iron biofortification. J Nutr 142:492–497

Petry N, Egli I, Campion B (2013) Genetic reduction of phytate in common bean (*Phaseolus vulgaris* L.) seeds increases iron absorption in young women. J Nutr 143:1219–1224

Petry N, Egli I, Gahutu JB, Tugirimana PL, Boy E, Hurrell R (2014) Phytic acid concentration influences iron bioavailability from biofortified beans in Rwandese women with low iron status. J Nutr 144:1681–1687

Petry N, Boy E, Wirth JP et al (2015) Review: the potential of the common bean (*Phaseolus vulgaris*) as a vehicle for iron biofortification. Nutrients 7:1144–1173

Petry N, Rohner F, Gahutu JB et al (2016) In Rwandese women with low iron status, iron absorption from low-phytic acid beans and biofortified beans is comparable, but low-phytic acid beans cause adverse gastrointestinal symptoms. J Nutr 146:970–975

Prasai N, Asare-Marfo D (2015) Biofortification Priority Index: a global strategy tool for investing in crop biofortification. http://www.harvestplus.org/knowledge-market/BPI.

Pueyo JJ, Hunt DC, Chrispeels MJ (1993) Activation of bean (*Phaseolus vulgaris*) α-amylase inhibitor requires proteolytic processing of the proprotein. Plant Physiol 101:1341–1348

Pusztai A (1991) Plant lectins. Cambridge University Press, Cambridge, pp 1–263

Pusztai A, Grant G, Spencer RJ et al (1993) Kidney bean lectin-induced Escherichia coli overgrowth in the small intestine is blocked by GNA, a mannose-specific lectin. J Appl Bacteriol 75:360–368

Quenzer NM, Huffman VL, Burns EE (1978) Some factors affecting pinto bean quality. J Food Sci 43:1059–1061

Reed S, Neuman H, Glahn RP et al (2017) Characterizing the gut (*Gallus gallus*) microbiota following the consumption of an iron biofortified Rwandan cream seeded carioca (*Phaseolus vulgaris* L.) bean-based diet. PLoS One 12:e0182431

Ribeiro ND, Jost E, Cerutti T et al (2008) Micromineral composition of common bean cultivars and its application in crop breeding. Bragantia 67:267–273

Romero-Andreas J, Yandell BS, Bliss FA (1986) Bean arcelin: 1. Inheritance of a novel seed protein of *Phaseolus vulgaris* L. and its effect on seed composition. Theor Appl Genet 72:123–128

Saltzman A, Birol E, Oparinde A et al (2017) Availability, production, and consumption of crops biofortified by plant breeding: current evidence and future potential. Ann N Y Acad Sci 1390:104–114

Santalla M, De Ron AM, Voysest O (2001) European bean market classes. In: Amurrio M, Santalla M, De Ron AM (eds) Catalogue of bean genetic resources. Fundación Pedro Barrié de la Maza/ PHASELIEU-FAIR3463/MBG-CSIC, Pontevedra, pp 77–94

Santimone M, Koukiekolo R, Moreau Y et al (2004) Porcine pancreatic alpha-amylase inhibition by the kidney bean (*Phaseolus vulgaris*) inhibitor (alpha-AI1) and structural changes in the alpha-amylase inhibitor complex. Biochim Biophys Acta 1696:181–190

Sarwar Gilani G, Wu Xiao C, Cockell KA (2012) Impact of antinutritional factors in food proteins on the digestibility of protein and the bioavailability of amino acids and on protein quality. Br J Nutr 108(Suppl 2):S315–S332

Schmutz J, McClean PE, Mamidi S et al (2014) A reference genome for common bean and genome-wide analysis of dual domestications. Nat Genet 46:707–713

Singh SP, Gepts P, Debouck DG (1991) Races of common bean (*Phaseolus vulgaris* Fabaceae). Econ Bot 45:379–396

Sparvoli F, Cominelli E (2015) Seed biofortification and phytic acid reduction: a conflict of interest for the plant? Plants 4:728–755

Sparvoli F, Daminati MG, Bollini R (1994) Biochemical and molecular characterisation of a Phaseolus vulgaris mutant lacking the major lectin related seed proteins. Ann Rep Bean Improv Coop 37:110

Sparvoli F, Bollini R, Cominelli E (2015) Nutritional value. In: De Ron AM (ed) Grain legumes, Handbook of plant breeding, vol 10. Springer, New York, NY, pp 291–325

Sparvoli F, Laureati M, Pilu R et al (2016) Exploitation of common bean flours with low antinutrient content for making nutritionally enhanced biscuits. Front Plant Sci 7:928

Sperotto RA, Ricachenevsky FK (2017) Common bean Fe biofortification using model species' lessons. Front Plant Sci 8:2187

Streit LG, Beach LR, Register JC et al (2001) Association of the Brazil nut protein gene and Kunitz trypsin inhibitor alleles with soybean protease inhibitor activity and agronomic traits. Crop Sci 41:1757–1760

Tabe LM, Droux M (2002) Limits to sulfur accumulation in transgenic lupin seeds expressing a foreign sulfur-rich protein. Plant Physiol 128:1137–1148

Taylor M, Chapman R, Beyaert R et al (2008) Seed storage protein deficiency improves sulphur amino acid content in common bean (*Phaseolus vulgaris*): redirection of sulphur from c-glutamyl-S-methylcysteine. J Agric Food Chem 56:5647–5654

The World Health Report (2002) Quantifying selected major risks to health. WHO, Geneva, p 4 http://www.who.int/whr/2002/chapter4/en/index3.html

Thompson MD, Brick MA, McGinley JN et al (2009) Chemical composition and mammary cancer inhibitory activity of dry bean. Crop Sci 49:179–186

Van Beem J, Kornegay J, Lareo L (1992) Nutritive value of the nuña popping bean. Econ Bot 46(2):164–170

Vasconcelos IM, Oliveira JTA (2004) Antinutritional properties of plant lectins. Toxicon 44:385–403

Vitale A, Bollini R (1995) Legume storage proteins. In: Kigel J, Galili G (eds) Seed development and germination. Dekker, New York, NY, pp 73–102

Vlasova A, Capella-Gutiérrez S, Rendón-Anaya M et al (2016) Genome and transcriptome analysis of the Mesoamerican common bean and the role of gene duplications in establishing tissue and temporal specialization of genes. Genome Biol 17:32

Voelker TA, Staswick P, Chrispeels MJ (1986) Molecular analysis of two phytohemagglutinin genes and their expression in *Phaseolus vulgaris* cv. Pinto, a lectin-deficient cultivar of the bean. EMBO J 5:3075–3082

Voysest O (2000) Mejoramiento genético del frijol (*Phaseolus vulgaris* L.). Legado de variedades de América Latina 1930–1999. CIAT, Cali

Welch RM, Graham RD (2004) Breeding for micronutrients in staple food crops from a human nutrition perspective. J Exp Bot 55:353–364

Welch RM, House WA (1984) Factors affecting the bioavailability of mineral nutrients in plant foods. In: Welch RM, Gabelman WH (eds) Crops as sources of nutrients for humans. American Society of Agronomy, Madison, WI, pp 37–54

Yin F, Pajak A, Chapman R et al (2011) Analysis of common bean expressed sequence tags identifies sulfur metabolic pathways active in seed and sulfur-rich proteins highly expressed in the absence of phaseolin and major lectins. BMC Genomics 12:268

Zaugg I, Magni C, Panzeri D et al (2013) QUES, a new *Phaseolus vulgaris* genotype resistant to common bean weevils, contains the Arcelin-8 allele coding for new lectin-related variants. Theor Appl Gen 126:647–661

Chapter 6
Marker-Assisted Breeding for Enrichment of Provitamin A in Maize

Hari Shanker Gupta, Firoz Hossain, Vignesh Muthusamy, and Rajkumar Uttamrao Zunjare

6.1 Introduction

Micronutrient malnutrition has emerged as a major public health problem worldwide in general, and underdeveloped and developing countries in particular (Bouis and Saltzman 2017). Around two billion people are known to suffer from micronutrient deficiency, whereas 815 million people are undernourished (Global Nutrition Report 2017). Among the 676 million children of less than five years of age, 155 million are stunted, while 52 million are affected due to wasting (UNICEF/WHO/World Bank 2017). Thirty five per cent of the world's poor live in South Asia and they are vulnerable to various health problems (IFPRI 2016). It is estimated that a country like India loses over $12 billion in GDP annually on account of vitamin and mineral deficiency alone (www.harvestplus.org). Among various micronutrients, vitamin A deficiency (VAD) is one of the most common problems among human populations (Giuliano 2014). Vitamin A is a group of organic compounds required for proper cell growth, eye vision, immune functions and reproductive system of human body (Sommer and West 1996). Since vitamin A cannot be synthesised in human body, it needs to be supplied externally either as supplement or through consumption of balanced diet. White maize, widely consumed as food in Africa, lacks provitamin A (proA), precursor of vitamin A. Yellow maize; however, contains very low level of proA. The human body is said to be vitamin A deficient when retinol reserve is <0.1 μmol/g (Tanumihardjo 2011). VAD affects approximately 4.4 million children and around 20 million women causing eyesight damage that leads to night blindness. Half of these cases with extreme VAD are reported from India (www.harvestplus.org). Keratomalacia, an inflammation that causes irreversible blindness, diarrhoea and respiratory diseases are also the results of VAD (Sommer and Davidson 2002; Mayer et al. 2008; Bouis and Saltzman 2017). Although

H. S. Gupta (✉) · F. Hossain · V. Muthusamy · R. U. Zunjare
ICAR-Indian Agricultural Research Institute, New Delhi, India

A. M. I. Qureshi et al. (eds.), *Quality Breeding in Field Crops*,
https://doi.org/10.1007/978-3-030-04609-5_6

fortification of food, supplementation and diet diversity have been the major means to address VAD, they were found non-viable in the long run due to issues like poverty, limited access to fortified foods, and poor infrastructure for regular distribution on one hand and low level of diversification and purchasing power on the other (Tanumihardjo 2011). By contrast, the breeding-based approach known as 'biofortification' is sustainable and cost-efficient for enhancing the density of micronutrients in edible portions of the crops (Giuliano 2014; Bouis and Saltzman 2017; Menkir et al. 2017) and ensures safe delivery after consumption. World Health Organisation (WHO) has recommended mean 250 RE (Retinol Equivalents) and 500 RE for children and adults per day, respectively (Bouis and Welch 2010).

Among cereals, maize (*Zea mays* L.) belongs to *Poaceae* family and occupies a significant place in the world agriculture (Gupta et al. 2015a; Hossain et al. 2016). It provides around 30% of the food calories to 4.5 billion people of 94 developing countries together with other cereals (Shiferaw et al. 2011). Global production of maize has reached about 1060 million tonnes from 187 million hectare area distributed in as many as 168 countries (FAOSTAT 2017). Livestock industry uses more than half of this global production, which has helped in rapid growth of economy (Gupta et al. 2015a). Yellow kernel maize is known to contain carotenoids predominant in lutein and zeaxanthin, which constitute non-proA fraction of carotenoids. In contrast, the quantity of proA carotenoids ranges merely from 0.25 to 2.50 μg/g, which is insufficient to meet the prescribed daily requirement of human being (Tanumihardjo 2011; Pixley et al. 2013; Zunjare et al. 2018a). Based on the factors such as bioavailability ratio (of 12:1), retention up to 50% after storage/processing, level of nutrients in the host, food matrix and food consumed in the meal, HarvestPlus, a plan of CGIAR (Consultative Group on International Agricultural Research), has fixed a target of 15 μg/g proA per unit of dry weight of maize kernel (Bouis et al. 2011; Pixley et al. 2013; Owens et al. 2014). In order to meet this target, researchers have been pursuing development of proA-rich maize hybrids through different approaches of genetic enhancement. The present review is a compilation of the global efforts in development, promotion and commercialisation of proA-rich maize that can help in alleviating VAD in human population.

6.2 Pathway for Carotenoid Biosynthesis

Genetic variation studies among maize genotypes across the countries reveal wide variations for total carotenoids, dominated by non-proA carotenoids (lutein and zeaxanthin), while narrow variation is observed for proA carotenoids (β-carotene, β-cryptoxanthin and α-carotene) (Ortiz-Monasterio et al. 2007; Burt et al. 2011; Vignesh et al. 2012; Tiwari et al. 2012; Sivaranjani et al. 2013, 2014; Suwarno et al. 2014; Muthusamy et al. 2015a, b, c). Higher influence of genotypes than genotype × environment interactions has been reported by Vignesh et al. (2012), Suwarno et al. (2014), and Muthusamy et al. (2015a, b, 2016). Apart from this, additive gene action for β-carotene, β-cryptoxanthin, and total carotenoids was reported (Egesel et al. 2003; Senete et al. 2011; Muthusamy et al. 2016).

Carotenoid metabolic pathway in maize has been well characterised (Fig. 6.1) (Hirschberg 2001; Aluru et al. 2008; Vallabhaneni and Wurtzel 2009; Yan et al. 2010; Burt et al. 2011) and the genes involved in the pathway are: *phytoene synthase1* (*psy1*) coded by *yellow endosperm1* (*y1*), (Buckner et al. 1996); *phytoene desaturase* (PDS) encoded by *viviparous5* (*vp5*) (Li et al. 1996) and *ζ-carotene desaturase* (ZDS) encoded by *viviparous9* (*vp9*) (Matthews et al. 2003); the *carotenoid isomerase* acting between PDS and ZDS (CRTISO) encoded by *yellow endosperm9* (*y9*) (Li et al. 2007); *lycopene beta cyclase* (*lcyB*) encoded by *pink scutellum1* (*ps1*) (Singh et al. 2003); *lycopene epsilon cyclase* (*lcyE*) (Harjes et al. 2008) and *β-carotene hydroxylase1* (*crtRB1* or *hyd3* or *BCH2*) (Menkir et al. 2008; Vallabhaneni et al. 2009; Yan et al. 2010).

The carotenoid biosynthesis pathway has two major branches viz., α-branch and β-branch that occur after the biosynthesis of linear all-trans-lycopene (DellaPenna and Pogson 2006; Menkir et al. 2008). Thereafter, lycopene may be cyclised either to form two β rings, as found in β-carotene and its derivatives or form one β ring and one ε ring, as found in α-carotene and its derivatives. The major proA carotenoids in maize, β-carotene and β-cryptoxanthin are produced in the β branch, whereas α-carotene is produced in α-branch of the carotenoid biosynthesis pathway (Menkir et al. 2008; Vallabhaneni and Wurtzel 2009).

6.3 Genes Responsible for Accumulation of proA

From the pathway shown above, three major genes are significant in the proA carotenoids accretion in the maize kernel: (1) *psy1* gene consisting of six exons spanning 5995 bp on chromosome 6 (bin: 6.01) involved in formation of phytoene from GGPP and shifting white coloured to yellow coloured maize (Buckler et al. 2006; Fu et al. 2013). (2) *lcyE* gene mapped to chromosome 8 (bin 8.05) consists of ten exons, spanning 3640 bp alters flux down α-carotene versus β-carotene branches of the carotenoid biosynthesis pathway (Harjes et al. 2008). The selection of favourable alleles for high β-carotene content was investigated through mapping strategy, and four natural *lcyE* polymorphisms were identified that explain 58% of the phenotypic variance and a threefold change in proA compounds. Identified four natural *lcyE* polymorphisms viz., *lcyE* 5′TE [Transposable Element; in 5′-untranslated region (UTR)], *lcyE* SNP216 (in exon 1), *lcyE* SNP2238 (in intron 4) and *lcyE* 3′InDel (in 3′-UTR) divert more lycopene to β-branch of the pathway, thereby increasing the concentration of proA. Of these, the favourable allele of *lcyE* 5′TE causes up to 30% reduction in the ratio of α- to β-branch carotenoids and increase in proA content (Harjes et al. 2008). Babu et al. (2013) also reported favourable

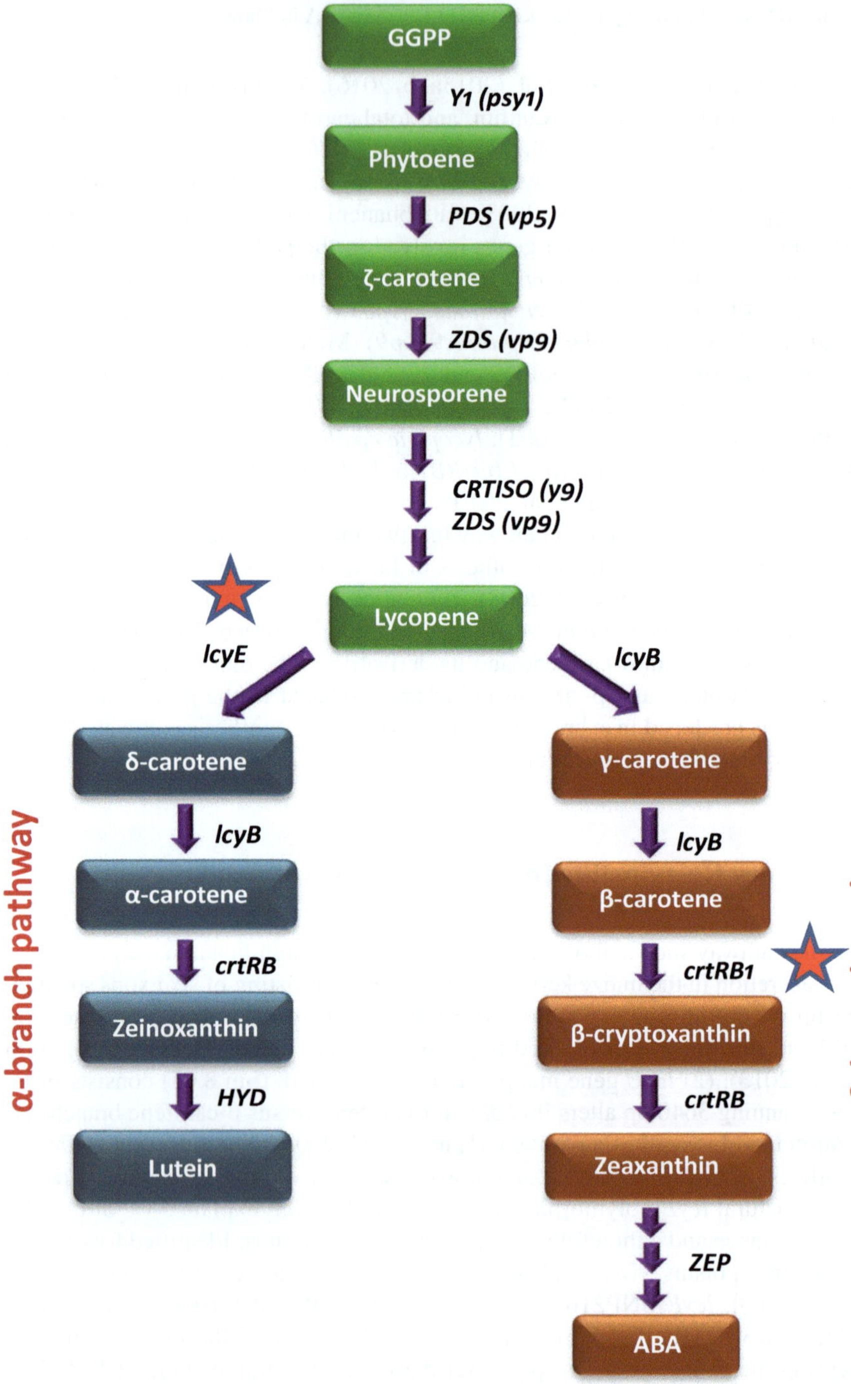

Fig. 6.1 Carotenoid biosynthesis pathway. *GGPP* geranyl geranyl pyrophosphate, *psy phytoene synthase*, *PDS phytoene desaturase*, *ZDS ζ-carotene desaturase*, *CRTISO carotene isomerase*, *lcyE lycopene ε-cyclase*, *lcyB lycopene β-cyclase*, *crtRB1 β-carotene hydroxylase*, *HYD α-carotene hydroxylase*, *ZEP zeaxanthin epoxidase*, *ABA* abscisic acid (Hirschberg 2001; Aluru et al. 2008; Vallabhaneni and Wurtzel 2009; Yan et al. 2010; Burt et al. 2011)

polymorphisms of *lcyE* that led to reduction 0–30% in the ratio of α- to β-branch carotenoids, and enhance proA concentration. (3) *crtRB1* (or *Hyd3/BCH2*) mapped on chromosome 10 (bin: 10.05) consists of 5455 bp that encodes *β-carotene hydroxylase* enzyme with three favourable alleles (Yan et al. 2010). *crtRB1* catalyses conversion of β-carotene into β-cryptoxanthin, β-cryptoxanthin to zeaxanthin and α-carotene to zenoxanthin. By association mapping, three polymorphisms viz., 5′TE (in the 5′UTR), InDel4 (in the coding region) and 3′TE (spanning the sixth exon and 3′UTR) in *crtRB1* limits the conversion of β-carotene into further components and lead to many fold increase in proA concentration of maize kernel (Vallabhaneni et al. 2009; Yan et al. 2010). The strong and statistically significant effect of *crtRB1*-favourable allele for enhanced β-carotene in maize is now very well established (Vallabhaneni et al. 2009; Yan et al. 2010; Vignesh et al. 2012; Zhang et al. 2012; Babu et al. 2013; Azmach et al. 2013; Kandianis et al. 2013; Muthusamy et al. 2014; Liu et al. 2015; Menkir et al. 2017; Zunjare et al. 2017a, 2018a).

6.3.1 Combined Effects of crtRB1 *and* lcyE

The accumulation of proA carotenoids regulated by *lcyE* and *crtRB1* genes has been reported and could be used for the enhancement of proA in the maize grain through breeding (Zhang et al. 2012; Babu et al. 2013; Kandianis et al. 2013; Muthusamy et al. 2015c; Zunjare et al. 2017a, 2018a). By developing 26 segregating populations for *crtRB1* and *lcyE* functional polymorphisms, Babu et al. (2013) demonstrated significant effect of *crtRB1* and *lcyE* interaction for proA accumulation in tropical maize. The interaction effect between *crtRB1* and *lcyE* was in agreement with that of Azmach et al. (2013) in yellow tropical maize inbreds. In their studies, it has been shown that the effect of *lcyE* alone was inconsistent for proA gain; however, it requires further validation. Recently, Zunjare et al. (2017a), while working on two digenic populations segregating for *crtRB1* and *lcyE* alleles, showed positive and favourable interaction of *crtRB1* and *lcyE* for proA, which has been corroborated by Gebremeskel et al. (2017). Although, positive interaction was observed, the favourable alleles of the both genes are rare and occur in low frequency in maize germplasm (Babu et al. 2013; Azmach et al. 2013; Muthusamy et al. 2015c; Zunjare et al. 2018a). Also, in the association mapping studies conducted by Harjes et al. (2008) and Yan et al. (2010) reported no genotypes possessing favourable alleles of both, *crtRB1* and *lcyE* genes. Hence, marker-assisted pyramiding would be useful to combine the favourable alleles in elite inbreds (Muthusamy et al. 2015c). Concerns have been raised earlier that reducing the amount of carotenoids may lead to compromised abiotic stress tolerance in crop plants (Tan et al. 1997). The transcript profiling efforts for *lcyE* by Harjes et al. (2008) and for *crtRB1* by Yan et al. (2010) revealed that the differences in expression levels were although very high in endosperm, it was not very different in embryos, and not at all different in leaves, which suggest tissue specific regulation of *lcyE* and *crtRB1*. The differential expression of *lcyE* in embryo and endosperm was also validated by Bai et al. (2009). Thus selecting for mutant allele of *lcyE* and/or *crtRB1*, whose expression is limited to endosperm, is unlikely to cause any undesirable effects in the carotenoid metabolism of leaves or other vegetative tissues (Babu et al. 2013).

6.3.2 Molecular Markers for crtRB1 *and* lcyE

DNA-based markers are recognised as important tool for accelerated breeding programmes (Collard et al. 2005). In contrast to phenotypic and biochemical markers, DNA markers are practically large in number and environmental- and plant stage-insensitive factors (Winter and Kahl 1995; Collard and Mackill 2008). In maize biofortification programme, the quantification of proA carotenoid content using HPLC is expensive as well as time-consuming. Therefore, marker-assisted selection (MAS) reduces the biochemical assay of large progenies by using gene-based markers (Gupta et al. 2013). Thus, by selecting for mutant alleles of *lcyE* and *crtRB1* genes, proA concentration can be significantly increased in the maize endosperm. A step forward in this direction was the development of the co-dominant markers for both *lcyE* and *crtRB1* by Babu et al. (2013) followed by preliminary screening of tropical maize germplasm using these PCR-based markers by breeders at CIMMYT (Pixley et al. 2013). Thus, individual PCR assays for *crtRB1* and *lcyE* given by Harjes et al. (2008) and Yan et al. (2010), respectively, were used in the marker-assisted introgression programme by Babu et al. (2013), Muthusamy et al. (2014) and Zunjare et al. (2017b). A multiplex protocol was developed in our laboratory for simultaneous foreground selection of *crtRB1* and *lcyE* in marker-assisted pyramiding programme, which in turn further accelerate proA biofortification programme by saving significant time and cost (Zunjare et al. 2017b). Further, sequencing studies identified SNP4, SNP13, InDel6 and InDel7 in 3′TE region of *crtRB1* (Vignesh et al. 2013); while four SNPs (SNP1, SNP2, SNP3 and SNP4) in 5′TE region of *lcyE* (Zunjare et al. 2018b) clearly differentiated the high and the low proA genotypes. These point mutations can, therefore, be utilised for development of cleaved amplified polymorphic sequence (CAPS) marker for their effective utilisation in MAS (Zunjare et al. 2018b).

6.4 Development of proA-Rich Maize Lines/Hybrids

To develop maize cultivars with higher proA concentration, two approaches have been adopted. The plant breeding approach assisted by MAS has led to incremental gains in proA. In contrast, genetic engineering approach has resulted in quantum jump in proA concentration, however, adoption of these lines face obstruction in release and commercial cultivation of genetically engineered line due to regulatory process in different countries. Results of both the approaches are presented below:

6.4.1 Molecular Marker-Assisted Breeding

MAS has greatly enhanced efficiency and effectiveness of breeding by offering several advantages such as reduction in time, selection at seedling stage, elimination of large scale phenotyping, combining multiple genes simultaneously, avoiding linkage drag and feasibility of improving traits of low heritability. To develop proA-rich hybrids, large scale experiment for elucidating the influence of marker polymorphisms in *lcyE* and *crtRB1* was carried out at CIMMYT, and the results suggest that by employing MAS for *lcyE* and *crtRB1*, it would be possible to develop tropical maize hybrids with 15 μg/g proA concentration, the level fixed by HarvestPlus for alleviating widespread VAD in humans (Babu et al. 2013). At ICAR-Indian Agricultural Research Institute (IARI), New Delhi, favourable allele of *crtRB1* gene (from CIMMYT-HarvestPlus genotypes) was introgressed in four hybrids viz., Vivek QPM9 and Vivek Hybrid27, HM4 and HM8 using MAS (Muthusamy et al. 2014). These single cross maize hybrids are popular in various agro-climatic zones of India. The improved version of Vivek QPM9 at harvest showed higher proA (21.5 μg/g), while the original hybrid contained only 2.1 μg/g (Fig. 6.2) (Muthusamy et al. 2014).

The proA level, measured after two months of harvest under normal storage conditions, was found to be as much as 8.1 μg/g in comparison to 2.2 μg/g in original hybrid. The improved hybrids showed grain characters, ear size, shape and yield similar to that of the original hybrids (Fig. 6.3).

Vivek Hybrid9, which was previously converted to quality protein maize (QPM) through MAS was released as Vivek QPM9 (Gupta et al. 2013) with high level of

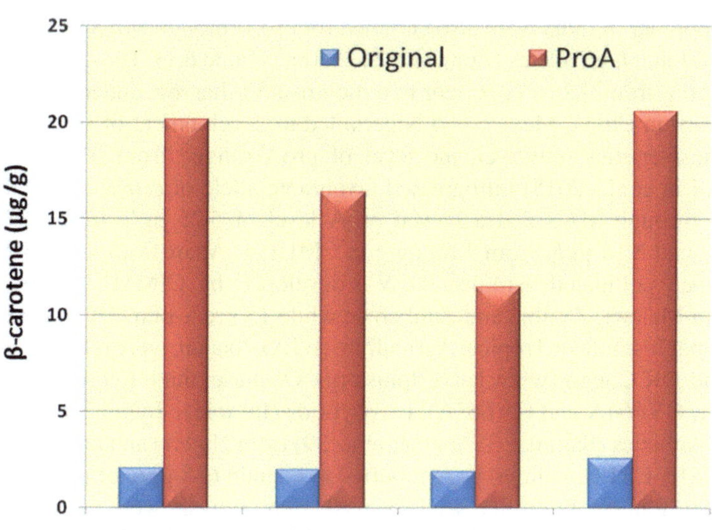

Fig. 6.2 Comparison of kernel proA concentration among original and improved hybrids

Fig. 6.3 Ear and grain characteristics of Vivek Hybrid9, Vivek QPM9 and Pusa Vivek QPM9 Improved

lysine and tryptophan-two essential amino acids that are not synthesised in human body. Lysine and tryptophan are the building blocks of protein, and are also involved as precursors for several neuro-transmitters and metabolic regulators (Gupta et al. 2015a). Normal maize is poor in these two essential amino acids in the endosperm but QPM possesses nearly two folds higher amount of these essential amino acids (Sarika et al. 2017, 2018; Hossain et al. 2018). The proA version of Vivek QPM9, developed by incorporating favourable allele of *crtRB1*, is thus a double biofortified hybrid (named as 'Pusa Vivek QPM9 Improved'), that was released in 2017 for commercial cultivation in India by the Central Sub-Committee on Standard and Notification of Varieties (CSCSNRV) earlier known as Central Variety Release Committee (Yadava et al. 2017). Likewise, four more QPM hybrids viz., HQPM1, HQPM4, HQPM5 and HQPM7 popular in India were also enriched for proA through MAS by incorporating both *crtRB1* and *lcyE* alleles (Zunjare et al. 2018a) (Table 6.1). These hybrids showed proA ranging from 9.2 to 12.9 μg/g two months after harvest under normal storage. Earlier work in China, Mexico and Nigeria led to development of several proA rich maize lines/varieties; however, the level of proA ranged from 5 to 8 μg/g only. Whereas, Liu et al. (2015) introgressed favourable allele of *crtRB1* in CML161 and CML171 through MAS and achieved proA level of 5.25 μg/g from 1.60 μg/g in CML161; and 8.14 μg/g from 1.80 μg/g in CML171. Apart from this, 11 proA-rich hybrids/open pollinated varieties (OPVs) developed by CIMMYT, Mexico, were released in Malawi, Zambia and Zimbabwe while 15 proA-rich OPVs, developed by International Institute of Tropical Agriculture (IITA), Ibadan, were released in Nigeria, Ghana and DR Congo (www.harvestplus.org). Of these, three hybrids from Zambia (GV662A, GV664A and GV665A), two hybrids (Ife maize hyb-3, Ife maize hyb-4) and two synthetics (Sammaz 38 and Sammaz 39) from Nigeria and one synthetic from Ghana (CSIR-CRI Honampa) were reported to contain 6–8 μg/g of proA (Table 6.1). Around 460 tonnes of certified seeds of proA-rich cultivars were produced for their cultivation by farmers (www.harvestplus.org). Furthermore, until 2016, 64 synthetics and 74 proA- rich hybrids were under extensive testing in 14 African countries (Manjeru et al. 2017).

Table 6.1 List of maize genotypes enriched with proA worldwide

S. No.	Genotype	Type	ProA (μg/g)	Country	Reference
1.	GV662A	Hybrid	7.0	Zambia	Simpungwe (2014)
2.	GV664A	Hybrid	7.0	Zambia	Simpungwe (2014)
3.	GV665A	Hybrid	8.0	Zambia	Simpungwe (2014)
4.	Ife maizehyb-3	Hybrid	8.0	Nigeria	Dhliwayo et al. (2014)
5.	Ife maizehyb-4	Hybrid	7.8	Nigeria	Dhliwayo et al. (2014)
6.	Sammaz 38	OPV	6.3	Nigeria	Dhliwayo et al. (2014)
7.	Sammaz 39	OPV	6.7	Nigeria	Dhliwayo et al. (2014)
8.	CSIR-CRI Honampa	OPV	6.2	Ghana	Dhliwayo et al. (2014)
9.	Pusa Vivek QPM9 Improved	QPM Hybrid	17.8–21.5[a]	India	Muthusamy et al. (2014)
10	HM4-ProA	Hybrid	10.5–12.5[a]	India	Muthusamy et al. (2014)
11.	HM8-ProA	Hybrid	19.4–21.7[a]	India	Muthusamy et al. (2014)
12.	Vivek Hybrid27-ProA	Hybrid	15.8–17.0[a]	India	Muthusamy et al. (2014)
13.	CML161	QPM Inbred	5.2	China	Liu et al. (2015)
14.	CML171	QPM Inbred	8.1	China	Liu et al. (2015)
15.	HQPM1-ProA	QPM Hybrid	9.8–10.1[b]	India	Zunjare et al. (2018a)
16.	HQPM4-ProA	QPM Hybrid	9.0–11.0[b]	India	Zunjare et al. (2018a)
17.	HQPM5-ProA	QPM Hybrid	9.2–10.0[b]	India	Zunjare et al. (2018a)
18.	HQPM7-ProA	QPM Hybrid	11.9–12.9[b]	India	Zunjare et al. (2018a)

[a]Two months after harvest and stored under normal condition
[b]Immediately after harvest

6.4.2 Genetic Engineering

Genetic engineering or transgenic technology has also been deployed to enrich carotenoids in maize (Aluru et al. 2008; Zhu et al. 2008; Naqvi et al. 2009). Overexpression bacterial (*Erwinia herbicola*) genes viz., *crtB* and *crtI* improved β-carotene content up to 10 μg/g in Hi-II maize line (Aluru et al. 2008). However, Zhu et al. (2008) and Naqvi et al. (2009) reported development of transgenic maize genotypes with very high amount of β-carotene (~60 μg/g) by stacking five genes (*psy1*, *crtI*, *lycb*, *bch* and *crtW*). White maize inbred, 'M37W' from South Africa was transformed, and showed higher amount of both proA and non-proA carotenoids (Zhu et al. 2008) but commercial use of these genotypes/lines is yet to come. Although transgenic approach is a rapid method for development of maize possessing high β-carotene, its successful adoption faces the challenges of (1) stringent regulatory mechanisms and (2) political and socio-economic factors prevailing in the country of release (Muthusamy et al. 2014).

6.5 Bioavailability of proA Carotenoids

An amount of 1 μg retinol is referred to, in the FAO/WHO recommendations, as 1 RE, which is equivalent to 12 μg of proA carotenoids. Accordingly, a conversion ratio, 12:1 was proposed by Institute of Medicine (2001). However, subsequent report by Howe and Tanumihardjo (2006a, b) proposed conversion ratio of 2.8:1. Li et al. (2010) estimated that biofortified maize porridge will have a vitamin A equivalence of 6.5:1 in North American women by measuring lipoprotein fraction of human blood. Subsequently, Muzhingi et al. (2011) measured intrinsically labelled β-carotene-rich yellow maize porridge and reported bioconversion ratio of 3.2:1 in Zimbabwean men. Both of these studies provide strong support to reducing VAD if proA-rich maize is consumed instead of normal maize. Furthermore, Gannon et al. (2014) in Zambia working on 140 rural Zambian children concluded that consumption of maize biofortified with β-carotene was efficacious and do not pose risk of hyper vitaminosis-A that was observed with the oral intake of pre-formed vitamin A from supplementation and food fortification. Bioavailability studies in animal system and in vitro simulated digestion/Caco-2 cell on biofortified maize also concluded efficient absorption of nutrients from biofortified maize (Howe and Tanumihardjo 2006b; Thakkar and Failla 2008). Neither study indicated a negative effect of the high carotenoids concentration of maize on absorption. A feeding study on Mongolian gerbil further showed that biofortified maize with high β-cryptoxanthin exhibits more efficient bioconversion (2.4:1) than β-carotene supplement (4.6:1) (Davis et al. 2008). Liu et al. (2012) reported hens fed with β-cryptoxanthin rich maize laid the eggs possessed higher proA equivalents. ProA carotenoids from biofortified maize are reported to be bioavailable to human as well as in poultry and therefore, proA-rich maize can also be effectively used as feed in the poultry diet (Diaz-Gomez et al. 2017a). The above reports suggest that conversion ratio can be reduced from 12:1 to at least 6:1 for calculating bioavailability in feeding experiments with proA-rich maize.

6.6 Stability and Retention of proA Carotenoids

The study on retention of maize carotenoids on storage is of critical importance for defining the availability of pro-A carotenoids in general and β-carotene in particular. Therefore, sound understanding of biosynthesis vis-a-vis degradation will be useful for managing to retain high carotenoid levels (Suwarno et al. 2015). ProA carotenoids are much more sensitive to degradation due to environmental factors like temperature, light, oxygen and acidic conditions (Gregory 1996; Kimura et al. 2007; De Moura et al. 2015). Several carotenoid cleavage dioxygenases (CCDs) are known to mediate degradation of carotenoids to apocarotenoids (group of cleavage products of carotenoids). A number of carotenoid cleavage genes have also been reported in maize (Vallabhaneni et al. 2010) and of these, *ZmCCD1* loci located on chromosome 9 (bin 9.07) efficiently degrades carotenoids resulting in reduction in total carotenoid content. Identification, mapping and introduction of favourable alleles of *ZmCCD* gene is expected to retain more carotenoids in kernels (Sun et al. 2008; Vallabhaneni et al. 2010). Suwarno et al. (2015) used association mapping panel of 380 inbred lines and high density genome-wide ~476,000 SNP markers platform for genome-wide association study for various carotenoids. It led to identification of genomic regions viz., *DXS1*, GGPS1 and *GGPS2* (along with *crtRB1* and *lcyE*) that play functional role in accumulation of precursor isoprenoids as well as downstream genes viz., *HYD5*, *CCD1* and *ZEP1*, which are involved in hydroxylation and degradation of carotenoids. A loci on chromosome 2, independent of *crtRB1* explains ~16% of the phenotypic contribution for β-carotene, and a mutant of *ZmCCD1* showed reduction of β-cryptoxanthin degradation (Suwarno et al. 2015). Most of the proA degradation in biofortified hybrids occurs during storage compared to cooking, while magnitude of this effect varies genotype to genotype (Mugode et al. 2014). Recently, Taleon et al. (2017) reported significant differences in retention of total proA after six months of storage under different storage conditions like aluminium bags, metal silos, multilayer polyethylene bags and common woven bags. The improved method of storage showed higher retention capacity for β-cryptoxanthin as compared to β-carotene. The storage studies indicate that with improved storage, the degradation of proA carotenoids can be slowed down, however, after six months of storage, on an average 50% reduction in β-carotene was recorded although degradation was faster during the first two weeks. The authors further reported that carotenoid degradation was similar in ears as well as shelled grains. Their investigation suggest that stored orange maize contribute 26.5% and 24.3% only for fulfilling estimated average requirement of children and women, respectively. By contrast, recent report of Dube et al. (2018) suggests that 200 g daily consumption of proA-rich maize supplies 52–64% of estimated average requirement of vitamin A. We observed in our laboratory that vacuum packaging significantly reduces the process of degradation of proA in maize grains. Further, Pillay et al. (2011) demonstrated that retention of proA in maize endosperm is affected by the cooking processes and therefore, it is important to identify the most efficient cooking method which ensures higher retention of proA carotenoids.

6.7 Consumer Preference for proA-Rich Maize

Few reports from African countries dealing with acceptability of proA maize are available (Muzhingi et al. 2008; De Groote et al. 2010; Pillay et al. 2011). Pillay et al. (2011) concluded that there is significant scope for promoting maize-based food products enriched with proA in South Africa. If proA-rich maize is cheaper and available in grocery stores, the people expressed readiness to consume yellow maize than white maize. Also, imparting education to women on the nutritional benefits of eating yellow maize will pay rich dividend (Pillay et al. 2011, 2014). The alternative would be to include yellow maize food products in pre-school feeding schemes (Pillay et al. 2011, 2014). Awobusuyi et al. (2016) studied acceptability of 'amahewu' (popular maize-based beverage in South Africa) processed using proA-rich maize. They showed that about 69% consumers liked proA-rich maize 'amahewu' either moderately or very much compared to white maize. The higher satisfaction from 'amahewu' prepared using proA-rich maize compared to that of white maize suggests that 'amahewu' holds potential to deliver proA nutrient to vulnerable population. Sensory acceptability evaluation study showed similar response to complementary food prepared from proA-rich maize and white maize (Amod et al. 2016). These results suggest that proA-rich maize has the potential to replace white maize in complementary feeding (Amod et al. 2016). In Zambia, at the end of year 2013, a sum total of 10,000 farming households had adapted proA-rich maize. It is estimated that approximately 500,000 farming households will have access to proA-rich maize by 2018 (Simpungwe 2014). These findings indicate that there is general acceptance of yellow maize biofortified with proA and there exists a potential to promote proA-rich maize consumption and its food products elsewhere.

6.8 Nutritional Impact of proA Maize

The importance of proA maize for health has been established well across the countries (Bouis and Saltzman 2017). ProA orange maize has been reported to be as efficacious as a vitamin A supplement (Gannon et al. 2014; Palmer et al. 2016). According to another study conducted in Zambia, the consumption of proA biofortified maize increased serum xanthophylls and ^{13}C-natural abundance of retinol in children (Sheftel et al. 2017). The consumption of proA-rich maize has conclusively been shown to reduce VAD in human population in several countries of Africa, however, in order to have better impact, ideal method of storage, milling and cooking methods needs to be identified that retains higher level of proA carotenoids.

Carotenoid biofortified maize has also emerged as an alternative to colour additives in poultry industry (Diaz-Gomez et al. 2017b). The consumption of proA-rich maize by chickens produced eggs rich in proA concentration (Liu et al. 2012; Moreno et al. 2016; Sowa et al. 2017). The chickens also possessed higher redness, yellowness and lower lightness in the meat and skin colour compared to white maize-fed chickens (Odunitan-wayas et al. 2016). Heying et al. (2014) showed that eating proA carotenoids daily by sows at the time of gestation and lactation enhanced liver retinol status in piglets. Hence, direct and indirect consumption of proA-rich maize would significantly contribute to nutritional security and proA-rich biofortified maize would be highly cost-effective strategy for reducing malnutrition (Lividini and Fiedler 2015).

6.9 Conclusion and Prospects

Malnutrition has turned out to be a major global public health problem especially in developing and underdeveloped world causing enormous economic loss. Deficiency of micronutrients in conjunction with low level of availability of better quality protein is known to be one of the major causes of malnutrition. Biofortified staple food crops have turned out to be the most effective way in supplying higher quantity and quality of proteins as well as micronutrients. Research conducted during the last decade has led to development and release of a large number of biofortified varieties of various food crops in general and cereals in particular. Maize, commonly used as food as well as feed, attracted attention of several researchers, who have come out with many biofortified maize hybrids/composites rich in vitamin A and/or better quality of protein. After the huge success of QPM in sub-Saharan Africa, biofortified yellow kernel maize rich in vitamin A was adopted by the farmers as well as consumers resulting in acceptance of proA-rich maize hybrids/composites/synthetics. Similar efforts in development, release and commercial cultivation of maize hybrids/composites in Asia is expected to help in meeting at least half of the recommended dietary allowance (RDA) of vitamin A in human population facing with VAD. Additionally, double biofortified proA-rich QPM will have dual advantage of providing not only vitamin A but better quality of protein (comparable to casein of milk) too. Vitamin E, measured by α-tocopherol, is known to be important for cardiovascular and neurological functions. In our laboratory, parental inbreds of proA-rich QPM hybrids have been enhanced with α-tocopherol. Iron and zinc are the two other micronutrients required in humans for basic cellular functions and are integral part of various enzymes (Gupta et al. 2015b). Widespread deficiency of these two micronutrients is known to be prevalent in the developing and underdeveloped countries. Low phytic acid mutants capable of enhancing the bioavailability of iron and zinc are now being introgressed into proA-rich QPM inbreds at ICAR-IARI, New Delhi. We aim to develop multi-nutrient maize by combining better quality of protein, high level of all the four essential micronutrients viz., vitamin A, vitamin E, iron and zinc. This multi-nutrient maize is expected to help in addressing the problem of malnutrition in a holistic manner.

References

Aluru M, Xu Y, Guo R et al (2008) Generation of transgenic maize with enhanced provitamin A content. J Exp Bot 59:3551–3562. https://doi.org/10.1093/jxb/ern212

Amod R, Pillay K, Siwela M et al (2016) Acceptance of a complementary food based on provitamin a- biofortified maize and chicken stew. J Hum Ecol 55:152–159

Awobusuyi TD, Siwela M, Kolanisi U et al (2016) Provitamin A retention and sensory acceptability of amahewu, a non-alcoholic cereal-based beverage made with provitamin A-biofortified maize. J Sci Food Agric 96:1356–1361

Azmach G, Gedil M, Menkir A et al (2013) Marker-trait association analysis of functional gene markers for provitamin A levels across diverse tropical yellow maize inbred lines. BMC Plant Biol 13:227. https://doi.org/10.1186/1471-2229-13-227

Babu R, Rojas NP, Gao S et al (2013) Validation of the effects of molecular marker polymorphisms in *LcyE* and *CrtRB1* on provitamin A concentrations for 26 tropical maize populations. Theor Appl Genet 126:389–399. https://doi.org/10.1007/s00122-012-1987-3

Bai L, Kim EH, Dellapenna D et al (2009) Novel *lycopene epsilon cyclase* activities in maize revealed through perturbation of carotenoid biosynthesis. Plant J 59:588–599. https://doi.org/10.1111/j.1365-313X.2009.03899.x

Bouis HE, Saltzman A (2017) Improving nutrition through biofortification: a review of evidence from HarvestPlus, 2003 through 2016. Glob Food Sec 12:49–58. https://doi.org/10.1016/j.gfs.2017.01.009

Bouis HE, Welch RM (2010) Biofortification—a sustainable agricultural strategy for reducing micronutrient malnutrition in the global south. Crop Sci 50:S20–S32. https://doi.org/10.2135/cropsci2009.09.0531

Bouis HE, Christine H, Bonnie M et al (2011) Biofortification: a new tool to reduce micronutrient malnutrition. Food Nutr Bull 32:202–215

Buckler ES, Gaut BA, McMullen MD (2006) Molecular and functional diversity of maize. Curr Opin Plant Biol 9:172–176

Buckner B, Miguel PS, Buckner DJ et al (1996) The *yl* gene of maize codes for *phytoene synthase*. Genetics 143:479–488

Burt AJ, Grainger CM, Smid MP et al (2011) Allele mining of exotic maize germplasm to enhance macular carotenoids. Crop Sci 51:991–1004. https://doi.org/10.2135/cropsci2010.06.0335

Collard BCY, Mackill DJ (2008) Marker-assisted selection: an approach for precision plant breeding in the twenty-first century. Philos Trans R Soc B 363:557–572. https://doi.org/10.1098/rstb.2007.2170

Collard BCY, Jahufer MZZ, Brouwer JB et al (2005) An introduction to markers, quantitative trait loci (QTL) mapping and marker-assisted selection for crop improvement: The basic concepts. Euphytica 142:169–196. https://doi.org/10.1007/s10681-005-1681-5

Davis C, Jing H, Howe JA et al (2008) β-Cryptoxanthin from supplements or carotenoid-enhanced maize maintains liver vitamin A in Mongolian gerbils (Meriones unguiculatus) better than or equal to β-carotene supplements. Br J Nutr 100:786–793. https://doi.org/10.1017/S0007114508944123

De Groote H, Tomlins K, Haleegoah J et al (2010) Assessing rural consumers' WTP for orange, biofortified maize in Ghana with experimental auctions and a simulated radio message. African Association of Agricultural Economists 48th Agricultural Economics Association of South Africa conference 25

De Moura FF, Miloff A, Boy E (2015) Retention of provitamin a carotenoids in staple crops targeted for biofortification in Africa: cassava, maize and sweet potato. Crit Rev Food Sci Nutr 55:1246–1269. https://doi.org/10.1080/10408398.2012.724477

DellaPenna D, Pogson BJ (2006) Vitamin synthesis in plants: tocopherols and carotenoids. Annu Rev Plant Biol 57:711–738. https://doi.org/10.1146/annurev.arplant.56.032604.144301

Dhliwayo T, Palacios-Rojas N, Babu R et al (2014) Vitamin A maize. Biofortification progress brief 5. International Food Policy Research Institute (IFPRI), Washington, DC

Diaz-Gomez J, Moreno JA, Angulo E et al (2017a) Provitamin A carotenoids from an engineered high carotenoid maize are bioavailable and zeaxanthin does not compromise β-carotene absorption in poultry. Transgenic Res 26:591–601. https://doi.org/10.1007/s11248-017-0029-y

Diaz-Gomez J, Moreno JA, Angulo E et al (2017b) High-carotenoid biofortified maize is an alternative to color additives in poultry feed. Anim Feed Sci Technol. https://doi.org/10.1016/j.anifeedsci.2017.06.007

Dube N, Mashurabad PC, Hossain F et al (2018) β-carotene bio-accessibility from biofortified 1 maize (*Zea mays* L.) is related to its density and is negatively influenced by lutein and zeaxanthin. Food Funct 9:379–388. https://doi.org/10.1039/c7fo01034f

Egesel CO, Wong JC, Lambert RJ et al (2003) Gene dosage effects on carotenoid concentration in maize grain. Maydica 48:183–190

FAOSTAT (2017). http://www.fao.org/faostat/en/#data/QC

Fu Z, Chai Y, Zhou Y et al (2013) Natural variation in the sequence of *PSY1* and frequency of favorable polymorphisms among tropical and temperate maize germplasm. Theor Appl Genet 126:923–935. https://doi.org/10.1007/s00122-012-2026-0

Gannon B, Kaliwile C, Arscott SA et al (2014) Biofortified orange maize is as efficacious as a vitamin A supplement in Zambian children even in the presence of high liver reserves of vitamin A: a community-based, randomized placebo-controlled trial. Am J Clin Nutr 100:1541–1550. https://doi.org/10.3945/ajcn.114.087379

Gebremeskel S, Garcia-oliveira AL, Menkir A et al (2017) Effectiveness of predictive markers for marker assisted selection of provitamin A carotenoids in medium-late maturing maize (*Zea mays* L.) inbred lines. J Cereal Sci 79:27–34. https://doi.org/10.1016/j.jcs.2017.09.001

Giuliano G (2014) Plant carotenoids: genomics meets multi-gene engineering. Curr Opin Plant Biol 19:111–117. https://doi.org/10.1016/j.pbi.2014.05.006

Global Nutrition Report (2017) https://www.globalnutritionreport.org

Gregory JF (1996) Vitamins. In: Fennema OR (ed) Food chemistry, 3rd edn. Marcel Dekker, New York, pp 545–546

Gupta HS, Babu R, Agrawal PK et al (2013) Accelerated development of quality protein maize hybrid through marker-assisted introgression of *opaque*-2 allele. Plant Breed 132:77–82. https://doi.org/10.1111/pbr.12009

Gupta HS, Hossain F, Muthusamy V (2015a) Biofortification of maize: an Indian perspective. Indian J Genet Plant Breed 75:1–22. https://doi.org/10.5958/0975-6906.2015.00001

Gupta HS, Hossain F, Nepolean T et al (2015b) Understanding genetic and molecular Bases of Fe and Zn accumulation towards development of micronutrient-enriched maize. In: Rakshit A et al (eds) Nutrient use efficiency: from basics to advances. Springer, New Delhi. https://doi.org/10.1007/978-81-322-2169-2_17

Harjes CE, Rocheford TR, Bai L et al (2008) Natural genetic variation in *Lycopene Epsilon Cyclase* tapped for maize biofortification. Science 319:330–333. https://doi.org/10.1126/science.1150255

Heying EK, Tanumihardjo JP, Vasic V et al (2014) Biofortified orange maize enhances b-cryptoxanthin concentrations in egg yolks of laying hens better than tangerine peel fortificant. J Agric Food Chem 62:11892–11900. https://doi.org/10.1021/jf5037195

Hirschberg J (2001) Carotenoid biosynthesis in flowering plants. Curr Opin Plant Biol 4:210–218

Hossain F, Muthusamy V, Bhat JS et al (2016) Maize. In: Singh M, Kumar S (eds) Broadening the genetic base of grain cereals. Springer, New York, NY. https://doi.org/10.1007/978-81-322-3613-9_1

Hossain F, Muthusamy V, Pandey N et al (2018) Marker-assisted introgression of *opaque2* allele for rapid conversion of elite hybrids into quality protein maize. J Genet. https://doi.org/10.1007/s12041-018-0914-z

Howe JA, Tanumihardjo SA (2006a) Evaluation of analytical methods for carotenoid extraction from biofortified maize (*Zea mays* sp.). J Agric Food Chem 54:7992–7997. https://doi.org/10.1021/jf062256f

Howe JA, Tanumihardjo SA (2006b) Carotenoid-biofortified maize maintains adequate vitamin a status in Mongolian gerbils. J Nutr 136:2562–2567

IFPRI (2016) Global food policy report. International Food Policy Research Institute, Washington, DC

Institute of Medicine (2001) Dietary reference intakes for vitamin A, vitamin K, arsenic, boron, chromium, copper, iodine, iron, manganese, molybdenum, nickel, silicon, vanadium, and zinc. Washington, DC, Panel on Micronutrients. National Academies Press

Kandianis CB, Stevens R, Liu W et al (2013) Genetic architecture controlling variation in grain carotenoid composition and concentrations in two maize populations. Theor Appl Genet 126:2879–2895. https://doi.org/10.1007/s00122-013-2179-5

Kimura M, Kobori CN, Rodriguez-Amaya DB et al (2007) Screening and HPLC methods for carotenoids in sweetpotato, cassava and maize for plant breeding trials. Food Chem 100:1734–1746. https://doi.org/10.1016/j.foodchem.2005.10.020

Li ZH, Matthews PD, Burr B et al (1996) Cloning and characterization of a maize cDNA encoding *phytoene desaturase*, an enzyme of the carotenoid biosynthetic pathway. Plant Mol Biol 30:269–279. https://doi.org/10.1007/BF00020113

Li F, Murillo C, Wurtzel ET (2007) Maize *Y9* encodes a product essential for 5-cis-carotene isomerization. Plant Physiol 144:1181–1189

Li S, Nugroho A, Rocheford T, White WS (2010) Vitamin A equivalence of the β-carotene in β-carotene- biofortified maize porridge consumed by women. Am J Clin Nutr 92:1105–1112. https://doi.org/10.3945/ajcn.2010.29802

Liu YQ, Davis CR, Schmaelzle ST et al (2012) β-Cryptoxanthin biofortified maize (*Zea mays*) increases β-cryptoxanthin concentration and enhances the color of chicken egg yolk. Poult Sci 91:432–438. https://doi.org/10.3382/ps.2011-01719

Liu L, Jeffers D, Zhang Y et al (2015) Introgression of the *crtRB1* gene into quality protein maize inbred lines using molecular markers. Mol Breed. https://doi.org/10.1007/s11032-015-0349-7

Lividini K, Fiedler JL (2015) Assessing the promise of biofortification: a case study of high provitamin A maize in Zambia. Food Policy 54:65–77. https://doi.org/10.1016/j.foodpol.2015.04.007

Manjeru P, Biljon AV, Labuschagne M (2017) The development and release of maize fortified with provitamin A carotenoids in developing countries. Crit Rev Food Sci Nutr. https://doi.org/10.1080/10408398.2017.1402751

Matthews PD, Luo RB, Wurtzel ET (2003) Maize *phytoene desaturase* and *ζ-carotene desaturase* catalyse a poly-Z desaturation pathway: implications for genetic engineering of carotenoid content among cereal crops. J Exp Bot 54:2215–2230. https://doi.org/10.1093/jxb/erg235

Mayer JE, Pfeiffer WH, Beyer P (2008) Biofortified crops to alleviate micronutrient malnutrition. Curr Opin Plant Biol 11:166–170. https://doi.org/10.1016/j.pbi.2008.01.007

Menkir A, White WS, Maziya-Dixon B et al (2008) Carotenoid diversity in tropical adapted yellow maize inbred lines. Food Chem 109:521–529

Menkir A, Maziya-Dixon B, Mengesha W et al (2017) Accruing genetic gain in provitamin A enrichment from harnessing diverse maize germplasm. Euphytica 213:1–12. https://doi.org/10.1007/s10681-017-1890-8

Moreno JA, Diaz-Gomez J, Nogareda C et al (2016) The distribution of carotenoids in hens fed on biofortified maize is influenced by feed composition, absorption, resource allocation and storage. Sci Rep 6:35346. https://doi.org/10.1038/srep35346

Mugode L, Ha B, Kaunda A et al (2014) Carotenoid retention of biofortified provitamin a maize (*Zea mays* L.) after Zambian traditional methods of milling, cooking and storage. J Agric Food Chem 62:6317–6325. https://doi.org/10.1021/jf501233f

Muthusamy V, Hossain F, Thirunavukkarasu N et al (2014) Development of β-carotene rich maize hybrids through marker-assisted introgression of *β-carotene hydroxylase* allele. PLoS One 9:1–22. https://doi.org/10.1371/journal.pone.0113583

Muthusamy V, Hossain F, Thirunavukkarasu N et al (2015a) Genetic variability and interrelationship of kernel carotenoids among indigenous and exotic maize (*Zea mays* L.) inbreds. Cereal Res Commun 43:567–578. https://doi.org/10.1556/0806.43.2015.012

Muthusamy V, Hossain F, Thirunavukkarasu N et al (2015b) Molecular characterization of exotic and indigenous maize inbreds for biofortification with kernel carotenoids. Food Biotechnol 29:276–295. https://doi.org/10.1080/08905436.2015.1059768

Muthusamy V, Hossain F, Thirunavukkarasu N et al (2015c) Allelic variations for *lycopene-ε-cyclase* and *β-carotene hydroxylase* genes in maize inbreds and their utilization in β-carotene enrichment programme. Cogent Food Agric 1:1033141–1033141. https://doi.org/10.1080/23311932.2015.1033141

Muthusamy V, Hossain F, Thirunavukkarasu N et al (2016) Genetic analyses of kernel carotenoids in novel maize genotypes possessing rare allele of *β-carotene hydroxylase* gene. Cereal Res Commun 44:669–680

Muzhingi T, Langyintuo AS, Malaba LC et al (2008) Consumer acceptability of yellow maize products in Zimbabwe. Food Policy 33:352–361. https://doi.org/10.1016/j.foodpol.2007.09.003

Muzhingi T, Gadaga TH, Siwela AH et al (2011) Yellow maize with high β-carotene is an effective source of vitamin A in healthy Zimbabwean men. Am J Clin Nutr 94:510–519. https://doi.org/10.3945/ajcn.110.006486

Naqvi S, Zhu C, Farre G et al (2009) Transgenic multivitamin corn through biofortification of endosperm with three vitamins representing three distinct metabolic pathways. Proc Natl Acad Sci U S A 106:7762–7767. https://doi.org/10.1073/pnas.0901412106

Odunitan-Wayas FA, Kolanisi U, Chimonyo M et al (2016) Effect of provitamin A biofortified maize inclusion on quality of meat from indigenous chickens. J Appl Poult Res 25:581–590

Ortiz-Monasterio JI, Palacios-Rojas N, Meng E et al (2007) Enhancing the mineral and vitamin content of wheat and maize through plant breeding. J Cereal Sci 46:293–307. https://doi.org/10.1016/j.jcs.2007.06.005

Owens BF, Gore MA, Magallanes-Lundback M et al (2014) A foundation for provitamin a biofortification of maize: genome-wide association and genomic prediction models of carotenoid levels. Genetics 198:1699–1716. https://doi.org/10.1534/genetics.114.169979

Palmer AC, Chileshe J, Hall AG et al (2016) Short term daily consumption of provitamin A carotenoid biofortified maize has limited impact on breast milk retinol concentration in Zambian women enrolled in a randomized controlled feeding trial. J Nutr. https://doi.org/10.3945/jn.116.233

Pillay K, Derera J, Siwela M et al (2011) Consumer acceptance of yellow, provitamin A-biofortified maize in KwaZulu-Natal. S Afr J Clin Nutr 24:186–191

Pillay K, Siwela M, Derera J et al (2014) Provitamin A carotenoids in biofortified maize and their retention during processing and preparation of South African maize foods. J Food Sci Technol 51:634–644. https://doi.org/10.1007/s13197-011-0559-x

Pixley K, Palacios-Rojas N, Raman B et al (2013) In: Tanumihardjo S (ed) Carotenoids and human health, chapter: biofortification of maize with provitamin a carotenoids. Springer, New York, NY, pp 271–292. https://doi.org/10.1007/978-1-62703-203-2_17

Sarika K, Hossain F, Muthusamy V et al (2017) Exploration of novel *opaque16* mutation as a source for high-lysine and -tryptophan in maize endosperm. Indian J Genet 77:59–64. https://doi.org/10.5958/0975-6906.2017.00008.6

Sarika K, Hossain F, Muthusamy V et al (2018) *opaque16*, a high lysine and tryptophan mutant, does not influence the key physico-biochemical characteristics in maize kernel. PLoS One 13(1):e0190945. https://doi.org/10.1371/journal

Senete CT, Guimaraes PED, Paes MCD et al (2011) Diallel analysis of maize inbred lines for carotenoids and grain yield. Euphytica 182:395–404

Sheftel J, Gannon BM, Davis CR (2017) Provitamin A-biofortified maize consumption increases seru xanthophylls and ^{13}C-natural abundance of retinol in Zambian children. Exp Biol Med 242(15):1508–1514. https://doi.org/10.1177/1535370217728500

Shiferaw B, Prasanna BM, Hellin J et al (2011) Crops that feed the world 6. Past successes and future challenges to the role played by maize in global food security. Food Secur 3:307–327. https://doi.org/10.1007/s12571-011-0140-5

Simpungwe E (2014) Delivery of vitamin A maize in Zambia. Biofortification progress brief 37. International Food Policy Research Institute (IFPRI), Washington, DC

Singh M, Lewis PE, Hardeman K et al (2003) Activator mutagenesis of the *pink scutellum1/viviparous7* locus of maize. Plant Cell 15:874–884. https://doi.org/10.1105/tpc.010249

Sivaranjani R, Prasanna BM, Hossain F et al (2013) Genetic variability for total carotenoid concentration in selected maize inbred lines. Indian J Agric Sci 83:431–436

Sivaranjani R, Santha IM, Pandey N et al (2014) Microsatellite-based genetic diversity in selected exotic and indigenous maize (*Zea mays* L.) inbred lines differing in total kernel carotenoids. Indian J Genet 74:34–41

Sommer A, Davidson FR (2002) Assessment and control of vitamin A deficiency: the Annecy Accords. J Nutr 132:S2845–S2850

Sommer A, West KP (1996) Vitamin A deficiency: health survival and vision. Oxford University Press, New York

Sowa M, Yu J, Palacios-Rojas N et al (2017) Retention of carotenoids in biofortified maize flour and b-cryptoxanthin-enhanced eggs after household cooking. ACS Omega 2:7320–7328. https://doi.org/10.1021/acsomega.7b01202

Sun Z, Hans J, Walter MH et al (2008) Cloning and characterisation of a maize *carotenoid cleavage dioxygenase* (*ZmCCD1*) and its involvement in the biosynthesis of apocarotenoids with various roles in mutualistic and parasitic interactions. Planta 228:789–801. https://doi.org/10.1007/s00425-008-0781-6

Suwarno WB, Pixley KV, Palacios-Rojas N et al (2014) Formation of heterotic groups and understanding genetic effects in a provitamin A biofortified maize breeding program. Crop Sci 54:14–24

Suwarno WB, Pixley KV, Palacios-Rojas N et al (2015) Genome-wide association analysis reveals new targets for carotenoid biofortification in maize. Theor Appl Genet 128:851–864. https://doi.org/10.1007/s00122-015-2475-3

Taleon V, Mugode L, Cabrera-Soto L et al (2017) Carotenoid retention in biofortified maize using different post-harvest storage and packaging methods. Food Chem 232:60–66. https://doi.org/10.1016/j.foodchem.2017.03.158

Tan BC, Schwartz SH, Zeevaart JA et al (1997) Genetic control of abscisic acid biosynthesis in maize. Proc Natl Acad Sci U S A 94:12235–12240

Tanumihardjo SA (2011) Vitamin A: biomarkers of nutrition for development. Am J Clin Nutr 94:658–665. https://doi.org/10.3945/ajcn.110.005777

Thakkar SK, Failla ML (2008) Micellarization of β-carotene during in vitro digestion of maize and uptake by Caco-2 intestinal cells is minimally affected by xanthophylls. FASEB J 22:1105–1106

Tiwari A, Prasanna BM, Hossain F et al (2012) Analysis of genetic variability for kernel carotenoid concentration in selected maize inbred lines. Indian J Genet 72:1–6

UNICEF/WHO/World Bank Group Joint Child Malnutrition Estimates (2017) Levels and trends in child malnutrition. UNICEF, Washington, DC

Vallabhaneni R, Wurtzel ET (2009) Timing and biosynthetic potential for carotenoid accumulation in genetically diverse germplasm of maize. Plant Physiol 150:562–572. https://doi.org/10.1104/pp.109.137042

Vallabhaneni R, Gallagher CE, Licciardello N et al (2009) Metabolite Sorting of a germplasm collection reveals the *hydroxylase3* locus as a new target for maize provitamin a biofortification. Plant Physiol 151:1635–1645. https://doi.org/10.1104/pp.109.145177

Vallabhaneni R, Bradbury LMT, Wurtzel ET et al (2010) The *carotenoid dioxygenase* gene family in maize, sorghum and rice. Arch Biochem Biophys 504:104–111

Vignesh M, Hossain F, Nepolean T et al (2012) Genetic variability for kernel beta -carotene and utilization of *crtRB1* 3′TE gene for biofortification in maize (*Zea mays* L.). Indian J Genet Plant Breed 72:189–194

Vignesh M, Nepolean T, Hossain F et al (2013) Sequence variation in 3′UTR region of *crtRB1* gene and its effect on β-carotene accumulation in maize kernel. J Plant Biochem Biotechnol 22:401–408. https://doi.org/10.1007/s13562-012-0168-4

Winter P, Kahl G (1995) Molecular marker technologies for plant improvement. World J Microbiol Biotechnol 11:438–448. https://doi.org/10.1007/BF00364619

Yadava DK, Choudhury PR, Hossain F et al (2017) Biofortified varieties: sustainable way to alleviate malnutrition. Indian Council of Agricultural Research, New Delhi, pp 1–32

Yan J, Kandianis CB, Harjes CE et al (2010) Rare genetic variation at *Zea mays crtRB1* increases β-carotene in maize grain. Nat Genet 42:322–327. https://doi.org/10.1038/ng.551

Zhang X, Pfei WH, Babu NPR et al (2012) Probability of success of breeding strategies for improving provitamin A content in maize. Theor Appl Genet 125:235–246. https://doi.org/10.1007/s00122-012-1828-4

Zhu C, Naqvi S, Breitenbach J et al (2008) Combinatorial genetic transformation generates a library of metabolic phenotypes for the carotenoid pathway in maize. Proc Natl Acad Sci U S A 105:18232–18237

Zunjare RU, Chhabra R, Hossain F et al (2017a) Development and validation of multiplex-PCR assay for simultaneous detection of rare alleles of *crtRB1* and *lcyE* governing higher accumulation of provitamin A in maize kernel. J Plant Biochem Biotechnol 27:1–7. https://doi.org/10.1007/s13562-017-0432-8

Zunjare RU, Hossain F, Muthusamy V et al (2017b) Influence of rare alleles of β-*carotene hydroxylase* and *lycopene epsilon cyclase* genes on accumulation of provitamin A carotenoids in maize kernels. Plant Breed 136:872–880. https://doi.org/10.1111/pbr.12548

Zunjare RU, Hossain F, Muthusamy V et al (2018a) Development of biofortified maize hybrids through marker-assisted stacking of β-*carotene hydroxylase*, *lycopene-ε cyclase* and *opaque2* genes. Front Plant Sci 9:178. https://doi.org/10.3389/fpls.2018.00178

Zunjare RU, Chhabra R, Hossain F et al (2018b) Molecular characterization of 5′ UTR of the *lycopene epsilon cyclase* (*lcyE*) gene among exotic and indigenous inbreds for its utilization in maize biofortification. 3 Biotech 8:75. https://doi.org/10.1007/s13205-018-1100-y

Chapter 7
Recent Advances in Breeding for Modified Fatty Acid Profile in Soybean Oil

Akshay Talukdar, M. Shivakumar, and Subhash Chandra

7.1 Introduction

Oils and fats are important sources of energy for the human diet. It also adds much to the sensory characteristics of food. About 80% of edible oils are derived from plant sources, of which about 60% are contributed by annual oilseeds viz., soybean, rapeseed and mustard, sunflower, and groundnut. Soybean oil occupies the lions share accounting for over half of the world vegetable oil production. However, soybean oil is highly unsaturated and oxidatively unstable. Therefore, majority of soybean oil undergoes partial chemical hydrogenation, which produces trans-fats as an unavoidable result. Dietary intake of trans-fats and most saturated fats are not advisable as it has negative impacts on cholesterol levels and cardiovascular health. Therefore, improving the functional and nutritional qualities of soybean oils has garnered much attention over the last 15 years or so (Edgar et al. 2009). The ideal oil should have about 7% saturates (palmitic + stearic acids), >55% oleic acid, and <3% linolenic acid (Jeong et al. 2007). Genetic modification of the fatty acid composition of soybean oil has been successful in better meeting the needs of end users than is possible with conventional oil (Fehr 2007). Reduced linolenic acid (18:3) content (say 1%) eliminates the need for chemical hydrogenation enhancing the stability and shelf life of the oil. Similarly, increased oleic acid (18:1) content (say 25 to >80%) also increases its stability and shelf life. Oil with palmitic acid (16:0) reduced from 11 to <4%, reduces saturated fatty acids, which is desirable for cardiovascular health.

A. Talukdar (✉)
Division of Genetics, ICAR-Indian Agricultural Research Institute, New Delhi, India

M. Shivakumar · S. Chandra
ICAR-Indian Institute of Soybean Research, Indore, Madhya Pradesh, India

A. M. I. Qureshi et al. (eds.), *Quality Breeding in Field Crops*,
https://doi.org/10.1007/978-3-030-04609-5_7

7.2 Breeding Approaches to Fatty Acid Modification

Great efforts are underway to manipulate soybean genetically to achieve oil with desired composition of saturated and unsaturated fatty acids. Successful results have been achieved by both conventional breeding and genetic engineering approaches. Enhancing oleic acid by exploiting natural variation among various sources of soybean germplasm has been made with great success (Takagi and Rahman 1996; Rahman et al. 2001; Alt et al. 2005a, b). Genotype with 'mid-oleic' phenotype generated through conventional approaches contains oleic acid in the range of 30–70%. However, this approach of developing the mid-oleic phenotype has some drawbacks. At the first place, pyramiding of multiple loci is a complex and difficult step (Alt et al. 2005a, b). Secondly, being polygenic in nature, the 'mid-oleic' phenotype gets affected by the growing environment; typically, it needs warmer climates for stability and maintenance of the elevated oleic acid trait. The temperature affects the desaturase activity and expression (Heppard et al. 1996; Tang et al. 2005). Besides relative amounts of polyunsaturated fatty acids, the content and composition of tocopherols in soybean oil also contribute to oil stability. Finally, the 'mid-oleic' trait from the germplasm comes with yield drag (Primomo et al. 2002). This may be due to the allelic variants selected for altered fatty acid profile during both vegetative and embryogenesis development, thereby increasing the chance of a negative agronomic effect associated with the mutant allele governing the novel oil phenotype (Tom and Edgar 2009). Further, the fatty acid profile of membranes in vegetative tissues is also gets altered. This may impact membrane fluidity and, in turn, affect the plant's ability to respond to environmental change, ultimately leading to reduced yield. The problems of yield drag and environmental instability of oleic acid levels in mid-oleic acid mutants have largely been addressed by down-regulation of FAD2-1A and -1B via seed-specific expression of post-transcriptional gene-silencing elements (Mazur et al. 1999; Buhr et al. 2002; Graef et al. 2009). Seeds produced through this approach found to contain oleic acid >80% of the total fatty acids, without alteration of the fatty acid composition of vegetative tissues. Moreover, the transgenic allele is inherited as a single dominant trait, thereby greatly facilitating breeding. High oleic acid mutants, with an oleic content ranging from 60% to 90%, have also been developed in corn, peanut, canola, and sunflower. Such mutants found to have a number of defective FAD2 genes (Perez-Vich et al. 2002; Patel et al. 2004; Tang et al. 2005; Hu et al. 2006; Belo et al. 2008). So, development of mutant lines with useful oleic acid contents would require combining several mutant loci (Mikkilineni and Rocheford 2003). Kim et al. (2015) compared a high oleic acid soybean line, JD11-0070 with three commercial soybean cultivars in four Korean environments for seed yield, protein, and oil concentrations. JD11-0070 had a stable oleic acid concentration across environments (around 79%). In addition, no significant differences were observed between JD11-0070 and other three cultivars for seed yield or protein concentration, although JD11-0070 had a lower oil concentration than others. This indicated possibility of developing high oleic acid soybean lines containing higher yields and favorable seed protein and quality oil. Chi et al. (2017) studied fatty acid composition of seeds from 319 wild soybean (*Glycine soja*) accessions in Japan. The distributions of palmitate (16:0), stearate (18:0), oleate (18:1), linoleate (18:2), and linolenate (18:3) in seeds were determined for each accession and correlations were studied. Significant inverse correlations were observed between the oleate and linolenate contents and the linoleate and linolenate contents. Moreover, a weak inverse correlation between the stearate and linolenate contents was indicated and an inverse correlation between the palmitate and linoleate contents was also found. The total palmitate and stearate contents were high in the seeds of *Glycine soja* collected from regions with high annual temperatures, while the total linolenate content was high in the seeds collected from regions with a low annual temperature. It highlighted the role of growing temperature on the concentration of various components of oil.

7.2.1 Mutation Breeding

Different breeding approaches have been utilized to change the fatty acid profile of soybean to make it more healthy and tasty. Among the conventional approaches, mutation breeding was used hugely to alter the fatty acid profile. Improvement in the level of palmitic, stearic, oleic, linoleic, and linolenic acid through mutation breeding approach is highlighted below:

White et al. (1961) studied genetic control of fatty acids in soybean seed oil to provide information to guide the selection of fatty acid traits in the breeding program. Linolenic acid was held primarily responsible for the off-flavor of the refined oil. Therefore, it caught attention of the breeders to initiate program for its reduction which got more intense after the mid-1980s.

In case of palmitic acid, until recent time, a total of seven loci (*Fap1*, *Fap2*, *Fap3*, *Fap4*, *Fap5*, *Fap6*, and *Fap7*) have been characterized and reported. Two allele viz., *fap1* and *fap3* reduce palmitate content, while *fap2*, *fap2-b*, *fap4*, *fap5*, *fap6*, and *fap7* contribute toward increased palmitate content (Erickson et al. 1988). The allele *fap5* is closely linked to *fap2-b* allele, and *fap7* is closely linked to *fap6* (Stoltzfus et al. 2000a, b). The large variation in palmitate content among lines with the same major allele is linked to the segregation of some minor genes or modifiers (Horejsi et al. 1994; Rebetzke et al. 1998).

The amount of linolenic acid in commodity soybean oil is roughly 10% of the total fatty acids, which reduces oxidative stability of the oil, leading to rancidity and decreased shelf-life of products. A number of low linolenic acid soybean oil (LLA) genotypes were developed through mutation breeding approach (Hammond and Fehr 1983; Wilcox and Cavins 1985; Fehr et al. 1992). The LLA soybean oil had <4% linolenic acid, with ultra-LLA <2%. Mutations within the *FAD*3 gene resulted in the low LLA oil phenotype (Chappell and Bilyeu 2006; Reinprecht et al. 2009). The *FAD*3 gene encodes for a D15 fatty acid desaturase that incorporates a third double bond into linoleic acid to produce linolenic acid. In soybean, three *FAD*3 genes have been identified viz., *GmFAD3A*, *GmFAD3B*, and *GmFAD3C*, of which *GmFAD3A* is highly expressed during embryogenesis and thus is the major determinant of the linolenic acid content of soybean oil (Bilyeu et al. 2003). Takagi and Rahman (1996) reported identification of two mutant lines, M23 and M11 which had elevated oleic acid content. Two different alleles at the *Ol* locus, *ol* in M23 and *ola* in M11 were responsible for the elevated oleic acid level (Istvan et al. 2008).

7.2.2 Biotechnological Approaches

Biotechnological approaches have been used to manipulate the target gene to achieve the goal in seed-specific fashion with least probability of agronomic performance being compromised. Targeted perturbation of FAD2 alleles in a seed-specific fashion in soybean has been shown to produce a high oleic acid (75–85%) phenotype in the seed oil (Kinney and Knowlton 1997; Mazur et al. 1999). Modulation of FAD2 expression using this transgenic approach was carried out by introducing transgenic elements designed to induce post-transcriptional gene silencing (Cerutti 2003). High oleic acid soybean derived from down-regulating FAD2 concurrently reduces polyunsaturated fatty acids to <6% (Kinney and Knowlton 1997); in addition, palmitic acid was reduced to about 7–8%. Seed-specific silencing of FATB genes in soybean leads to a major reduction in total saturated fatty acids (15% to <6%) (Kinney 1996). When this gene was combined with a silenced FAD2, oleic acid contents got elevated to >90% (Buhr et al. 2002). By effectively suppressing the expression of all three *GmFAD3* genes with a single RNA interference construct, Flores et al. (2008) were able to generate an ultra LLA phenotype in soybean. William et al. (2014) generated soybean lines that are low in polyunsaturated fats by introducing mutations in two fatty acid desaturase 2 genes (*FAD2-1A* and *FAD2-1B*) through TALEN technology. The fatty acid profile of the seed was dramatically changed in plants homozygous for mutations in both *FAD2-1A* and *FAD2-1B*: oleic acid increased from 20% to 80% and linoleic acid decreased from 50% to <4%. It will enable development of soybean genotype with higher oleic acid, high stearic, and low linolenic acid oils. Such oils will be free from undesirable saturated fatty acids and trans fatty acids and would facilitate improvement in cardiovascular health (Korver and Katan 2006; Lichtenstein et al. 2006; Mozaffarian and Willett 2007).

7.2.3 Molecular Breeding

Molecular markers have been used for genetic improvement of various traits in soybean including oil. Publication of the complete genome sequence of soybean has made it easy for all to develop molecular as per need and choice paving the way for genomics-assisted improvement of soybean. With the availability of cheap and quick sequencing facility, attention has now been given to utilize more of the sequence-based information for addressing issues related to crop improvement including enhanced oxidative stability of soybean oil. Masum et al. (2014) identified QTL for protein, oil, and fatty acids content in a set of $F_{5:8}$ RILs derived from a cross between lines, 'MD 96-5722' and 'Spencer' using 5376 Single Nucleotide Polymorphism (SNP) markers. One protein content QTL on Chr.14, 11 QTLs associated with oil content on six chromosomes (Chr.3, Chr.5, Chr.9, Chr.13, Chr.14, and Chr.16), and 16 QTLs for five major fatty acids (palmitic, stearic, oleic, linoleic, and linolenic acids) on Chr.3, Chr.13, Chr.14, Chr.15, Chr.16, and Chr.18 were mapped. The SNP markers closely linked to the QTL will be useful for development of cultivars with altered oil and fatty acid compositions in soybean. Zhou et al. (2015) re-sequenced 302 wild-type genotypes, landraces and improved soybean accessions at >11× depth and identified 21 fatty acid biosynthesis genes. Using GBS approach, Crystal and Jason (2017) identified and mapped 21 minor and major effects QTL for six seed oil-related traits and plant height (Table 7.1). They identified mutants that had elevated seed stearic acid, a saturated fat which has no negative impacts on cardiovascular health, from 3% to 4% in typical cultivars to 20% of the seed oil. The inheritance of a large genomic deletion on Chr.14 was the basis for largest effect QTL, resulting in 18% enhancement in seed stearic acid contents. The deleted region of the chromosome contained *SACPD-C* and another gene; loss of both the genes boosted seed stearic acid levels to about 18%. Although the deletion has enhanced the level of stearic acid, its impact on yield needs to be seen. Some reports have shown the deletion to be inextricably correlated with reduced seed yield. Molecular markers linked to stearic (18:0), oleic (18:1), linoleic (18:2), and linolenic (18:3) have been reported by various workers some of which are presented in Table 7.2. Bin et al. (2017) constructed 2534.42 cM long genetic map using 3541 markers with an average distance of 0.72 cM between adjacent markers. Inclusive composite interval mapping (ICIM) revealed 26 stable QTLs for five fatty acids, explaining 0.4–37.0% of the phenotypic variance for individual fatty acids across environments. Of these QTLs, nine were novel and stable (Table 7.3). Deployment of such QTLs through marker-assisted selection would improve the quality of oil.

Table 7.1 List of QTLs detected in a mapping population raised from the cross A6 × 194D (Crystal and Jackson 2017)

Trait	QTL	Chr. No.	SNP marker nucleotide	Position (cM)	Interval	LOD	R^2(%)	QTL effect
Palmitic acid	q.2p	2	Gm2:7138451	44.58	38–128	5.55	10.49	+0.36%
	q.5p	5	Gm5:1420686	2.48	0–8	6.79	13.06	+0.44%
	q.14p	14	Gm14:12506615	44.91	42–58	6.89	13.26	+0.44%
Stearic acid	q.2.1 s	2	Gm2:5946912	39.0	38–42	14.63	13.75	21.16%
	q.2.2 s	2	Gm2:15552879	79.79	78–80	9.05	7.86	20.52%
	q.4 s	4	Gm4:18312993	75.99	73–78	7.23	6.12	+1.24%
	q.14 s	14	Gm14:42206409	58.0	57–59	40.84	56.76	25.00%
Oleic acid	q.4o	4	Gm4:19496796	76.38	74–80	6.57	12.17	+1.37%
	q.14o	14	Gm14:32290452	51.0	49–54	13.7	28.05	+2.9%
Linoleic acid	q.4e	4	Gm4:32351007	78.88	73–84	6.74	14.75	22.31%
	q.14e	14	Gm14:34918500	54.0	45–63	4.48	9.49	+2.02%
Linolenic acid	q.2n	2	Gm2:15552879	79.79	78–81	7.05	10.4	+0.33%
	q.4n	4	Gm4:27332180	78.12	76–84	7.04	10.39	20.21%
	q.8n	8	Gm8:44116750	150.0	148–152	6.93	10.2	20.15%
	q.10n	10	Gm10:45310798	89.68	81–98	8.13	12.16	20.42%
	q.13n	13	Gm13:41141355	118.19	108–125	6.48	9.48	20.17%
	q.14n	14	Gm14:32372635	51.27	44–53	5.52	7.97	20.25%
Oil	q.9oi	9	Gm9:6801513	16.0	15–17	6.85	14.43	+0.28%
	q.14oi	14	Gm14:42206409	62.0	49–68	9.43	20.58	+0.71%

Table 7.2 QTL position, linked markers and its parental source associated with stearic, oleic, linoleic, and linolenic fatty acids in soybean oil

Fatty acid	Chr. No.	Map position (cM)	Marker	R^2 (%)	Parent 1	Parent 2	References
Stearic Fatty acid (18:0)	14	55.2	Satt168	18	N87-984-16	TN93-99	Panthee et al. (2006)
	14	72.8-75.3	Satt070	61	Dare	FAM94-41	Spencer et al. (2003)
	6	121.3	–	13	Essex	Williams	Hyten et al. (2004)
	13	27.1	Sat_090	10	RG10	OX948	Reinprecht et al. (2006)
	18	76.7	Satt288	10-19	RG10	OX948	Reinprecht et al. (2006)
	16	83.3	A233_1	19	A81-365022	PI468916	Diers and Shoemaker (1992)
	16	12.3	Satt249	11	N87-984-16	TN93-99	Panthee et al. (2006)
	19	58.2	–	16	Essex	Williams	Hyten et al. (2004)
Oleic Fatty acid (18:1)	5	92.3	A104_1	26	A81-365022	PI468916	Diers and Shoemaker (1992)
	5	92.6	A170_1	23	A81-365022	PI468916	Diers and Shoemaker (1992)
	5	96.0	Satt211	4	G99-G725	N00-3350	Monteros et al. (2008)
	14	–	A619_1	19	A81-365022	PI468916	Diers and Shoemaker (1992)
	17	79.2	Satt389	6	G99-G725	N00-3350	Monteros et al. (2008)
	15	11.0	A242_2	20	A81-365022	PI468916	Diers and Shoemaker (1992)
	15	13.6	Pb	21	A81-365022	PI468916	Diers and Shoemaker (1992)
	15	45.4	Satt263	10	N87-984-16	TN93-99	Panthee et al. (2006)
	18	0.0	Satt163	10	RG10	OX948	Reinprecht et al. (2006)
	18	76.7	Satt288	10	RG10	OX948	Reinprecht et al. (2006)
	18	43.4	Satt394	13	G99-G725	N00-3350	Monteros et al. (2008)
	18	96.6	Satt191	7	G99-G725	N00-3350	Monteros et al. (2008)
	19	82.5	–	35	Essex	Williams	Hyten et al. (2004)
	19	30.9	Satt418	9	G99-G725	N00-3350	Monteros et al. (2008)
	19	71.4	Satt561	0.25	G99-G725	N00-3350	Monteros et al. (2008)

(continued)

Table 7.2 (continued)

Fatty acid	Chr. No.	Map position (cM)	Marker	R^2 (%)	Parent 1	Parent 2	References
Linoleic Fatty acid (18:2)	5	92.3	A104_1	33	A81-365022	PI468916	Diers and Shoemaker (1992)
	5	92.6	A170_1	30	A81-365022	PI468916	Diers and Shoemaker (1992)
	5	102.3	A082_1	38	A81-365022	PI468916	Diers and Shoemaker (1992)
	11	58.9	A118_1	20	A81-365022	PI468916	Diers and Shoemaker (1992)
	14	–	Fad3i6	70-75	RG10	OX948	Reinprecht et al. (2006)
	15	11.0	A242_2	21	A81-365022	PI468916	Diers and Shoemaker (1992)
	15	13.6	Pb	20	A81-365022	PI468916	Diers and Shoemaker (1992)
	15	44.8	Satt185	13	N87-984-16	TN93-99	Panthee et al. (2006)
	13	93.7	–	10	Essex	Williams	Hyten et al. (2004)
	19	74.5	–	50	Essex	Williams	Hyten et al. (2004)
Linolenic fatty acid (18:3)	14	–	pB194-1 pB124	85	C1640	PI479750	Brummer et al. (1995)
	14	87.5	Satt534 Fad3i6	72-78	RG10	OX948	Reinprecht et al. (2006)
	15	6.3	SAC7_1	31	A81-365022	PI468916	Diers and Shoemaker (1992)
	15	11.0	A242_2	23	A81-365022	PI468916	Diers and Shoemaker (1992)
	15	28.3	K229_1	20	A81-365022	PI468916	Diers and Shoemaker (1992)
	15	30.9	A454_1	22	A81-365022	PI468916	Diers and Shoemaker (1992)
	15	34.6	A203_1	22	A81-365022	PI468916	Diers and Shoemaker (1992)
	15	45.4	Satt263	12	N87-984-16	TN93-99	Panthee et al. (2006)
	18	21.9	Satt235	22	N87-984-16	TN93-99	Panthee et al. (2006)
	9	–	A065_3	20	A81-365022	PI468916	Diers and Shoemaker (1992)
	19	36.7	A023_1	26	A81-365022	PI468916	Diers and Shoemaker (1992)
	19	50.6	–	13	Essex	Williams	Hyten et al. (2004)
	19	82.5	–	24	Essex	Williams	Hyten et al. (2004)

Table 7.3 Stable QTL for specific fatty acids in soybean seed across multiple environments

Trait	QTL	Environment	Chr. No.	Marker interval	LOD	PVE (%)	Novelty	Reported QTL or QTN by	Gene annotation
Palmitic acid	*qPA4_1*	2009/2011	4	Gm04_47098–Gm04_44762	5.1–16.3	1.7–7.7	Novel		
	qPA6_1	2009/2010	6	Gm06_8091–Gm06_2457	8.0–14.4	4.1–6.1	Novel		
	qPA8_1	2010/2011	8	Gm08_9171–Gm08_38849	43.8–48.5	28.0–37.0		Bachlava et al. (2009)	
	qPA16_1	2009/2010	16	Gm16_28849–Gm16_52098	5.9–13.5	2.2–7.5	Novel		
Stearic acid	*qSA8_1*	2009/2011	8	Gm08_53938–Gm08_3252	34.2–39.8	22.4–28.7		Li et al. (2015)	
	qSA12_1	2009/2011	12	Gm12_1782–Gm12_30904	18.0–32.6	9.9–11.9	Novel		*GmFabG*
	qSA14_1	2009/2010	14	Gm14_21994–Gm14_55450	3.8–28.8	3.0–7.0		Xie et al. (2012), Bachlava et al. (2009)	
	qSA18_1	2009/2011	18	Gm18_84824–Gm18_77518	6.0–32.0	2.8–11.6		Fan et al. (2015)	

Oleic acid	*qOA7_1*	2010/2011	7	Gm07_58581–Gm07_29390	13.0–54.0	1.3–33.6		Fan et al. (2015)	
	qOA8_1	2009/2010	8	Gm08_13258–Gm08_43422	6.0–6.1	1.1–6.5		Bachlava et al. (2009)	
	qOA9_1	2009/2010	9	Gm09_45996–Gm09_5152	6.9–23.1	1.3–14.9		Fan et al. (2015), Li et al. (2015), Xie et al. (2012), Reinprecht et al. (2006), Wang et al. (2012)	*GmACP*
	qOA13_1	2009/2011	13	Gm13_25899–Gm13_9722	11.9–28.9	2.3–3.5		Fan et al. (2015), Wang et al. (2012)	
	qOA17_1	2009/2010	17	Gm17_33278–Gm17_7573	8.9–15.0	1.5–1.7		Xie et al. (2012)	
	qOA18_1	2009/2011	18	Gm18_79781–Gm18_53964	52.1–68.1	8.6–27.4	Novel		
	qFA19_1	2009/2011	19	Gm19_67930–Gm19_42890	10.4–20.5	1.0–4.4	Novel		
	qFA19_2	2009/2010	19	Gm19_58422–Gm19_16080	3.6–20.9	1.8–4.4		Fan et al. (2015)	
Linoleic acid	*qFA19_1*	2009/2010	19	Gm19_67930–Gm19_42890	7.9–26.0	0.7–12.0	Novel		
	qFA19_2	2009/2010	19	Gm19_58422–Gm19_16080	28.3–40.3	13.6–19.3		Fan et al. (2015)	
	qLA1_1	2009/2011	1	Gm01_79467–Gm01_15147	6.7–25.9	3.0–4.4	Novel		
	qLA4_1	2009/2010	4	Gm04_67180–Gm04_61504	7.6–27.4	2.7–4.8	Novel		
	qLA5_1	2010/2011	5	Gm05_27463–Gm05_53236	3.9–4.1	0.4–1.3		Bachlava et al. (2009)	
	qLA8_1	2009/2010/2011	8	Gm08_47647–Gm08_8866	7.9–33.6	1.3–16.9		Fan et al. (2015)	

(continued)

Table 7.3 (continued)

Trait	QTL	Environment	Chr. No.	Marker interval	LOD	PVE (%)	Novelty	Reported QTL or QTN by	Gene annotation
Linolenic acid	*qLNA2_1*	2009/2011	2	Gm02_57581–Gm02_69905	6.2–9.1	2.5–3.6	Novel		
	qLNA3_1	2010/2011	3	Gm03_40606–Gm03_10118	13.1–42.1	3.7–21.7		Fan et al. (2015)	*GmFAD8*
	qLNA15_1	2009/2010	15	Gm15_4886–Gm15_16198	7.9–35.2	3.0–17.6		Li et al. (2015)	
	qLNA16_10	2010/2011	16	Gm16_11545–Gm16_47706	8.6–9.9	2.8–3.0		Diers and Shoemaker (1992)	
	qLNA19_1	2010/2011	19	Gm19_45448–Gm19_57039	48.3–50.1	28.2–32.3		Fan et al. (2015)	
	qLNA20_1	2009/2010/2011	20	Gm20_19811–Gm20_41673	14.9–44.4	3.6–26.0		Fan et al. (2015)	

7.3 Future Prospects

A lot of progress has been made in manipulating the fatty acid profile in soybean oil. However, lots more to be done to achieve the final goal. Future achievements in this direction depend largely on application of the genomics tools available for soybean (Shoemaker et al. 2003; Jackson et al. 2006), including the complete genome sequences available (www.phytozome.net/soybean). The genomic resources coupled with proteomics (Hajduch et al. 2005), metabolomics (Bino et al. 2004), and flux map analysis (Iyer et al. 2008) would allow the researchers to gain insight into carbon flow during embryogenesis. This, in turn, will allow for the designing of experiments that target manipulation of the metabolic control points. Such studies will address questions underlying the relative partitioning of carbon between seed storage components and the intricacies of fatty acid metabolism in soybean seeds. Such information will constitute the blueprints for developing the designer soybean with target traits in the near future.

References

Alt JL, Fehr WR, Welke GA, Sandu D (2005a) Phenotypic and molecular analysis of oleate content in the mutant soybean line M23. Crop Sci 45:1997–2000

Alt JL, Fehr WR, Welke GA, Shannon JG (2005b) Transgressive segregation for oleate content in three soybean populations. Crop Sci 45:2005–2007

Bachlava E, Dewey RE, Burton JW, Cardinal AJ (2009) Mapping and comparison of quantitative trait loci for oleic acid seed content in two segregating soybean populations. Crop Sci 49:433–444

Belo A, Zheng P, Luck S, Shen B, Meyer DJ, Li B, Tingey S, Rafalski A (2008) Whole genome scan detects an allelic variant of fad2 associated with increased oleic acid levels in maize. Mol Gen Genomics 279:1–10

Bilyeu KD, Palavalli L, Sleper DA, Beuselinck PR (2003) Three microsomal omega-3 fatty-acid desaturase genes contribute to soybean linolenic acid levels. Crop Sci 43:1833–1838

Bin L, Shengxü F, Fukuan Y, Ying C, Shengrui Z, Fenxia H, Shurong Y, Lianzheng W (2017) Junming sun high-resolution mapping of QTL for fatty acid composition in soybean using specific-locus amplified fragment sequencing. Theor Appl Genet 130:1467–1479

Bino RJ, Hall RD, Fiehn O, Kopka J, Saito K, Draper J, Nikolau BJ, Mendes P, Roessner-Tunali U, Beale MH et al (2004) Potential of metabolomics as a functional genomics tool. Trends Plant Sci 9:418–425

Brummer EC, Nickell AD, Wilcox JR, Shoemaker RC (1995) Mapping the *Fan* locus controlling linolenic acid in soybean oil. J Hered 86:245–247

Buhr T, Sato S, Ebrahim F, Xing A, Zhou Y, Mathiesen M, Schweiger B, Kinney AJ, Staswick P, Clemente T (2002) Ribozyme termination of RNA transcripts down-regulate seed fatty acid genes in transgenic soybean. Plant J 30:155–163

Cerutti H (2003) RNA interference: traveling in the cell and gaining functions? Trends Genet 19:39–46

Chappell AS, Bilyeu KD (2006) A GmFAD3A mutation in the low linolenic acid mutant C1640. Plant Breed 125:535–536

Chi DW, Masatsugu H, Toyoaki A, Akihiro S, Ryo A (2017) Fatty acid composition and distribution in wild soybean (Glycine soja) seeds collected in Japan. Asian J Plant Sci 16(2):52–64. https://doi.org/10.3923/ajps.2017.52.64

Crystal BH, Jason DG (2017) Genotyping-by-sequencing-based investigation of the genetic architecture responsible for a sevenfold increase in soybean seed stearic acid. Genes Genomes Genet 7:299–308

Diers BW, Shoemaker RC (1992) Restriction fragment length polymorphism of soybean fatty acid content. J Am Oil Chem Soc 69:1242–1244

Edgar BC, Thomas TE, Howard GD, Anthony JK (2009) Modifying vegetable oils for food and non-food purposes. In: Vollmann J, Rajcan I (eds) Oil crops, handbook of plant breeding, 4th edn. Springer, New York, NY. https://doi.org/10.1007/978-0-387-77594-4_2

Erickson EA, Wilcox JR, Cavins JF (1988) Inheritance of altered palmitic acid percentage in two soybean mutants. Heredity 79(6):465–468

Fan S, Li B, Yu F, Han F, Yan S, Wang L, Sun J (2015) Analysis of additive and epistatic quantitative trait loci underlying fatty acid concentrations in soybean seeds across multiple environments. Euphytica 206:689–700

Fehr WR (2007) Breeding for modified fatty acid composition in soybean. Crop Sci 47(3):72–87. https://doi.org/10.2135/cropsci2007.04.0004IPBS

Fehr WR, Welke GA, Hammond EG, Duvick DN, Cianzio SR (1992) Inheritance of reduced linolenic acid content in soybean genotypes A16 and A17. Crop Sci 32:903–906

Flores T, Karpova O, Su X, Zheng P, Bilyeu K, Sleper DA, Nguyen HT, Zhang ZJ (2008) Silencing of the GmFAD3 gene by siRNA leads to low a-linolenic acids (18:3) of fad3-mutant phenotype in soybean Glycine max (Merr.). Transgenic Res 17:839–850

Graef G, LaVallee BJ, Tenopir P, Tat ME, Schweiger BJ, Kinney AJ, Van Gerpen J, Clemente TE (2009) A high oleic acid and low palmitic acid soybean: agronomic performance and evaluation as a feedstock for biodiesel. Plant Biotechnol J 7:411–421

Hajduch M, Ganapathy A, Stein JW, Thelen JJ (2005) A systematic proteomic study of seed filling in soybean: establishment of highresolution two-dimensional reference maps, expression profiles, and an interactive proteome database. Plant Physiol 137:1397–1419

Hammond EG, Fehr WR (1983) Registration of A5 germplasm line of soybean. Crop Sci 23:192

Heppard EP, Ki nney AJ, Stecca KL, Miao GH (1996) Developmental and growth temperature regulation of two different microsomal o-6 desaturase genes in soybeans. Plant Physiol 110:311–319

Horejsi TF, Fehr WR, Welke GA, Duvick DN, Hammond EG, Cianzio SR (1994) Genetic control of reduced palmitate content in soybean. Crop Sci 34:331–334

Hu X, Sullivan-Gilbert M, Gupta M, Thompson SA (2006) Mapping of the loci controlling oleic and linolenic acid contents and development of fad2 and fad3 allele-specific markers in canola (Brassica napus L.). Theor Appl Genet 113:497–507

Hyten DL, Pantalone VR, Saxton AM, Schmidt ME, Sams CE (2004) Molecular mapping and identification of soybean fatty acid modifier quantitative trait loci. J Am Oil Chem Soc 81:1115–1118

Istvan R, Guangyun H, Aron DW (2008) Advances in breeding of seed-quality traits in soybean. J Crop Improv 14(1-2):145–174

Iyer VV, Sriram G, Fulton DB, Zhou R, Westgate ME, Shanks JV (2008) Metabolic flux maps comparing the effect of temperature on protein and oil biosynthesis in developing soybean cotyledons. Plant Cell Environ 31:506–517

Jackson SA, Rokhsar D, Stacey G, Shoemaker R, Schmutz J, Grimwood J (2006) Toward a reference sequence of the soybean genome: a multiagency effort. Plant Genome 46:S-55–S-61

Jeong DL, Kristin DB, GS J (2007) Genetics and breeding for modified fatty acid profile in soybean seed oil. J Crop Sci Biotechnol 10(4):201–210

Kim HJ, Ha B-K, Ha K-S, Chae J-H, Park J-H, Kim M-S, Asekova S, Grover Shannon J, Son C-K, Lee J-D (2015) Comparison of a high oleic acid soybean line to cultivated cultivars for seed yield, protein and oil concentrations. Euphytica 201:285–292. https://doi.org/10.1007/s10681-014-1210-5

Kinney AJ (1996) Development of genetically engineered soybean oils for food applications. J Food Lipids 3 273–292

Kinney AJ, Knowlton S (1997) Designer oils: the high oleic soybean. In: Harander S, Roller S (eds) Genetic engineering for food industry: a strategy for food quality improvement. Blackie Academic, London, pp 193–213

Korver O, Katan MB (2006) The elimination of trans fats from spreads: how science helped to turn an industry around. Nutr Rev 64:275–279

Li YH, Reif JC, Ma YS, Hong HL, Liu ZX, Chang RZ, Qiu LJ (2015) Targeted association mapping demonstrating the complex molecular genetics of fatty acid formation in soybean. BMC Genomics 16:841

Lichtenstein AH, Matthan NR, Jalbert SM, Resteghini NA, Schaefer EJ, Ausman LM (2006) Novel soybean oils with different fatty acid profiles alter cardiovascular disease risk factors in moderately hyperlipidemic subjects. Am J Clin Nutr 84:497–504

Masum A, Shiming L, Melanie B, Stella KK, Khalid M, Nacer B, David AL, Abdelmajid K (2014) Identification of quantitative trait loci (QTL) underlying protein, oil, and five major fatty acids' contents in soybean. Am J Plant Sci 5:158–167

Mazur B, Krebbers E, Tingey S (1999) Gene discovery and product development for grain quality traits. Science 285:372–375

Mikkilineni V, Rocheford TR (2003) Sequence variation and genomic organization of fatty acid desaturase-2 (fad2) and fatty acid desaturase-6 (fad6) cDNAs in maize. Theor Appl Genet 106:1326–1332

Monteros MJ, Burton JW, Boerma HR (2008) Molecular mapping and confirmation of QTLs associated with oleic acid content in N00-3350 soybean. Crop Sci 48:2223–2234

Mozaffarian D, Willett WC (2007) Trans fatty acids and cardiovascular risk: a unique cardiometabolic imprint? Curr Atheroscler Rep 9:486–493

Panthee DR, Pantalone VR, Saxton AM (2006) Modifier QTL for fatty acid composition in soybean oil. Euphytica 152:67–73

Patel M, Jung S, Moore K, Powell G, Ainsworth C, Abbott A (2004) High-oleate peanut mutants result from a MITE insertion into the FAD2 gene. Theor Appl Genet 108:1492–1502

Perez-Vich B, Ferna ndez-Martınez JM, Grondona M, Knapp SJ, Berry ST (2002) Stearoyl- ACP and oleoyl-PC desaturase genes cosegregate with quantitative trait loci underlying high stearic and high oleic acid mutant phenotypes in sunflower. Theor Appl Genet 104:338–349

Primomo VS, Falk DE, Ablett GR, Tanner JW, Rajcan I (2002) Genotype X environment interactions, stability, and agronomic performance of soybean with altered fatty acid profiles. Crop Sci 42:37–44

Rahman SM, Kinoshita T, Anai T, Takagi Y (2001) Combining ability in loci for high oleic and low linolenic acids in soybean. Crop Sci 41:26–29

Rebetzke GJ, Burton JW, Carter TE Jr, Wilson RF (1998) Genetic variation for modifiers controlling reduced saturated fatty acid content in soybean. Crop Sci 38:303–308

Reinprecht Y, Poysa VW, Yu K, Rajcan I, Ablett GR, Pauls KP (2006) Seed and agronomic QTL in low linolenic acid, lipoxygenase-free soybean (*Glycine max* (L.) Merrill) germplasm. Genome 49:1510–1527

Reinprecht Y, Luk-Labey SY, Larsen J, Poysa VW, Yu K, Rajcan I, Ablett GR, Pauls KP (2009) Molecular basis of the low linolenic acid trait on soybean EMS mutant line RG10. Plant Breed 128:253–258

Shoemaker RC, Schlueter JA, Cregan P, Vodkinn L (2003) The status of soybean genomics and its role in the development of soybean biotechnologies. Ag Bio Forum 6(1&2):4–7

Spencer MM, Pantalone VR, Meyer EJ, Landau-Ellis D, Hyten DL Jr (2003) Mapping the *Fas* locus controlling stearic acid contents in soybean. Theor Appl Genet 106:615–619

Stoltzfus DL, Fehr WR, Welke GA, Hammond EG, Cianzio SR (2000a) A fap5 allele for elevated palmiatate in soybean. Crop Sci 40:647–650

Stoltzfus DL, Fehr WR, Welke GA, Hammond EG, Cianzio SR (2000b) A fap7 allele for elevated palmiatate in soybean. Crop Sci 40:1538–1542

Takagi Y, Rahman SM (1996) Inheritance of high oleic acid content in the seed oil of soybean mutant M23. Theor Appl Genet 92:179–182

Tang GQ, Novitzky WP, Griffin HC, Huber SC, Dewey RE (2005) Oleate desaturase enzymes of soybean: evidence of regulation through differential stability and phosphorylation. Plant J 44:433–446

Tom EC, Edgar BC (2009) Soybean oil: genetic approaches for modification of functionality and total content. Plant Physiol 151:1030–1040

Wang X, Jiang GL, Green M, Scott RA, Hyten DL, Cregan PB (2012) Quantitative trait locus analysis of saturated fatty acids in a population of recombinant inbred lines of soybean. Mol Breed 30:1163–1179

White HBJ, Quackenbush FW, Probst AH (1961) Occurrence and inheritance of linolenic and linoleic acids in soybean seeds. J Am Oil Chem Soc 38:113–117

Wilcox JR, Cavins JF (1985) Inheritance of low linolenic acid content of the seed oil of a mutant Glycine max. Theor Appl Genet 71:74–78

William H, Andrew C, Benjamin MC, Zachary LD, Anita O, Erin R, Adam R, Thomas S, Alexandre J, Frederic C, Luc M, Daniel FV, Feng Z (2014) Improved soybean oil quality by targeted mutagenesis of the fatty acid desaturase 2 gene family. Plant Biotechnol J 12:934–940

Xie D, Han Y, Zeng Y, Chang W, Teng W, Li W (2012) SSR-and SNPrelated QTL underlying linolenic acid and other fatty acid contents in soybean seeds across multiple environments. Mol Breed 30:169–179

Zhou Z, Jiang Y, Wang Z, Gou Z et al (2015) Resequencing 302 wild and cultivated accessions identifies genes related to domestication and improvement in soybean. Nat Biotechnol 33(4):408–414

Chapter 8
Breeding Dry Beans (*Phaseolus vulqaris* L.) with Improved Cooking and Canning Quality Traits

Asif M. Iqbal Qureshi and Rie Sadohara

8.1 Introduction

Dry beans (*Phaseolus vulgaris* L.) are the most produced pulse crop with a worldwide production of 26 million tons in 2016 (Food and Agriculture Organization of the United Nations 2016), serving as a staple food in parts of Africa, Latin America, and east Asia, where they are often a primary source of protein, calories, and minerals (Wortmann et al. 1998; Broughton et al. 2003; Watts 2011). A large genetic diversity exists in dry beans (also known as common beans) germplasm presenting potential to be exploited in breeding programs. Dry beans were domesticated separately in Mesoamerican (Mexico) and Andean regions, which resulted in two distinct gene pools. Within each gene pool exist three races of common beans differing in plant and seed morphology, growth habit, and biological markers (Singh et al. 1991). As a consequence, there is a rich diversity in the shape, size, and color of seeds, forming distinct market classes. An example set of market classes found in the U.S. is shown in Fig. 8.1. There are cultural and regional preferences toward particular market classes to be grown and consumed in a given region (Castellanos et al. 1996; Scott and Maideni 1998). Therefore, local preferences should be taken into consideration when developing new bean varieties exploiting a large diversity in dry bean germplasm.

A. M. I. Qureshi (✉)
Genetics, Plant Breeding & Biotechnology, MRCFC, Khudwani, Sher-e-Kashmir University of Agricultural Sciences & Technology of Kashmir, Srinagar, Jammu and Kashmir, India

R. Sadohara
Department of Plant, Soil and Microbial Sciences, Michigan State University, East Lansing, MI, USA

A. M. I. Qureshi et al. (eds.), *Quality Breeding in Field Crops*,
https://doi.org/10.1007/978-3-030-04609-5_8

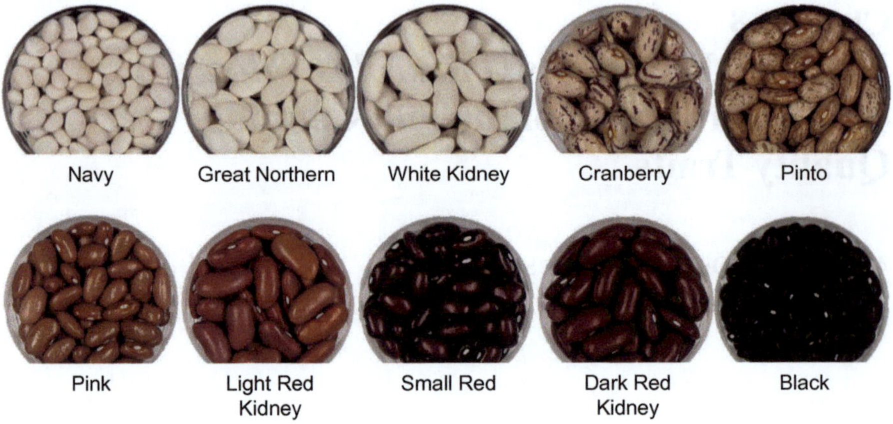

Fig. 8.1 Diverse market classes of dry beans found in the U.S.

Dry beans offer numerous health-promoting effects: the inclusion of dry beans in the diet leads to the prevention and control of diabetes, cardiovascular diseases, and certain types of cancer (Hayat et al. 2014; Messina 2014; Suárez-Martínez et al. 2016). Thus, the consumption of pulses including beans is recommended by many health organizations (Leterme 2002; Havemeier et al. 2017). Moreover, cultivation of dry beans, as with other legumes, contributes to sustainable agriculture because of their nitrogen fixation capacity (Foyer et al. 2016). Despite those benefits, dry beans are much less utilized compared to major cereal crops: the world production of dry beans is 3% of the production of wheat or rice (Food and Agriculture Organization of the United Nations 2016). Improving quality traits will be key to promoting dry bean consumption, including anti-nutritional factors, consequent poor digestibility, long cooking times, and sensory attributes. It is a challenge, however, to evaluate those quality traits because it often involves laborious processes such as quantification of chemical compounds, efficacy trials, soaking and cooking, and sensory evaluations by a trained panel of assessors. Marker-assisted selection (MAS) would be a useful tool to select bean breeding lines for a quality trait, bypassing the need of phenotyping. This technology is becoming more feasible with the constantly evolving DNA sequencing technologies and the whole genome sequence of *Phaseolus vulgaris* (Schmutz et al. 2014). In addition, indirect measurement of some quality traits using prediction models would enable highly efficient screening (Hacisalihoglu et al. 2010; Ngatchou et al. 2012; Plans et al. 2012; Boukar 2013; Plans et al. 2013; Mendoza et al. 2017). The combination of rapid genotyping and phenotyping tools will together accelerate genetic investigations as well as screening processes of dry bean breeding. In this chapter, important quality traits, their implications, evaluation methods, and improvement efforts will be discussed.

8.2 Cooking Quality

8.2.1 Protein Quality and Digestibility

Common beans are a major source of protein to those who do not have regular access to animal proteins or who follow vegetarian diets. The major storage protein of dry beans, phaseolin, accounts for 40–50% of the total protein in the seed (Ma and Bliss 1978). Phaseolin is deficient in sulfur-amino acids such as methionine and cysteine, thus protein quality of dry beans is suboptimal compared to animal-derived proteins (Salunkhe 1982). One practical solution is to consume dry beans with cereals, which are low in lysine; due to the different limiting amino acids in beans and cereals, the composite meal will provide better overall amino acid profiles. As an approach through breeding, there have been investigations to improve methionine content in dry beans. Monti and Grillo (1983) summarized previous studies which reported the correlation of protein content and sulfur-amino acids to be either insignificant or negative. They argued that the enhancement of sulfur-amino acid content should be possible without adversely affecting the protein content, but increasing the ratio of the albumin fraction, which have a high sulfur-amino acid content, may be more efficient. Kelly and Bliss (1975) discussed the percentage of available methionine in total protein of dry beans, microbiologically determined. The available methionine content was estimated to be moderately heritable and was positively correlated with protein content, presenting a possibility of improving both the protein quantity and sulfur amino acid content (Kelly and Bliss 1975). Despite the existence of some high-methionine lines (Gepts and Bliss 1984), variability in the sulfur-containing amino acids is low among cultivated lines and wild accessions, posing a difficulty in improvement through breeding (Monti and Grillo 1983; Montoya et al. 2008). A recent study, however, found that an alteration in seed protein composition resulted in an increase in methionine and cysteine, which will be discussed later in this section (Taylor et al. 2008).

The sulfur-amino acid deficiency is not the sole problem in the bean protein that should be addressed. A feeding trial showed that enrichment of precooked beans with methionine had minimal improvement on the protein utilization by malnourished children and that the low digestibility of bean protein is the major limiting factor (Graham et al. 1979). Indeed, the protein digestibility of dry beans is 55–77%, lower than that of meat, which is 82–86% (Navarrete and Bressani 1981). Therefore, increasing the digestibility and utilization efficiency may be more effective in improving overall nutritional quality of dry beans. Although dry beans are highly nutritious, the presence of anti-nutrients decreases the digestibility and availability of nutrients. They include lectins (Vasconcelos and Oliveira 2004), phenolic compounds especially tannins, and phytic acid (Reddy et al. 1985; Sandberg 2002; Umar Lule and Xia 2005; Kumar et al. 2010).

8.2.1.1 Reduction of Lectins

Lectins are closely related proteins that are considered to be a defense mechanism against insect pests and also show toxicity to higher animals (Lajolo and Genovese 2002). Lectins can be classified into three major constituents: phytohemagglutinin (PHA), arcelin, and α-amylase inhibitors (AI). PHA and AI are commonly found in dry bean cultivars and accessions. On the other hand, arcelin is present in wild Mesoamerican lines (Osborn et al. 1986), but the genes for this protein have been introgressed to some cultivated lines in order to confer resistance to bruchids (Osborni et al. 1988). Although heat treatment (i.e. cooking) deactivates most of the lectins, small residual activity may be present depending on the cooking temperature and time (Bender and Reaidi 1982; McPherson 1990); therefore, from nutritional perspective, decreasing or alternating lectin contents is likely to improve digestion and utilization of nutrients present in beans. Interestingly, a study showed that the ingestion of bean seeds with high arcelin and low PHA contents showed less toxicity to rats compared to a cultivar with a high PHA content (Cardona et al. 1990; Pusztai et al. 1993).

In an effort to develop bean lines with lower anti-nutrients, a line with altered protein composition with no phaseolin, PHA, and arcelin accumulation was developed. This line, SMARC1N-PN1, did not differ from the parental lines in protein content and days to maturity (Hartweck and Osborn 1997; Osborn et al. 2003). Interestingly, the elimination of those storage protein led to an increase in methionine, cysteine, and soluble protein in the mature seeds, resulting in a bean line with two desirable traits combined (Taylor et al. 2008). More recently, lectin-free (*lf*) lines were generated by introgressing mutant alleles of the lectin genes from a wild accession into existing cultivars. The resulting *lf* lines were devoid of PHA, arcelin, and AI, and some lines showed higher dry seed yield than the high-yielding parents (Campion et al. 2009a). The authors mentioned that they never observed bruchid infestations in the *lf* lines, but the susceptibility to insect damages should be evaluated by an appropriate method (Campion et al. 2009a). Taken together, those breeding efforts showed the possibility of developing bean lines without lectin accumulation, which will not only increase the nutritional quality of dry beans but also open up a possibility of utilizing raw beans as livestock feed. It is of importance to note that breeding nutritionally superior bean lines may result in a compromise in agronomic performance, thus the two traits should both be at a satisfactory level.

8.2.2 Iron Concentration and Bioavailability

Dry beans serve as a good source of minerals, especially iron and zinc (Beebe et al. 2000). Iron deficiency is a worldwide health concern especially in women and children in South East Asia, Africa, and the Eastern Mediterranean region (WHO 2015). Improvement of those micronutrient profiles of beans will have a positive impact especially in Africa, where beans are a major source of iron and other minerals (Broughton et al. 2003). Although beans are high in minerals, the presence of anti-nutrients such as polyphenols and phytic acid are known to hinder the absorption of the micronutrients (Sandberg 2002). The fractional absorption rate of iron from beans is low, ranging from 1% to 7% depending on the concentrations of iron and anti-nutritional factors, the iron status of the subject, and the study design (Donangelo et al. 2003; Petry et al. 2010, 2012). Since not all the mineral contents are absorbed by human body, the concept of bioavailability is used to better characterize the potential of food as a source of minerals. It is an important trait that should be taken into consideration when breeding dry beans for alleviating iron deficiency (White and Broadley 2005; Vaz-Tostes et al. 2016). Iron bioavailability in vivo, however, is cumbersome and expensive to measure; thus, an alternative strategy would be more useful for screening promising breeding lines. For this purpose, Glahn et al. (1998) developed a human Caco-2 cell model, which mimics the digestion in the gastrointestinal system and uses resulting ferritin formation as an indicator of iron absorption in the body. This method has been used to study genetic diversity and predict in vivo iron bioavailability of beans and other crops (Glahn et al. 2002; Oikeh et al. 2003; Tako et al. 2015; Wiesinger et al. 2016). In order to increase iron bioavailability of beans, breeding efforts should be directed not only to increase the iron concentration in the seeds but also to reduce anti-nutrients that inhibit iron absorption (Petry et al. 2015; Vaz-Tostes et al. 2016).

8.2.2.1 Increment of Mineral Concentrations

There are some instances of plant breeding aiming at increasing the iron concentration in dry beans, including High-Iron Bean (HIB) varieties developed by HarvestPlus and its collaborators. Ten HIB varieties were released to tackle anemia in Rwanda, one of the highest bean-consuming countries (Broughton et al. 2003). The HIB varieties included eight climber types and two bush types with seed coat color ranging from white, cream, red to dark red (Asare-Marfo et al. 2016). The consumption of cream-colored HIB for 18 weeks significantly improved the hemoglobin and total body iron of college-age women compared to the control group that had conventional beans of the same color for the same period of time (Haas et al. 2016). Five years after the release of the HIB varieties, a follow-up nationwide interview revealed that the HIB varieties were adopted by 21% of households in Rwanda (Asare-Marfo et al. 2016). This report also revealed distinctive local preferences; among the ten HIB varieties released, by far the most preferred was a bush type bean because it is high-yielding. On the other hand, the other bush type variety was not popular potentially because of its cream-colored seeds (Asare-Marfo et al. 2016). It presented a challenge in promoting high iron (and high iron bioavailability) varieties that are not the usual type that the locals consume. Therefore, new varieties with improved nutritional benefits need to have competitive agronomic yield and meet the local preferences and practices in the target region to be accepted by consumers.

Large genetic variability has been reported in mineral concentrations of dry beans, presenting an opportunity to enhance their mineral concentrations (White and Broadley 2005; Hacisalihoglu and Settles 2013); however, quantification of minerals in beans is time-consuming. Thus, a predictive screening tool such as near-infrared spectroscopy (NIRS) will be useful. Plans et al. (2012) used NIRS to determine calcium and magnesium content. Prediction was more accurate when ground seed coat was used, but the methodology using intact beans will be more efficient when screening large populations. Further work will be necessary to increase prediction accuracy and to simplify the sample preparation procedures. Some genetic analyses were conducted to find quantitative trait loci (QTL) associated with mineral concentrations. For instance, QTL for seed iron and zinc were detected using a population derived from an inter-genepool cross (Blair et al. 2009). A single dominant gene model was proposed on zinc accumulation in navy bean, and breeding high-zinc cultivar was deemed possible (Cichy et al. 2005). The use of DNA markers will facilitate the initial screening for high concentration of minerals before further testing for their bioavailability.

8.2.2.2 Reduction of Anti-nutrient Factors for High Iron Bioavailability

A feeding trial using a high iron bean with black seed coat resulted in no increased iron absorption in women in Rwanda, reiterating the importance of decreasing phytic acid and phenolic compounds to enhance iron absorption (Petry et al. 2012). In light of this, low phytic acid (*lpa*) mutants via mutagenesis using EMS were developed, and the *lpa* lines showed 90% reduction in phytic acid and increase in the concentration of free iron and inorganic phosphorus (Campion et al. 2009b). The *lpa* lines exhibited comparable germination rate and dry seed yield to the control lines (Campion et al. 2009b). In a succeeding study, the *lpa* trait was combined with the aforementioned *lf* trait, and the three resulting lectin-free and low phytic acid (*lf* + *lpa*) lines exhibited comparable agronomic performance as the control cultivars, demonstrating the possibility of improving the iron profile of bean cultivars without compromising agronomic traits, which in practice will be the most important for wide adaptation (Campion et al. 2013). However, the iron bioavailability increased only in the white-seeded line, and not in black and brown lines because of tannins present in the seed coat (Campion et al. 2013), similar to the result of the study conducted by Petry et al. (2012). Tannins strongly bind to metal cations hindering their absorption (Reddy et al. 1985), and therefore the authors concluded that the benefit of reduction in phytic acid would only be achieved when the seed does not contain tannins (Campion et al. 2013). A similar observation was made where iron absorption was only improved when both polyphenols and phytic acid were removed (Petry et al. 2010). Taken together, it would be necessary to decrease both polyphenols and phytic acids to improve the bioavailability of micronutrients in dark-colored seeds. A wide genetic variability for polyphenol contents has been observed in landraces and RILs from an inter-genepool cross, which could be used as a source of low anti-nutrient trait (Díaz et al. 2010; Doria et al. 2012). More genetic studies on polyphenols and phytic acid will pave the way to the development of bean varieties with high mineral bioavailability and with a seed coat color that caters to the needs in the target region.

Cooking Time

As beans are most commonly prepared by boiling, cooking time of dry beans is an important quality trait. Beans typically take 1–2 h to be cooked (Borchgrevink 2012). The long looking time is likely one of the major contributors to the limited consumption of dry beans in countries where beans are not a staple food. Long cooking times not only cause inconvenience but also affect various aspects of life in the bean-consuming countries. First, long-cooking beans require more fuel use. In sub-Saharan Africa, the majority of households use fuelwood and charcoal as primary energy resource (IEA 2014), and a longer time will need to be allocated to collect fuelwood to cook beans. Menéndez and Curt (2013) reported that women and children spend 5 h per week on average to gather fuelwood, sacrificing their time for household chores and other activities in Tanzania. Moreover, prolonged exposure to smoke generated by burning fuelwood indoors may cause health problems to those who are responsible for cooking, mostly women (Shellie-Dessert and Hosfield 1990; Maes and Verbist 2012; Menéndez and Curt 2013). Second, long cooking times have nutritional significance. As firewood becomes more and more scarce, some households may have to give up the consumption of beans, often the primary source of protein and minerals in those regions (Makungwa et al. 2013). In addition, long cooking time is associated with less favorable nutritional profiles in terms of protein digestibility and iron bioavailability (Iyer et al. 1980; Wiesinger et al. 2016). Therefore, improving the cooking time of dry beans will have a positive impact on the quality of life and nutritional well-being of dry bean consumers.

Cooking time can be affected by numerous factors such as genotype, growing environments, storage conditions, and the type of soaking solutions (Nleya et al. 2000; Coelho et al. 2007; Cichy et al. 2015; Kinyanjui et al. 2015). When stored at a high temperature and high humidity, cooking time tends to be extended, and this phenomenon is termed as "Hard to Cook", or HTC (Liu and Bourne 1995). Due to its high negative impact on the economy and nutrition, extensive research has been conducted to elucidate its mechanism. HTC beans show difficulty in the cotyledon cell separation during cooking, and the proposed causes for this phenomenon include divalent cations cross-linking pectin (Jones and Boulter 1983), and polyphenolic compounds (Srisuma et al. 1989; Stanley 1992) and lignification (Hincks and Stanley 1987) strengthening cell wall and middle lamella. The reader is referred to a comprehensive review by Reyes-Moreno et al. (1993) for details on the HTC defects. The severity of HTC varies among genotypes; thus, selection toward less HTC-prone lines will be desirable if the resultant variety is expected to be stored under suboptimal conditions. Aside from HTC, bean seeds with low moisture content may exhibit the "hard shell" defect, where seeds fail to absorb sufficient amount of water after long time of soaking, typically 18–24 h. A long cooking time may be due to HTC or hard shell defect, or both (Jackson and Varriano-Marston 1981). Castellanos et al. (1995) proposed using less than 9% of the initial moisture content of the seeds and poor water uptake after 18 h of soaking as a means of screening against the hard-shell defect.

8.2.2.3 Measurement and Genetic Analyses of Cooking Time

Cooking time is most commonly measured using automated Mattson cooker (Wang and Daun 2005). It consists of a platform where individual bean seeds are placed and weighted plungers which sit on each seed. The lower part of the cooker is placed in boiling water, and when a seed becomes sufficiently soft, the plunger above it will penetrate the seed, and the movement is recorded by the sensor connected to the apparatus (Fig. 8.2). Each pin drop time will be recorded, and the cooking time is defined as the time when the majority of bean seeds (50–80%) are cooked. Using this method, the heritability of cooking time was estimated, ranging from 0.46 to 0.97 (Elia et al. 1997; Jacinto-Hernandez et al. 2003; Garcia et al. 2012). Recently, Carvalho et al. (2017) utilized the Mattson cooker to select fast-cooking lines by cooking bean seeds for a fixed amount of time and expressing cookability as the percentage of seeds fully cooked (i.e. penetrated by Mattson cooker pins), demonstrating the possibility of faster phenotypic screening for cooking time. Although the automated Mattson cooker has significantly reduced the labor required to measure cooking time, it still involves actual cooking, which is a bottleneck when thousands of lines are to be tested. In addition, the Mattson apparatus and the specific software may not be accessible to all breeders. Therefore, establishing an MAS scheme would accelerate the breeding and screening process for shortening cooking time.

Several genetic analyses have been conducted to identify genomic loci associated with cooking time of dry beans. Jacinto-hernandez et al. (2003) identified one RAPD marker associated with cooking time using three generations of recombinant

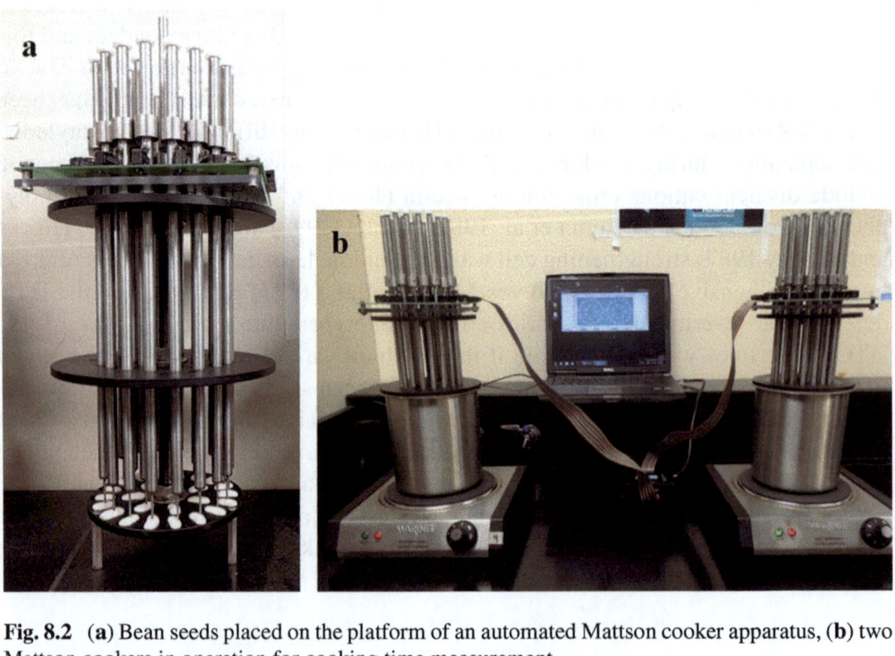

Fig. 8.2 (**a**) Bean seeds placed on the platform of an automated Mattson cooker apparatus, (**b**) two Mattson cookers in operation for cooking time measurement

inbred lines from a cross of a long- and a short-cooking lines, and proposed a two-gene model that controls the cooking time based on the observation of 3:1 ratio of cooking time distribution in the progeny. Garcia et al. (2012) used SSR markers and a bi-parental progeny population to detect QTL for cooking time. As a result, one significant QTL was detected on chromosome 1 constantly in the two environments. The QTL explained up to 20% of the phenotypic variance with significant environment × genotype effect. The environmental effect on cooking time has been indicated by other studies (Proctor and Watts 1987; Nleya et al. 2000). Cichy et al. (2015) identified a few accessions that cooked in less than 25 min from a 206-genotype panel with 5.5-fold cooking time variability, and conducted a genome-wide association study (GWAS). They detected significant SNP markers located on chromosome Pv02, Pv03, and Pv06, and suggested some candidate genes which will be useful in elucidating the mechanism of the fast-cooking trait. Further analyses would be necessary to validate those QTLs to utilize them in MAS schemes in breeding programs.

8.2.3 Sensory Properties

Sensory characteristics are another cooking quality attribute of dry beans to which consumers attach importance (Castellanos et al. 1996; Scott and Maideni 1998). Sensory attributes include flavor, aroma, texture, and appearance. Attempts have been made to explain consumer acceptance scores by using descriptive sensory attributes. For example, Arvanitoyannis et al. (2007) evaluated five landraces for sensory properties, and the acceptability scores were influenced by hardness, odor, and brightness of the seeds, indicating that not only flavor but also texture, smell, and appearance are important factors. Mkanda et al. (2007) conducted sensory evaluation by a trained panel and consumer acceptance tests on six varieties grown at two locations, and reported that highly preferred beans had soft texture and cooked-bean flavor. In other words, less preferred samples had hard texture, and it may have been because the beans were not sufficiently cooked to the extent that consumers would regard as cooked. Since cooking time is a highly variable trait among genotypes, cooking each sample for an optimum length of time will be more appropriate for sensory evaluation of the cooked beans. In this regard, Romero del Castillo et al. (2012) recently proposed a standardized method for cooking beans for sensory evaluation, which calls for a trained taster to determine when the bean seeds are cooked, thereby providing assessors with more optimally cooked samples. The authors stated that the results were highly reproducible and that two replications per sample will be sufficient. Location effects and genotype × environment interactions on various sensory attributes have been reported (Sanz-Calvo and Atienza-del-Rey 1999; Mkanda et al. 2007).

Sensory characteristics were considered in selecting landrace inbred lines in the work of Casañas et al. (1999), in which agricultural traits such as disease resistance and seed size, and sensory properties were measured on seven pure lines of landrace cultivars, two of which were selected for their superior overall performance. It serves as an example of a breeding program taking various traits into consideration including sensory attributes. The drawback of sensory analysis, however, is that it requires time, labor, and cost because of the need of training assessors and a large number of replications and/or assessors. It is expensive and cumbersome to arrange training sessions and evaluations, and only a small number of samples can be tested per session to prevent sensory fatigue. To speed up the evaluation process, prediction models were developed to estimate the flavor, mealiness, seed coat roughness, and seed coat brightness using near infrared spectroscopy (NIRS) technology (Plans et al. 2014). The sample preparation still requires actual cooking and a few more steps for better prediction, but it is more efficient and less expensive to use as an initial screening of a large number of lines for sensory quality. Further refinement of this technology by accumulation and incorporation of data will improve the accuracy of the prediction.

8.3 Canning Quality

Dry beans are often subjected to various treatments, such as cooking at normal or elevated pressure, storage under different environmental conditions, soaking in water or salt solutions, frying after cooking preceding to consumption or they are used as germinated and cooked beans (Reddy et al. 1984). Among the different physical treatments used to process common beans, canning is one of the methods which not only enhances the shelf stability of the bean products, but also improves the palatability and saporous of beans and reduces flatulence factors (raffinose oligosaccharides) (Xu and Chang 2009). Improving composition traits for canning quality in common bean are key for the processing industry, as these traits affect the profitability of a canning operation.

Among the different beans, black beans are usually sold as canned products in the U.S. market and their color and appearance are important attributes for both processors and consumers (Cichy et al. 2014). Well-trained sensory panel usually assesses the back bean canning quality and the widely used parameters in the evaluation of black beans are color, visual appearance, texture, hydration coefficient, and percent washed–drained solids (Merwe et al. 2006; Khanal et al. 2015). Canning quality traits are paramount to bean consumers and processors, and improving them would contribute to the profitability of canned bean market (Khanal et al. 2015). However, dry bean types that show darker colors such as black, small red, and red kidney are more likely to lose their color during the canning process. The quality ratings for color and appearance images are evaluated with sensory scores using multistatistical models (Mendoza et al. 2014). The canned bean color retention is an important attribute for processors and consumers and the degree of color retention is genetically controlled trait and depends on a cultivar's genetic makeup. The quantitative trait loci (QTLs) for color retention and seed anthocyanin concentration following canning in black beans are co-localized on chromosome and interestingly one QTL was also detected on Pv04 and Pv05 associated with L* trait and had R2 of 3.3 and 9.8 respectively (Cichy et al. 2014). They were able to determine a small number of candidate genes for marker-assisted selection (MAS) for the two traits on Pv05 and on Pv11 for the color retention trait. Similarly, two QTLs on Pv05 were also found associated with the trait of color with R2 of 0.10 and 0.13, and carried different effect on the trait (Wright and Kelly 2011).

Furthermore, the beans lose many of their anti-nutritional factors during the canning process (Xu and Chang 2009).

Canning quality of dry bean are known to be enormously influenced by the growing environment (Balasubramanian et al. 1999; Lange and Labuschagne 2001) and canning processing procedures (Jackson and Wiese 1993), and it has been widely accepted that the bean genotype and its interaction with environmental and processing factors determine the final canning quality of dry bean (Hosfield and Uebersax 1980). The higher variation due to location and year as compared to genotype for washed–drained solids in navy bean were reported by Walters et al. (1997) and they attributed and revealed the significant year effect to large differences in temperature and soil moisture conditions during the growing seasons. In three Mesoamerican market classes: navy, black, and pinto bean, genotype and genotype × environment interaction significantly influence washed drain solids (Balasubramanian et al. 1999). The canning quality varies significantly between years and that genetic variation accounts for a significant portion of the total trait variation; notwithstanding, in a quantitative trait loci (QTL) analysis, very few QTLs were found to be repeated in multiple years and across multiple environments. Posa-Macalincag et al. (2002) and Wright and Kelly (2011) found canning quality traits in dark red kidney beans. The canning quality in black beans is influenced by genetics, production environment, and seed handling (Wright and Kelly 2011).

8.3.1 Canning Quality Traits

The breeding materials are often evaluated by bean breeders for a series of phenotypic parameters, through which they can predict the final canning quality of bean genotypes. These quality parameters range from various pre-cooking characteristics to physical parameters measured after canning and may also involve sensory and organoleptic evaluations of the canned products.

The hydration coefficient is measured as the ratio of the weight of the blanched bean to the dry bean, which implies the number of cans obtained from 1 kg of dry bean seeds and it potentially increases the can yield and therefore the economic return of a bean variety for the bean processor. The significant variation among years for all physical, texture, and chemical parameters of canning quality except for hydration coefficient and can yield in both navy and large seeded bean were revealed by Hosfield (1991). A value of 1.8–2.0 for hydration coefficient is often considered optimum for well-soaked bean and bean processors may reject a seed lot if the hydration coefficient is significantly <1.8 due to the added cost per can, as a higher dry weight of bean seeds will be required to fill each can. Texture readings, measured by the specialized equipment called as texture analyzer, are used to quantify the firmness or softness of a bean sample after canning, and is another key determinant of palatability of a canned product and is often used as a measure of consumer acceptance of canned bean (Ghaderi et al. 1984; Hosfield 1991). The Visible/NIRS and Hypers have plausible and potential for predicting the texture of canned beans; the robustness and tenacity of the models is impacted by genotypic diversity, planting year, and phenotypic variability for canned bean texture (Mendoza et al. 2018).

The processors in general seek bean with high hydration coefficient that expand rapidly and uniformly in the can during processing, leave no clump at the bottom of the can, have acceptable percent washed–drained solids and texture after cooking, and cook to tenderness rapidly and uniformly (Deshpande et al. 1984). The strong associations were seen between hydration coefficient and texture with higher water uptake in black beans and are being associated with a softer canned product (Cichy et al. 2014; Mendoza et al. 2014), and the washed–drained weight was negatively associated with firmness of canned bean (Lu and Chang 1996).

There is another parameter called percent washed–drained solids which refers to the mass of processed bean, rinsed and drained and is widely used in the breeding programs and by the processing industry (Hosfield et al. 1984). A lower value of washed–drained solids may indicate excessive loss of solids during processing and lead to an increased degree of clumping of bean at the bottom of the can after processing which ultimately may lead to rejection of a cultivar by the processor because of high degree of clumping is seen as an unpalatable quality. The associations between different canning quality parameters are of paramount importance for selecting genotypes for superior canning quality (Walters et al. 1997), and a clear indication from industry for a preferred canning phenotype would help guide future selection efforts in bean breeding programs.

8.3.2 Processing and Quality Standards

Canning of dry beans involves several unit operations from cleaning to packaging which are presented in Fig. 8.1. Initially to remove extraneous matter such as dirt and small pieces of foreign material, dry beans may be passed through a winnowing mill or clipper cleaner. After then split and defective (discolored, moldy, broken, and off size) dried beans are inspected manually or using the machine. Dry beans are typically soaked in stainless steel tanks and the initial moisture content of raw beans to be processed should be about 12–16%. The main objective of soaking is to provide beans with a final moisture content of approximately 53–57%.

Generally the soaked beans are commonly run over a water riffle for removal of small stones, however advanced systems are also now available that can combine cleaning, destoning, and washing steps into one operation. The blanching is a thermal process and the main aim is to evacuate gases and to inactivate enzymes in the beans and provide a more uniform product during filling. The beans are initially introduced into 62.7–73.9 °C water, then increased to 76.7–82.2 °C and finally increased to 90.6–96.1 °C. Blanching temperatures of 50 °C, 70 °C, and 88 °C pro-

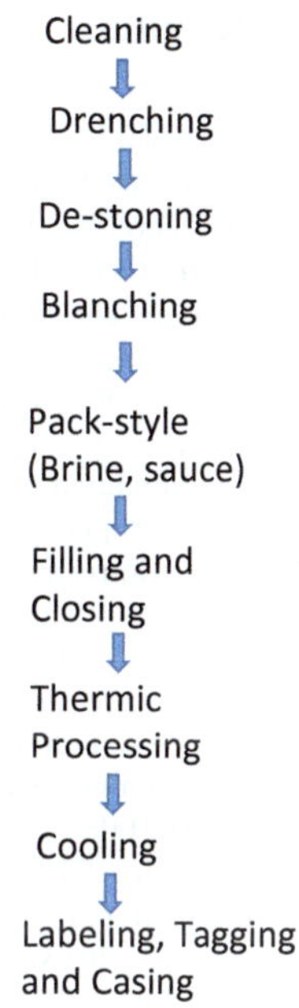

Fig. 8.3 Diagram of canned bean processing

duced no significant effects on canning quality traits of navy, black, and pinto beans (Balasubramanian et al. 2000). Regardless of blanching method employed, the beans frequently are washed in cold water following blanching. The main aim of soaking and blanching is to provide hydration of the bean cotyledon so sufficient softening can occur during cooking. The conditions for soaking and blanching vary throughout the industry with respect to time, temperature, and water composition. At 100 °C sauce/brine should be added to maintain a high closing temperature and generally, the fill is about 52% beans and 48% sauce by weight (Fig. 8.3).

8.3.2.1 Quality Standards

Different standards have been established by USDA-AMS for the US Standards of Grades of canned dried beans, canned pork and beans, and canned baked beans (USDA 1976). For products receiving "grade A" quality must score 90 points or higher on a 100-point scale with the absence of defects and character weighted 40 points each and color weighted 20 points (score at least 17 points for color), for dried canned beans packed in brine, and a reasonably good consistency if packed in tomato or sweetened sauce. Quality of canned pork and beans for "grade A" product must score 90 points or higher on 100-point scale with absence of defects and character weighted 40 points each and consistency weighted 20 points. The canned baked beans must score 90 points or higher on a 100-point scale with absence of defects weighed 25 points, consistency, flavor and character weighted 20 points each and color weighted 15 points for to receive a grade "A" quality and the product must score at least 17 points for consistency, character and flavor, at least 13 points for color, have similar varietal characteristics, and be practically free from defects.

References

Arvanitoyannis IS, Mavromatis A, Rodiatis A, Goulas C (2007) Physicochemical and sensory analysis of dry bean landraces in conjunction with multivariate analysis: An exploratory approach. Int J Food Sci Technol 42:819–826. https://doi.org/10.1111/j.1365-2621.2007.01289.x

Asare-Marfo D, Herrington C, Alwang J et al (2016) Assessing the adoption of high iron bean varieties and their impact on iron intakes and other livelihood outcomes in Rwanda. IFPRI, Washington, DC

Balasubramanian P, Slinkard A, Tyler R, Vandenberg A (1999) Genotype and environment effect on canning quality of dry bean grown in Saskatchewan. Can J Plant Sci 79:335–342

Balasubramanian P, Slinkard A, Tyler R, Vandenberg A (2000) A modified laboratory canning protocol for quality evaluation of dry bean (*Phaseolus vulgaris L.*). J Sci Food Agric 80:732–738

Beebe S, Gonzalez AV, Rengifo J (2000) Research on trace minerals in the common bean. Food Nutr Bull 21:387–391. https://doi.org/10.1177/156482650002100408

Bender AE, Reaidi GB (1982) Toxicity of Kidney Beans (*Phaseolus vulgaris*) with particular reference to lectins. J Plant Foods 4:15–22. https://doi.org/10.1080/0142968X.1982.11904243

Blair MW, Astudillo C, Grusak MA et al (2009) Inheritance of seed iron and zinc concentrations in common bean (*Phaseolus vulgaris* L.). Mol Breed 23:197–207. https://doi.org/10.1007/s11032-008-9225-z

Borchgrevink CP (2012) Culinary perspective of dry beans and pulses. In: Dry beans and pulses production, processing and nutrition. Blackwell Publishing Ltd., Oxford, pp 311–334

Boukar O (2013) Nondestructive determination of beans water absorption capacity using CFA images analysis for Hard-To-Cook evaluation. Int J Electr Comput Eng 3. https://doi.org/10.11591/ijece.v3i2.2149

Broughton WJ, Hernández G, Blair M et al (2003) Beans (*Phaseolus* spp.) – model food legumes. Plant Soil 252:55–128. https://doi.org/10.1023/A:1024146710611

Campion B, Perrone D, Galasso I, Bollini R (2009a) Common bean (*Phaseolus vulgaris* L.) lines devoid of major lectin proteins. Plant Breed 128:199–204. https://doi.org/10.1111/j.1439-0523.2008.01569.x

Campion B, Sparvoli F, Doria E et al (2009b) Isolation and characterisation of an *lpa* (low phytic acid) mutant in common bean (*Phaseolus vulgaris* L.). Theor Appl Genet 118:1211–1221. https://doi.org/10.1007/s00122-009-0975-8

Campion B, Glahn RP, Tava A et al (2013) Genetic reduction of antinutrients in common bean (*Phaseolus vulgaris* L.) seed, increases nutrients and *in vitro* iron bioavailability without depressing main agronomic traits. F Crop Res 141:27–37. https://doi.org/10.1016/j.fcr.2012.10.015

Cardona C, Kornegay J, Posso CE et al (1990) Comparative value of four arcelin variants in the development of dry bean lines resistant to the Mexican bean weevil. Entomol Exp Appl 56:197–206. https://doi.org/10.1111/j.1570-7458.1990.tb01397.x

Carvalho BL, Ramalho MAP, Vieira Júnior IC, Abreu Â de FB (2017) New strategy for evaluating grain cooking quality of progenies in dry bean breeding programs. Crop Breed Appl Biotechnol 17:115–123. https://doi.org/10.1590/1984-70332017v17n2a18

Casañas F, Bosch L, Pujolà M et al (1999) Characteristics of a common bean landrace (*Phaseolus vulgaris* L) of great culinary value and selection of a commercial inbred line. J Sci Food Agric 79:693–698. https://doi.org/10.1002/(SICI)1097-0010(199904)79:5<693::AID-JSFA239>3.0.CO;2-S

Castellanos JZ, Guzmán-Maldonado H, Acosta-Gallegos JA, Kelly JD (1995) Effects of hardshell character on cooking time of common beans grown in the semiarid highlands of Mexico. J Sci Food Agric 69:437–443. https://doi.org/10.1002/jsfa.2740690406

Castellanos JZ, Guzman-Maldonado SH, Jimenez A, Acosta-Gallegos JA (1996) Preferential habits of common bean (*Phaseolus vulgaris* L.) consumers in Mexico. In: Reports of bean improvement cooperative and National Dry Bean Council research conference. National Dry Bean Council, Vienna, VA, pp 182–183

Cichy KA, Forster S, Grafton KF, Hosfield GL (2005) Inheritance of seed zinc accumulation in navy bean. Crop Sci 45:864. https://doi.org/10.2135/cropsci2004.0104

Cichy KA, Fernandez A, Kilian A, Kelly JD, Galeano CH, Shaw S, Brick B, Hodkinson D, Troxtell E (2014) QTL analysis of canning quality and color retention in black beans (Phaseolus vulgaris L.). Mol Breed. https://doi.org/10.1007/s11032-013-9940-y

Cichy KA, Wiesinger JA, Mendoza FA (2015) Genetic diversity and genome-wide association analysis of cooking time in dry bean (*Phaseolus vulgaris* L.). Theor Appl Genet 128:1555–1567. https://doi.org/10.1007/s00122-015-2531-z

Coelho CMM, de Mattos Bellato C, Santos JCP et al (2007) Effect of phytate and storage conditions on the development of the 'hard-to-cook' phenomenon in common beans. J Sci Food Agric 87:1237–1243. https://doi.org/10.1002/jsfa.2822

Deshpande SS, Sathe SK, Salunkhe DK (1984) Interrelationships between certain physical and chemical properties of dry bean (*Phaseolus vulgaris* L.). Plant Foods Hum Nutr 34:53–65. https://doi.org/10.1007/BF01095072

Díaz AM, Caldas GV, Blair MW (2010) Concentrations of condensed tannins and anthocyanins in common bean seed coats. Food Res Int 43:595–601. https://doi.org/10.1016/j.foodres.2009.07.014

Donangelo CM, Woodhouse LR, King SM et al (2003) Iron and zinc absorption from two bean (*Phaseolus vulgaris* L.) genotypes in young women. J Agric Food Chem 51:5137–5143. https://doi.org/10.1021/jf030151w

Doria E, Campion B, Sparvoli F et al (2012) Anti-nutrient components and metabolites with health implications in seeds of 10 common bean (*Phaseolus vulgaris* L. and *Phaseolus lunatus* L.) landraces cultivated in southern Italy. J Food Compos Anal 26:72–80. https://doi.org/10.1016/j.jfca.2012.03.005

Elia FM, Hosfield GL, Kelly JD, Uebersax M (1997) Genetic analysis and interrelationships between traits for cooking time, water absorption, and protein and tannin content of andean dry beans. J Am Soc Hort Sci 122:512–518

Food and Agriculture Organization of the United Nations (2016) FAOSTAT. Crop Stat. http://www.fao.org/faostat/en/#data/QC. Accessed 24 Aug 2018

Foyer CH, Lam H-M, Nguyen HT et al (2016) Neglecting legumes has compromised human health and sustainable food production. Nat Plants 2:16112. https://doi.org/10.1038/nplants.2016.112

Garcia RAV, Rangel PN, Bassinello PZ et al (2012) QTL mapping for the cooking time of common beans. Euphytica 186:779–792. https://doi.org/10.1007/s10681-011-0587-7

Gepts P, Bliss FA (1984) Enhanced available methionine concentration associated with higher phaseolin levels in common bean seeds. Theor Appl Genet 69. https://doi.org/10.1007/BF00262537

Ghaderi A, Hosfield GL, Adams MW, Uebersax MA (1984) Variability in culinary quality, component interrelationships, and breeding implications in navy and pinto beans. J Am Soc Hortic Sci 109:85–90

Glahn RP, Lee OA, Yeung A et al (1998) Caco-2 cell ferritin formation predicts nonradiolabeled food iron availability in an *in vitro* digestion/Caco-2 cell culture model. J Nutr 128:1555–1561

Glahn RP, Cheng Z, Welch RM, Gregorio GB (2002) Comparison of iron bioavailability from 15 rice genotypes: studies using an *in vitro* digestion/Caco-2 cell culture model. J Agric Food Chem 50:3586–3591. https://doi.org/10.1021/jf0116496

Graham GG, Morales E, Placko RP, MacLean WCJ (1979) Nutritive value of brown and black beans for infants and small children. Am J Clin Nutr 32:2362–2366

Haas JD, Luna SV, Lung'aho MG et al (2016) Consuming iron biofortified beans increases iron status in Rwandan women after 128 days in a randomized controlled feeding trial. J Nutr 146:1586–1592. https://doi.org/10.3945/jn.115.224741

Hacisalihoglu G, Settles AM (2013) Natural variation in seed composition of 91 common bean genotypes and their possible association with seed coat color. J Plant Nutr 36:772–780. https://doi.org/10.1080/01904167.2012.754041

Hacisalihoglu G, Larbi B, Settles AM (2010) Near-infrared reflectance spectroscopy predicts protein, starch, and seed weight in intact seeds of common bean (*Phaseolus vulgaris* L.). J Agric Food Chem 58:702–706. https://doi.org/10.1021/jf9019294

Hartweck LM, Osborn TC (1997) Altering protein composition by genetically removing phaseolin from common bean seeds containing arcelin or phytohemagglutinin. TAG Theor Appl Genet 95:1012–1017. https://doi.org/10.1007/s001220050655

Havemeier S, Erickson J, Slavin J (2017) Dietary guidance for pulses: the challenge and opportunity to be part of both the vegetable and protein food groups. Ann N Y Acad Sci 1(9). https://doi.org/10.1111/nyas.13308

Hayat I, Ahmad A, Masud T et al (2014) Nutritional and health perspectives of beans (*Phaseolus vulgaris* L.): an overview. Crit Rev Food Sci Nutr 54:580–592. https://doi.org/10.1080/10408398.2011.596639

Hincks MJ, Stanley DW (1987) Lignification: evidence for a role in hard-to-cook beans. J Food Biochem 11:41–58

Hosfield G, Ghaderi A, Uebersax M (1984) A factor analysis of yield and sensory and physicochemical data from tests used to measure culinary quality in dry edible beans. Can J Plant Sci 64:285–293. https://doi.org/10.4141/cjps84-042

Hosfield GL (1991) Genetic control of production and food quality factors in dry bean. Food Technol 45(9):98

Hosfield GL, Uebersax MA (1980) Variability in physicochemical properties and nutritional components of tropical and domestic dry bean germplasm. J Am Soc Hortic Sci 105(2):246–252

IEA (2014) WEO-2014 special report: Africa energy outlook. IEA, Paris

Iyer V, Salunkhe DK, Sathe SK, Rockland LB (1980) Quick-cooking beans (*Phaseolus vulgaris* L.): I. Investigations on quality. Qual Plant Plant Foods Hum Nutr 30:27–43. https://doi.org/10.1007/BF01112102

Jacinto-Hernandez C, Azpiroz-Rivero S, Acosta-Gallegos JA et al (2003) Genetic analysis and random amplified polymorphic DNA markers associated with cooking time in common bean. Crop Sci 43:329–332

Jackson GM, Varriano-Marston E (1981) Hard-to-Cook phenomenon in beans: Effects of accelerated storage on water absorption and cooking time. J Food Sci 46:799–803. https://doi.org/10.1111/j.1365-2621.1981.tb15351.x

Jackson ER, Wiese KL (1993) Changes in calculated process times and drained weight based on soaking and blanching of kidney and navy beans. J Food Prot 56:239–242

Jones PMB, Boulter D (1983) The analysis of development of hardbean during storage of black beans (*Phaseolus vulgaris* L). Plant Foods Hum Nutr 33:77–85. https://doi.org/10.1007/BF01093740

Kelly JD, Bliss FA (1975) Heritability estimates of percentage seed protein and available methionine and correlations with yield in dry beans. Crop Sci 15:753. https://doi.org/10.2135/cropsci1975.0011183X001500060004x

Khanal R, Burt AJ, Woodrow L, Balasubramanian P, Navabi A (2015) Genotypic association of parameters commonly used to predict canning quality of dry bean. Crop Sci 54:2564–2573

Kinyanjui PK, Njoroge DM, Makokha AO et al (2015) Hydration properties and texture fingerprints of easy- and hard-to-cook bean varieties. Food Sci Nutr 3:39–47. https://doi.org/10.1002/fsn3.188

Kumar V, Sinha AK, Makkar HPS, Becker K (2010) Dietary roles of phytate and phytase in human nutrition: a review. Food Chem 120:945–959. https://doi.org/10.1016/j.foodchem.2009.11.052

Lajolo FM, Genovese MI (2002) Nutritional significance of lectins and enzyme inhibitors from legumes. J Agric Food Chem 50:6592–6598. https://doi.org/10.1021/jf020191k

Lange AD, Labuschagne M (2001) Multivariate assessment of canning quality, chemical characteristics and yield of small white canning beans (*Phaseolus vulgaris* L.) in South Afric. J Sci Food Agric 81:30–35

Leterme P (2002) Recommendations by health organizations for pulse consumption. Br J Nutr 88:S239–S242. https://doi.org/10.1079/BJN2002712

Liu K, Bourne MC (1995) Cellular, biological, and physicochemical basis for the hard-to-cook defect in legume seeds. Crit Rev Food Sci Nutr 35:263–298. https://doi.org/10.1080/10408399509527702

Lu W, Chang KC (1996) Correlations between chemical composition and canning quality attributes of navy bean (*Phaseolus vulgaris* L.). Cereal Chem 73:785–787

Ma Y, Bliss FA (1978) Seed proteins of common bean. Crop Sci 18:431. https://doi.org/10.2135/cropsci1978.0011183X001800030018x

Maes WH, Verbist B (2012) Increasing the sustainability of household cooking in developing countries: policy implications. Renew Sust Energ Rev 16:4204–4221. https://doi.org/10.1016/j.rser.2012.03.031

Makungwa SD, Epulani F, Woodhouse IH (2013) Fuelwood supply: a missed essential component in a food security equation. J Food Secur 1:49–51. https://doi.org/10.12691/jfs-1-2-6

McPherson LL (1990) The effect of the consumption of red kidney beans (*Phaseolus vulgaris*) on the growth of rats and the implications for human populations. J R Soc Health 110:222–226

Mendoza FA, Cichy K, Lu R, Kelly J (2014) Evaluation of canning quality traits in black beans (*Phaseolus vulgaris* L.) by visible/near-infrared spectroscopy. Food Bioprocess Technol 7(9):2666–2678

Mendoza FA, Kelly JD, Cichy KA (2017) Automated prediction of sensory scores for color and appearance in canned black beans (*Phaseolus vulgaris* L.) using machine vision. Int J Food Prop 20:83–99. https://doi.org/10.1080/10942912.2015.1136939

Mendoza FA, Cichy KA, Sprague C, Amanda C, Goffnett A, Luc R, Kelly JD (2018) Prediction of canned black bean texture (Phaseolus vulgaris L) fromintact dry seeds using visible/near infrared spectroscopy and hyperspectral imaging data. J Sci Food Agric 98:283–290

Menéndez A, Curt MD (2013) Energy and socio-economic profile of a small rural community in the highlands of central Tanzania: a case study. Energy Sustain Dev 17:201–209. https://doi.org/10.1016/j.esd.2012.12.002

Merwe D, Osthoff G, Pretorius AJ (2006) Evaluation and standardization of small scale canning methods for small white beans (Phaseolus vulgaris L.) canned in tomato sauce. J Sci Food Agric 86:1115

Messina V (2014) Nutritional and health benefits of dried beans. Am J Clin Nutr 100:437S–442S. https://doi.org/10.3945/ajcn.113.071472.2

Mkanda AV, Minnaar A, de Kock HL (2007) Relating consumer preferences to sensory and physicochemical properties of dry beans (*Phaseolus vulgaris*). J Sci Food Agric 87:2868–2879. https://doi.org/10.1002/jsfa.3046

Monti LM, Grillo S (1983) Legume seed improvement for protein content and quality. Qual Plant Plant Foods Hum Nutr 32:253–266. https://doi.org/10.1007/BF01091190

Montoya CA, Leterme P, Victoria NF et al (2008) Susceptibility of phaseolin to in vitro proteolysis is highly variable across common bean varieties (*Phaseolus vulgaris*). J Agric Food Chem 56:2183–2191. https://doi.org/10.1021/jf072576e

Navarrete DA, Bressani R (1981) Protein digestibility and protein quality of common beans (*Phaseolus vulgaris*) fed alone and with maize, in adult humans using a short-term nitrogen balance assay. Am J Clin Nutr 34:1893–1898

Ngatchou A, Bitjoka L, Boukar O, Tonye E (2012) Color and texture information processing to improve storage beans. Br J Appl Sci Technol 2:96–111. https://doi.org/10.9734/BJAST/2012/796

Nleya T, Vandenberg A, Araganosa G et al (2000) Produce quality of food legumes: genotype (G), environment (E) and (G×E) considerations. In: Knight R (ed) Linking research and marketing opportunities for pulses in the 21st century. Springer, Dordrecht, pp 173–182

Oikeh SO, Menkir A, Maziya-Dixon B et al (2003) Assessment of concentrations of iron and zinc and bioavailable iron in grains of early-maturing tropical maize varieties. J Agric Food Chem 51:3688–3694. https://doi.org/10.1021/jf0261708

Osborn TC, Blake T, Gepts P, Bliss FA (1986) Bean arcelin. Theor Appl Genet 71:847–855. https://doi.org/10.1007/BF00276428

Osborn TC, Hartweck LM, Harmsen RH et al (2003) Registration of *Phaseolus vulgaris* genetic stocks with altered seed protein compositions. Crop Sci 43:1570

Osborni TC, Alexander DC, Sun SSM et al (1988) Insecticidal activity and lectin homology of arcelin seed protein. Science 240:207–210. https://doi.org/10.1126/science.240.4849.207

Petry N, Egli I, Zeder C et al (2010) Polyphenols and phytic acid contribute to the low iron bioavailability from common beans in young women. J Nutr 140:1977–1982. https://doi.org/10.3945/jn.110.125369

Petry N, Egli I, Gahutu JB et al (2012) Stable iron isotope studies in Rwandese women indicate that the common bean has limited potential as a vehicle for iron biofortification. J Nutr 142:492–497. https://doi.org/10.3945/jn.111.149286

Petry N, Boy E, Wirth JP, Hurrell RF (2015) Review: the potential of the common bean (*Phaseolus vulgaris*) as a vehicle for iron biofortification. Nutrients 7:1144–1173. https://doi.org/10.3390/nu7021144

Plans M, Simó J, Casañas F, Sabaté J (2012) Near-infrared spectroscopy analysis of seed coats of common beans (*Phaseolus vulgaris* L.): a potential tool for breeding and quality evaluation. J Agric Food Chem 60:706–712. https://doi.org/10.1021/jf204110k

Plans M, Simó J, Casañas F et al (2013) Characterization of common beans (*Phaseolus vulgaris* L.) by infrared spectroscopy: comparison of MIR, FT-NIR and dispersive NIR using portable and benchtop instruments. Food Res Int 54:1643–1651. https://doi.org/10.1016/j.foodres.2013.09.003

Plans M, Simó J, Casañas F et al (2014) Estimating sensory properties of common beans (*Phaseolus vulgaris* L.) by near infrared spectroscopy. Food Res Int 56:55–62. https://doi.org/10.1016/j.foodres.2013.12.003

Posa-Macalincag MT, Hosfield GL, Grafton KF, Uebersax MA, Kelly JD (2002) Quantitative trait loci (QTL) analysis of canning quality traits in kidney bean (Phaseolus vulgaris L.). J Am Soc Hortic Sci 127:608–615

Proctor JP, Watts BM (1987) Effect of cultivar, growing location, moisture and phytate content on the cooking times of freshly harvested navy beans. Can J Plant Sci 67:923–926

Pusztai A, Grant G, Stewart JC et al (1993) Nutritional evaluation of RAZ-2, a new *Phaseolus vulgaris* bean cultivar containing high levels of the natural insecticidal protein arcelin 1. J Agric Food Chem 41:436–440. https://doi.org/10.1021/jf00027a017

Reddy NR, Pierson MD, Sathe SK, Salunkhe DK (1984) Chemical, nutritional and physiological aspects of dry bean carbohydrates. A review. Food Chem 13:25–68

Reddy NR, Pierson MD, Sathe SK, Salunkhe DK (1985) Dry bean tannins: a review of nutritional implications. J Am Oil Chem Soc 62:541–549. https://doi.org/10.1007/BF02542329

Reyes-Moreno C, Paredes-López O, Gonzalez E (1993) Hard-to-cook phenomenon in common beans – a review. Crit Rev Food Sci Nutr 33:227–286. https://doi.org/10.1080/10408399309527621

Romero del Castillo R, Costell E, Plans M et al (2012) A standardized method of preparing common beans (*Phaseolus vulgaris* L.) for sensory analysis. J Sens Stud 27:188–195. https://doi.org/10.1111/j.1745-459X.2012.00381.x

Salunkhe DK (1982) Legumes in human nutriton: current status and future research needs. Curr Sci 51:387–394

Sandberg A-S (2002) Bioavailability of minerals in legumes. Br J Nutr 88:S281–S285. https://doi.org/10.1079/BJN/2002718

Sanz-Calvo M, Atienza-del-Rey J (1999) Sensory analysis of beans (*Phaseolus vulgaris*). Biotechnol Agron Soc Environ 3:201–204

Schmutz J, McClean PE, Mamidi S et al (2014) A reference genome for common bean and genome-wide analysis of dual domestications. Nat Genet 46:707–713. https://doi.org/10.1038/ng.3008

Scott J, Maideni M (1998) Socio-economic survey of three bean growing areas of Malawi. Netw Bean Res Africa Occas Pap Ser 24:1–48

Shellie-Dessert KC, Hosfield GL (1990) Implications of genetic variability for dry bean cooking time and novel cooking methods for fuelwood conservation in Rwanda. Ecol Food Nutr 24:195–211

Singh SP, Gepts P, Debouck DG (1991) Races of common bean (*Phaseolus vulgaris*, Fabaceae). Econ Bot 45:379–396

Srisuma N, Hammerschmidt R, Uebersax MA et al (1989) Storage induced changes of phenolic acids and the development of hard-to-cook in dry beans (*Phaseolus vulgaris*, var. Seafarer). J Food Sci 54:311–314. https://doi.org/10.1111/j.1365-2621.1989.tb03069.x

Stanley DW (1992) A possible role for condensed tannins in bean hardening. Food Res Int 25:187–192. https://doi.org/10.1016/0963-9969(92)90136-S

Suárez-Martínez SE, Ferriz-Martínez RA, Campos-Vega R et al (2016) Bean seeds: leading nutraceutical source for human health. CyTA J Food 14:131–137. https://doi.org/10.1080/19476337.2015.1063548

Tako E, Reed S, Anandaraman A et al (2015) Studies of cream seeded carioca beans (*Phaseolus vulgaris* L.) from a Rwandan efficacy trial: *in vitro* and *in vivo* screening tools reflect human studies and predict beneficial results from iron biofortified beans. PLoS One 10:1–15. https://doi.org/10.1371/journal.pone.0138479

Taylor M, Chapman R, Beyaert R et al (2008) Seed storage protein deficiency improves sulfur amino acid content in common bean (*Phaseolus vulgaris* L.): redirection of sulfur from γ-glutamyl-S- methyl-cysteine. J Agric Food Chem 56:5647–5654. https://doi.org/10.1021/jf800787y

Umar Lule S, Xia W (2005) Food phenolics, pros and cons: a review. Food Rev Int 21:367–388. https://doi.org/10.1080/87559120500222862

Vasconcelos IM, Oliveira JTA (2004) Antinutritional properties of plant lectins. Toxicon 44:385–403. https://doi.org/10.1016/j.toxicon.2004.05.005

USDA (1976) U.S. Standards Canned Beans, Dried, Pork, Baked

Vaz-Tostes M das G, Verediano TA, de Mejia EG, Brunoro Costa NM (2016) Evaluation of iron and zinc bioavailability of beans targeted for biofortification using *in vitro* and *in vivo* models and their effect on the nutritional status of preschool children. J Sci Food Agric 96:1326–1332. https://doi.org/10.1002/jsfa.7226

Walters KJ, Hosfield GL, Uebersax MA, Kelly JD (1997) Navy bean canning quality: correlations, heritability estimates, and randomly amplified polymorphic DNA markers associated with component traits. J Am Soc Hortic Sci 122:338–343

Wang N, Daun JK (2005) Determination of cooking times of pulses using an automated Mattson cooker apparatus. J Sci Food Agric 85:1631–1635. https://doi.org/10.1002/jsfa.2134

Watts P (2011) Global pulse industry: state of production, consumption and trade; marketing challenges and opportunities. In: Tiwari BK, Gowen A, McKenna BBT-PF (eds) Pulse foods: processing, quality and nutraceutical applications. Academic, San Diego, CA, pp 437–464

White PJ, Broadley MR (2005) Biofortifying crops with essential mineral elements. Trends Plant Sci 10:586–593. https://doi.org/10.1016/j.tplants.2005.10.001

WHO (2015) The global prevalence of anaemia in 2011

Wiesinger JA, Cichy KA, Glahn RP et al (2016) Demonstrating a nutritional advantage to the fast-cooking dry bean (*Phaseolus vulgaris* L.). J Agric Food Chem 64:8592–8603. https://doi.org/10.1021/acs.jafc.6b03100

Wortmann CS, Kirkby RA, Eledu CA, Allen DJ (1998) Atlas of common bean (*Phaseolus vulgaris* L.) production in Africa. CIAT Publ 297:131

Wright EM, Kelly JD (2011) Mapping QTL for seed yield and canning quality following processing of black bean (*Phaseolus vulgaris* L.). Euphytica 179(3):471–484. https://doi.org/10.1007/s10681-011-0369-2

Xu BJ, Chang SK (2009) Total phenolic, phenolic acid, anthocyanin, flavan-3-ol, and flavonol profiles and antioxidant properties of pinto and black beans (*Phaseolus vulgaris* L.) as affected by thermal processing. J Agric Food Chem 57:4754–4764

Chapter 9
Genomic Approaches for Biofortification of Grain Zinc and Iron in Wheat

Govindan Velu and Ravi P. Singh

Micronutrient malnutrition, resulting from diets primarily deficient in zinc (Zn) and iron (Fe), has been widely recognized as a major health problem affecting almost three billion people worldwide, especially in countries with a high consumption of cereals (Black et al. 2013). Zn is an essential trace element for all organisms and its role has been thoroughly reviewed in both plant and human health (Cakmak 2000; Graham et al. 2012). About 17% of the world's population risks micronutrient deficiency due to inadequate Zn intake (Wessells and Brown 2012), and annually more than 100,000 deaths of children under age five are attributed to Zn deficiency (Black et al. 2013).

The biofortification approach of improving the nutritional quality of staple food crops through breeding offers a cost-effective and sustainable solution to the global malnutrition problems. Wheat is the second most produced cereal crop after rice and constitutes about 28% of dietary energy and 20% protein to consumers in many parts of the world (Braun et al. 2010). Improving the nutritional levels of wheat is therefore of paramount importance. The wheat biofortification breeding program at the International Maize and Wheat Improvement Center (CIMMYT) is leading the global partnership to develop and disseminate competitive wheat varieties with high grain Zn and other essential agronomic features to target regions in South Asia and beyond. The emphasis of this breeding program is to introduce novel sources of genetic diversity from wild species and landraces, into the adapted wheat background which resulted in the development of widely adapted, durable rust and foliar disease-resistant, high Zn wheat varieties (Velu et al. 2014). The high zinc wheat varieties with 20–40% superior in grain Zn concentration over the baseline commercial cultivars are being adapted by small-holder farmers in target regions (Velu et al. 2015).

G. Velu (✉) · R. P. Singh
Global Wheat Program, International Maize and Wheat Improvement Center, Texcoco, Estado de México, Mexico
e-mail: velu@cgiar.org; r.singh@cgiar.org

A. M. I. Qureshi et al. (eds.), *Quality Breeding in Field Crops*,
https://doi.org/10.1007/978-3-030-04609-5_9

Identification of molecular markers linked to nutritional traits would be of great interest as nutritional elements are rather difficult and extensive to phenotype. QTL mapping is a highly useful tool for the discovery of markers to use in breeding programs, especially in the post genomics era genotyping costs getting cheaper and application of molecular markers in wheat breeding have been increased. One way to implement molecular markers in breeding programs is by the identification of linkage between DNA markers and the loci that control the traits of interest, to make selections in segregating progenies based on those marker-trait associations. The marker-assistant selection (MAS) procedures can greatly facilitate the breeding programs by identifying genomic regions associated with higher grain Zn concentrations. To date, twenty-one QTLs for increased grain zinc content (GZnC) have been reported on ten chromosomes in diploid (*T. monococcum* and *T. boeoticum*), tetraploid (*T. dicoccoides*), or hexaploid wheat (*T. aestivum*) (Xu et al. 2011; Velu et al. 2014) (Table 9.1).

We initiated the marker discovery for GZnC in wheat at CIMMYT using a recombinant inbred line (RIL) population from the cross between PBW343 and Kenya Swara. Two novel QTLs of large effect were stably detected for increasing GZnC on chromosomes 2Bc (centromeric region) and 3AL (long arm) (Table 9.2). The two QTLs individually explained about 10–15% of the total phenotypic variation (Hao et al. 2014). The 2Bc QTL from PBW343 has pleiotropic effect and can increase thousand-kernel weight at significant level. The flanking markers associated for these QTLs (Table 9.2) were converted into usable farm (SNP) to be able to use in early generation MAS or MABC program.

Another QTL mapping study conducted in Seri M82 × Synthetic Hexaploid Wheat population revealed two major QTLs for GZnC on chromosome 4B and 6B, that appears to have pleiotropic effects for GFeC (Crespo-Herrera et al. 2016). The QTL on chromosome 4B fully linked with the marker TP81797 and the QTL on

Table 9.1 Summary of QTLs detected for GZnC in diploid, tetraploid and hexaploid wheat and source cultivars for GZnC from our mapping studies and from literature review

GZnC	1	2	3	4	5	6	7
A	*T. monococcum* ID-362	*T. durum* Langdon, *T. dicoccoides* G18-16	*T. aestivum* Kenya Swara	*T. aestivum* Hanxuan 10	*T. aestivum* Hanxuan 10, *T. dicoccoides* G18-16, *T. aestivum* Xiaoyan 54, *T. monococcum* ID-362	*T. spelta* H+ 26 (PI 348449), *T. aestivum* BV2010-13	*T. aestivum* Lumai 14, *T. dicoccoides* G18-16, *T. boeoticum* pau5088, *T. monococcum* pau14087, *T. aestivum* RAC875-2
B	*T. aestivum* BV2010-48	*T. aestivum* PBW343, *T.aestivum* HUW 234		*T. aestivum* Jing 411, *T. aestivum* RAC875-2	*T. aestivum* BV2010-13	*T. dicoccoides* G18-16, *T. aestivum* Cascades, *T. dicoccoides* LDN(DIC-6B)	*T. dicoccoides* G18-16
D			*T. aestivum* RAC875-2, *T. aestivum* BV2010-48	*T. aestivum* Lumai 14			

Note: The QTL detected in diploid, tetraploid, and hexaploid levels of wheat were colored in red, brown and blue, respectively, followed by variety or accession names as sources; the boxes in grey shading are QTL we have identified in PBW343/Kenya Swara population; the boxes in pink represent chromosomal location of QTL detected in the Picus…Francolin x Croc/Ae. Squarrosa population.

Table 9.2 Position and QTL effect associated with high grain zinc content (GZnC) in the PBW343 × Kenya Swara RIL population

Environ.	QTL name	Marker interval	Marker position	Peak LOD	Peak position (cM)	R^2 (%)[a]	Additive effect[b]
GZnC-2012	*QGzncpk.cimmyt-1BS*	*wPt-8622*	68.3	5.6***	68.3	10	−2.16
	QGzncpk.cimmyt-2Bc	*wPt-6174–wPt-2430*	37.6–41.9	4.9**	38.6	9	1.96
	QGzncpk.cimmyt-3AL	*wPt-0286*	79.0	7.9***	79.0	15	−2.43
	QGzncpk.cimmyt-4AS	*wPt-7191–wPt-8007*	24.0–26.3	3.4*	26.1	7	−1.70
	QGzncpk.cimmyt-5BL	*Xcfp393–tPt-3144*	220.8–225.4	3.4*	221.9	8	−1.49
GZnC-2013	*QGzncpk.cimmyt-1BS*	*wPt-8622*	68.3	3.1	68.3	5	−1.65
	QGzncpk.cimmyt-1BL	*Xwmc44–wPt2861*	245.8–249.7	3.0	248.9	5	1.46
	QGzncpk.cimmyt-2Bc	*wPt-6174–wPt-2430*	37.6–41.9	6.6***	37.7	11	2.52
	QGzncpk.cimmyt-2D	*wPt-6847–tPt-6105*	8.9–49.0	5.2***	22.5	26	3.35
	QGzncpk.cimmyt-3AL	*wPt-0286–Xwmc222*	79.0–87.3	4.7**	80.0	11	−2.34
	QGzncpk.cimmyt-6AL	*wPt-667,817–tPt-6278*	100.6–103.0	4.0**	101.7	7	−2.16
	QGzncpk.cimmyt-7DL	*wPt-671,530*	142.9	3.4	142.9	5	1.36
GZnC-Mean	*QGzncpk.cimmyt-1BS*	*wPt-3103–wPt-8622*	68.0–68.3	7.0***	68.1	11	−2.47
	QGzncpk.cimmyt-2Bc	*wPt-6174*	37.6	6.6***	37.6	10	2.09
	QGzncpk.cimmyt-3AL	*wPt-0286–Xwmc222*	79.0–87.3	9.0***	79.1	15	−2.56
	QGzncpk.cimmyt-4AS	*wPt-7191–wPt-8007*	24.0–26.3	3.1	26.2	5	−1.52
	QGzncpk.cimmyt-6AL	*wPt-667,817–tPt-6278*	100.6–103.0	3.0	101.7	6	−1.68

*P < 0.05; ** P < 0.01; *** P < 0.001.

[a] R^2, phenotypic variation associated with the QTL.

[b] Positive and negative value indicate that the alleles were inherited from PBW343 and Kenya Swara, respectively; the unit is mg/kg

chromosome 6B is also rather close to marker TP29689 (0.9 cM). This marker is being fine mapped and validated in other breeding populations.

Another mapping study involving RILs derived from a cross between Picus/3/..* Francolin (low Zn parent) and Reh/Hare//2*BCN/3/Croc_1/Ae.squarrosa (213)//.../ Huites/7/Mutus (high Zn parent) showed three major QTLs on chromosomes 1B,

Table 9.3 Position and effect of QTLs associated with high grain zinc concentration in a RIL from PICUS/3/..*FRANCOLIN and REH/HARE//2*BCN/3/CROC_1/AE.SQUARROSA(213)//.../HUITES/7/MUTUS

QTL name	Interval	Range	Peak position (cM)	Peak LOD	R^2 (%)[a]	Additive effect[b]
QGzncpr.cmt-1B[c]	*4,663,991–wPt-10,518*	48.8–49.0	48.9	8.8**	15	−2.02
QGzncpr.cmt-3D	*1,007,328*	12.4	12.4	3.4	6	−1.23
QGzncpr.cmt-5B	*wPt-8163–1,139,328*	84.3–84.5	84.4	6.5**	11	1.75
QGzncpr.cmt-6A	*4,990,410*	109.1	109.1	4.7**	8	1.36

**Significant at the 0.01 probability level.
[a] R^2, phenotypic variation associated with the QTL
[b] Positive and negative value indicate that the all eles were inherited from low and high Zn parents, respectively; the unit is mg/kg

5B, and 6A, respectively. The 1B QTL from high Zn parent explains about 15% of phenotypic variation, and the 5B and 6A QTL from low Zn parent contribute about 11% and 8% of total variation, respectively. Major QTLs contributed from both parents explain the trangressive segregation pattern observed in the RIL population. The three QTLs were closely linked to markers *wPt-10,518*, *wPt-8163*, and *4,990,410*, respectively (Table 9.3). Both the 1B and 5B QTL should represent novel loci for increasing grain zinc concentration based on our literature review (Table 9.1).

9.1 Molecular Marker-Assisted Selection

During the 2014–2015 crop season as a proof-of-concept strategy, marker assisted back-crossing strategy has been applied using selected RILs that showed significantly enhanced GZnC than either of the parental lines from PBW 343 × Kenya Swara population to transfer QTL of interest. Selected RIL lines high in GZnC and best bet lines adapted to South Asia were crossed and F1s were backcrossed to adapted parent. The marker-assisted selection was begun with BC1 plants to select plants with favorable GZnC alleles. DNA samples from individual BC1 plants were genotyped, and PCR-based probes for these QTLs were used to identify plants which have favorable donor alleles before the pollination time. The plants positive for donor allele were backcrossed again to the recurrent parent. The resultant BC2 families were advanced through conventional selection schemes of shuttle breeding between Ciudad Obregon and Toluca valley in Mexico. Promising, agronomically superior plants were selected which were screened for presence of favorable alleles for GZnC. This strategy will serve to move the desirable alleles quickly and more precisely into the adapted background.

The rich genetic diversity for Zn and Fe in wheat from diverse wild species and landraces provides novel alleles for genetic enhancement of Zn and Fe in wheat. Over the years through traditional breeding approach several novel alleles have been incorporated for grain Zn into elite germplasm which has led to development and release of biofortified wheats such as 'Zinc Shakti (Chitra)', WB-02, HPBW 01 (PBW 1 Zn), Zincol-2016, and BARI-Gom 33 (Singh and Velu 2017). These were developed from crosses between high-yielding elite wheat lines and the *T. dicoccon* based synthetic hexaploid wheats or *T. spelta* accessions and the derivative had about 40% higher grain Zn over local checks (Velu et al. 2015). This indicates significant contribution of alleles from ancient wheats and landraces for genetic enhancement of grain Zn in wheat. The biofortification breeding program at CIMMYT has demonstrated that targeted crosses with increased population sizes and selection for agronomic traits in the early generation could facilitates incorporation of several novel alleles for grain Zn in elite, high-yielding germplasm (Velu et al. 2014).

Grain Zn concentration, though under quantitative genetic control and influenced by soil and other associated factors, can further be improved through pyramiding of multiple high Zn QTLs in high-yielding wheats. High-throughput, non-destructive phenotyping for grain Zn and Fe using the X-ray fluorescence (XRF) analysis has facilitated the selection dramatically. Gene discovery and mapping studies led to the utilization of markers to further improve the breeding efficiency.

In a recent genetic study using the diverse panel of 330 CIMMYT wheat lines phenotyped in a range of environments in India and in Mexico and genotyped using the high-density iselect 90K SNP assay disclosed promising candidate genes for high Zn accumulation in wheat grain. Genome-wide association study (GWAS) identified 39 marker-trait associations (MTA) for Zn, and the associated candidate genes for grain Zn in wheat has been localized using GWAS approach have identified two major effect stable QTL regions on chromosome 2 and 7. Earlier studies have shown large number of QTLs for high grain Fe and Zn mapped on two and seven chromosomes of wheat (Srinivasa et al. 2014; Hao et al. 2014; Velu et al. 2016; Crespo-Herrera et al. 2017). These results confirm that the group 2 and 7 chromosomes hold genes for nutrient uptake, translocation and sequestration of mineral in wheat plant (Tiwari et al. 2009; Krishnappa et al. 2017; Crespo-Herrera et al. 2017).

In conclusion, we identified some of the promising genetic loci associated with grain Zn and Fe in wheat through QTL mapping analyses. Among the several genomic regions, the novel candidate regions identified in chromosome groups 2 and 7 offer potential candidate regions for fine mapping and candidate genes for marker-assisted breeding.

Acknowledgments The authors acknowledge the financial support from the HarvestPlus challenge program and CGIAR research program on Agriculture for Nutrition and Health. Special thanks to Susanne Dreisigacker, CIMMYT Wheat Genomics Laboratory, Mexico.

References

Black RE, Victora CG, Walker SP, Bhutta ZA, Christian P, de Onis M, Ezzati M, Grantham-McGregor S, Katz J, Martorell R, Uauy R (2013) Maternal and child undernutrition and overweight in low-income and middle-income countries. Lancet 382:427–451

Braun HJ, Atlin G, Payne T (2010) Multi-location testing as a tool to identify plant response to global climate change. In: Reynolds MP (ed) Climate change and crop production. CABI Press, Oxford, pp 115–138

Cakmak I (2000) Possible roles of zinc in protecting plant cells from damage by reactive oxygen species. New Phytol 146:185–205

Crespo-Herrera LA, Velu G, Singh RP (2016) QTL mapping reveals pleiotropic effect for grain iron and zinc concentrations in wheat. Ann Appl Biol 169(1):27–35

Crespo-Herrera LA, Velu G, Singh RP, Hao Y, Stangoulis J (2017) QTL mapping of grain Zn and Fe concentrations in two hexaploid wheat RIL populations with ample transgressive segregation in wheat. Front Plant Sci 8:1800

Graham R, Knez M, Welch R (2012) How much nutritional iron deficiency in humans globally is due to an underlying zinc deficiency? Adv Agron 115:1–40

Hao Y, Velu G, Pena RJ, Singh S, Singh RP (2014) Genetic loci associated with high grain zinc concentration and pleiotropic effect on kernel weight in wheat (Triticum aestivum L.). Mol Breed 34:1893–1902

Krishnappa G, Singh AM, Chaudhary S, Ahlawat AK, Singh SK, Shukla RB et al (2017) Molecular mapping of the grain iron and zinc concentration, protein content and thousand kernel weight in wheat (Triticum aestivum L.). PLoS One 12. https://doi.org/10.1371/journal.pone.0174972

Singh RP, Velu G. (2017) Zinc-biofortified wheat: harnessing genetic diversity for improved nutritional quality. Science brief: biofortification no. 1–4. Crop Trust online portal

Srinivasa J, Arun B, Mishra VK, Singh GP, Velu G, Babu R et al (2014) Zinc and iron concentration QTL mapped in a Triticum spelta × T. aestivum cross. Theor Appl Genet 127:1643–1651

Tiwari VK, Rawat N, Chhuneja P, Neelam K, Aggarwal R, Randhawa GS et al (2009) Mapping of quantitative trait Loci for grain iron and zinc concentration in diploid A genome wheat. J Hered 100:771–776

Velu G, Ortiz-Monasterio I, Cakmak I, Hao Y, Singh RP (2014) Biofortification strategies to increase grain zinc and iron concentrations in wheat. J Cereal Sci 59:365–372

Velu G, Singh RP, Arun B, Mishra VK, Tiwari C, Joshi A, Cherian B, Virk P, Pfeiffer WH (2015) Reaching out to farmers with high zinc wheat varieties through public-private partnerships – an experience from eastern-Gangetic plains of India. Adv Food Technol Nutr Sci 1(3):73–75

Velu G, Tutus Y, Gomez-Becerra HF, Hao Y, Demir L, Kara R et al (2016) QTL mapping for grain zinc and iron concentrations and zinc efficiency in a tetraploid and hexaploid wheat mapping populations. Plant Soil 411:81–99

Wessells KR, Brown KH (2012) Estimating the global prevalence of zinc deficiency: results based on zinc availability in national food supplies and the prevalence of stunting. PLoS One 7:e50568

Xu Y, An D, Li H, Xu H (2011) Breeding wheat for enhanced micronutrients. Can J Plant Sci 91:231–237

Chapter 10
Current Breeding Approaches for Developing Rice with Improved Grain and Nutritional Qualities

Wendy Chui Phing Lau and Mohammad Abdul Latif

10.1 Introduction

Rice is one of the most important food crops as it is the staple food that supports the livelihood for nearly half of the world's population (Global Rice Science Partnership (GRiSP) 2013). About 90% of the rice produced and consumed in year 2014 is in Asia (Food and Agriculture Organization of the United Nations (FAO) 2017). Despite the influence of urbanization, income growth, and westernization of Asian diets, rice consumption in Asia is projected to increase by 19.8%, from 388 million tons in 2010 to 465 million tons in 2035 (Pingali 2007; GRiSP 2013; Kelly 2016). Furthermore, rice would remain as the developing world's most important food crop for the next 30 years (Timmer et al. 2010). A study also showed that rice is still the primary choice and main dish in everyday meals for consumers in both urban and rural areas of South-East Asian countries, particularly Philippines, Malaysia and Indonesia (Lipoeto et al. 2012). The young Asians may be consuming less rice than their parents but rice consumption will not cease at the younger generation (Timmer et al. 2010). In addition, there is also a notable increase in rice consumption in western countries (Ferrero and Nguyen 2004). The increase in rice consumption in western countries is due to raising awareness of rice as healthy food (Braun and Bos 2005; Suwannaporn and Linnemann 2008).

In fact, with the growing income, consumers are inclining towards high-quality rice (Pingali 2007; Timmer et al. 2010). The shift towards high-quality rice and the increase in demand and consumption of high-quality rice are apparent in some customarily rice-consuming countries such as China and Bangladesh (Hsu et al. 2001; Peng et al. 2009; Minten et al. 2011). Nowadays, the consumers are discriminating

W. C. P. Lau (✉)
Independent researcher, Kuching, Sarawak, Malaysia

M. A. Latif
Bangladesh Rice Research Institute, Gazipur, Bangladesh

A. M. I. Qureshi et al. (eds.), *Quality Breeding in Field Crops*,
https://doi.org/10.1007/978-3-030-04609-5_10

rice based on the grain quality attributes and nutritional quality, such as fragrance, texture and protein content (Hsu et al. 2001; Juliano 2005; Farah et al. 2011). Besides protein content, micronutrients and vitamins are also important nutritional qualities that greatly emphasized to curb essential nutrient deficiencies in developing countries that are subsisting on rice (Arsenault et al. 2010; Pinkaew et al. 2013; Htet et al. 2016). Therefore, rice is not merely a daily staple to fill one's stomach, but its grain and nutritional qualities also must be able to satisfy the consumers' palate and to fulfil their daily intake of nutrients.

Rice grain quality is judged by a variety of attributes. Preference of rice grain quality varies by region and culture, and among individuals. Grain appearance, such as length, width, breakage, colour and chalkiness are some grain quality attributes that would affect the consumers' purchasing decision as those are the first impression to the consumers (Tomlins et al. 2007). Fragrance, cooked kernel elongation, gel consistency, gelatinization temperature, amylose content and protein content are examples of cooking and eating qualities, which also make rice grain quality an important determining factor of consumers' preference, because rice is eaten as whole, milled and cooked, usually with some or without seasoning (Khush and Juliano 1985; Champagne 2008). Furthermore, fragrant rice fetches higher premium in local and export market compared to non-fragrant rice. Nutritional qualities in rice includes micronutrients and vitamins, some of the intensively studied are vitamin A, iron (Fe) and zinc (Zn) (Brar et al. 2012).

Besides yield, rice grain and nutritional qualities must keep up with the burgeoning demand and changing lifestyle of consumers. Rice breeders have to tailor the rice grain and nutritional qualities to suit the preference of discerning consumers and also nourishing the people of their respective countries. Each country has a remarkable rice breeding journey that produces the rice varieties that are widely accepted by their local farmers and consumers. Some of the high-quality rice varieties represent their nation's culture and pride, and also highly priced in domestic and international market, such as Basmati rice varieties of India, Pakistan and Bangladesh, Khao Dawk Mali of Thailand, Azucena of the Philippines, and Sadri of Iran (Lorieux et al. 1996; Garris et al. 2005; Calingacion et al. 2014; Ashfaq et al. 2015). These varieties were developed by conventional breeding and some of the varieties are still used as parents in the breeding programme nowadays.

The advances in DNA markers, sequencing technology and gene editing become pragmatic tools for plant breeders, allowing plant breeders to improve traits that cannot be achieved by conventional breeding alone. In addition to the phenotype evaluation and selection, DNA markers have been integrated in plant selection and breeding strategy. Following the advancement of DNA sequencing technology, rice genome is fully sequenced and thereby facilitates in gene annotation (IRGSP 2005). The up-to-date information on rice gene annotations available from the Rice Annotation Project Database (RAP-DB) opens up the possibilities of rice improvement via genomic research (Sakai et al. 2013). Apart from that, the advances of nucleases in gene editing which has been applied in other crops would provide alternative method to rice breeder for improving rice quality. Hence, these recent advances would facilitate rice breeders in the endeavour to improve rice grain and nutritional qualities.

10.2 Marker-Assisted Selection

Marker-assisted selection (MAS) approach has been around for more than a decade. MAS is a method of choice by most of the rice breeders. It can be easily incorporated into breeding schemes, such as marker-assisted recurrent selection (MARS), marker-assisted backcrossing (MABC) and gene pyramiding. MAS had been widely applied by breeders to improve many traits in rice (Salgotra et al. 2012; Luo et al. 2014a, 2016).

The advantage of MAS is that it allows detection and selection of desired genotype at early breeding stage. In comparison with conventional breeding, plants have to be grown to a certain stage and had to undergo phenotypic screening that is time and labour consuming. In MAS, the plants with desirable genotype can be identified at seedling stage using DNA markers linked to the trait and thus, no time or effort is wasted on the undesirable plants (Collard et al. 2008).

The markers utilized in MAS changes following the advancement of molecular marker and sequencing technologies. For instance, restriction fragment length polymorphism (RFLP) and randomly amplified polymorphic DNA (RAPD) markers were used in linkage mapping of the gene for fragrance in chromosome 8 (Ahn et al. 1992; Lorieux et al. 1996). The previous gene mapping studies have then initiated further studies that utilized simple sequence repeats (SSR) markers in quantitative trait loci (QTL) mapping to map the SSR markers that linked to the fragrance gene (Garland et al. 2000; Cordeiro et al. 2002; Amarawathi et al. 2008). From the flanking SSR markers, many researchers have conducted fine mapping and revealed the single nucleotide polymorphisms (SNPs) and insertions or deletions (INDELs) that lead to loss of function of gene encoding betaine aldehyde dehydrogenase (BADH2) to metabolize the precursor of 2-acetyl-1-pyrroline (Bradbury et al. 2005a; Chen et al. 2006; Vanavichit et al. 2008). As a result, the accumulation of 2-acetyl-1-pyrroline produces the fragrance that often described as "pandan"-like or popcorn-like in the fragrant rice varieties (Buttery et al. 1983; Chen et al. 2006; Vanavichit et al. 2008; Bradbury et al. 2008). The extensive studies on the fragrance gene have developed many useful markers for use in MAB, from linked SSR markers to functional markers (Garland et al. 2000; Cordeiro et al. 2002; Bradbury et al. 2005b; Shi et al. 2008; Amarawathi et al. 2008; Sakthivel et al. 2009). Functional markers are developed from the polymorphic sites of the genes that affect the trait (Andersen and Lübberstedt 2003; Lau et al. 2015). Hence, prior to MAS, there are intensive studies on the QTL effects or genes of the traits to develop markers that can be used in MAS effectively.

MAS is particularly useful for improving rice grain and nutritional qualities of breeding programs, such that MAS allow early identification of plant with desirable genotype, thus time is saved from evaluating grain and nutritional qualities which can only be conducted after harvesting the mature grain at every plant generations (Dreher et al. 2003). For instance, Jin et al. (2010) introgressed three grain quality genes, namely *Wx*, *SSIIa* and *fgr* genes which control apparent amylose content, gelatinization temperature and fragrance, respectively, from Yixiang B into II-32B, a maintainer line used in hybrid breeding in China. The functional markers used in their study were *Wx*-$(CT)_{17}$ microsatellite marker for apparent amylose content, *SSIIa*-TT SNP marker for gelatinization temperature and *fgr-E7* marker for fragrance. Among the 300 F_2 plants, five plants that were homozygous for the three genes as indicated by the markers were selected and used in the following backcross generation. These plants were also evaluated for their grain quality traits and were similar to Yixiang B. Therefore, time, labour and cost are saved from assessing the grain quality of all the F_2 plants that has to be conducted after harvesting the mature grains. The functional markers were also used to select heterozygous genotypes for the genes in their BC_1F_1 and BC_2F_1 generations. This also shows the advantage of MAS where heterozygous genotype can be identified, which cannot be distinguished at phenotypic level. In addition, the improved II-32B variety was generated in two backcrosses and three selfings. On the contrary, conventional backcrossing would have taken up six to eight backcrosses to achieve similar results (Boopathi 2012). Therefore, MAS is able to accelerate the breeding and development of quality rice by reducing the number of breeding generation. General MAS steps are shown in Fig. 10.1.

The initial cost of establishment of genotyping platform for MAS may be expensive; however, in the long run, MAS is time, labour and cost efficient (Dreher et al. 2003). There are many high-throughput methods that are adaptable into the MAS

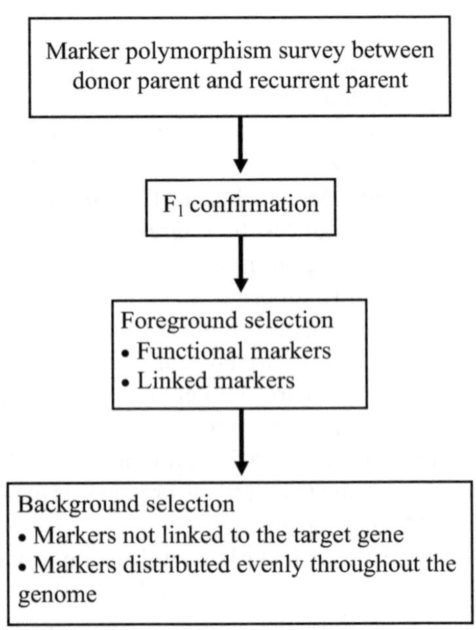

Fig. 10.1 General steps of MAS in a breeding programme

programme, such as DNA extraction for large number of samples, multiplex PCR and capillary electrophoresis system for genotyping many markers and samples would expedite the process in generating marker data with less input of cost and labour (Salgotra et al. 2011; Kim et al. 2016). Thus far, MAS has delivered many fruitful outcomes in improving rice grain quality (Yi et al. 2009; Wang et al. 2010a; Ni et al. 2011; Lau et al. 2017).

10.3 Genomic Selection

MAS is effective for traits that are controlled by a few major QTLs. For complex quantitative traits that are controlled by many minor QTLs, genomic selection (GS) would be a better alternative compared to MAS. Despite that the genes or major QTLs of rice grain and nutritional qualities are known, there are inevitable quantitative effects on the traits (Battenfield et al. 2016). For example, the *BADH2* gene on chromosome 8 alone is unable to explain the variation of fragrance in rice because there are also other minor QTLs that contribute to the overall fragrance in rice (Lorieux et al. 1996; Amarawathi et al. 2008; Lau et al. 2017). GS utilizes a large number of markers covering the whole genome such that all genes are expected to be in linkage disequilibrium (LD) with at least some of the markers to generate the genotypic data of the individuals in the study in order to prevent bias or information loss, whereas in MAS, the individual is selected from the studied population based on markers that linked to the phenotype (Meuwissen 2007; Spindel et al. 2015; Bhat et al. 2016).

The evaluation on rice grain and nutritional qualities has to be conducted after the harvesting stage, and also requiring that each individual has the sufficient amount of grains for the evaluation. Besides that, to evaluate the quality traits for all of the individuals is expensive, time and labour consuming. GS allows the prediction of the grain and nutritional qualities at early breeding stage, even before going through selfing to produce sufficient amount of seeds for evaluation over years or environments, and also allowing more individuals to be predicted than it would be in phenotypic evaluation (Battenfield et al. 2016; Bhat et al. 2016).

GS develops a prediction model with a training population having both genotypic and phenotypic data (Meuwissen et al. 2001) (Fig. 10.2). The generated prediction model is then applied to a test population that is genotyped to derive the genomic-estimated breeding values (GEBV) of each individuals in the test population (Meuwissen et al. 2001; Bhat et al. 2016). Based on the GEBVs, the potential individuals are selected from the test population for use as parent for crossing or next generation of the breeding programme, despite that the individuals have no phenotypic data and no progeny (Meuwissen et al. 2001). GS has shown promising results in dairy cattle breeding (Hayes et al. 2009; Nguyen et al. 2016). GS has also been applied in wheat breeding to improve quality traits (Battenfield et al. 2016). Battenfield et al. (2016) have genotyped and phenotyped 5520 lines from year 2010 to 2015 and 3075 SNPs were used in their GS models. As their study progressed

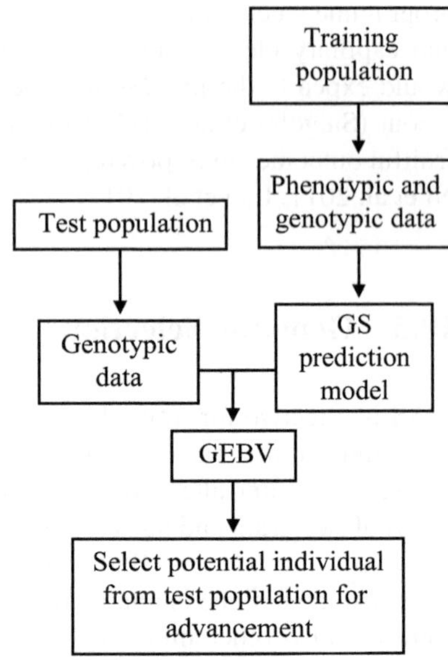

Fig. 10.2 General steps in GS in a breeding programme

over the years, more data on the lines and environment were available to train their GS models, and the GS prediction accuracy increased over time, ranging from 0.32 for grain hardness and 0.62 for mixing time (Battenfield et al. 2016).

Recently, GS has also been applied in rice, for improving grain yield, plant height and flowering time (Spindel et al. 2015). The study by Spindel et al. (2015) used genotyping-by-sequencing (GBS) protocol, which applies next generation sequencing (NGS) technology for discovering and genotyping SNPs in large population with methylation sensitive restriction enzymes that simplify library preparation and also utilizes unphosphorylated adapters to destabilize the formation of adaptor dimers for library preparation (Elshire et al. 2011). Spindel et al. (2015) conducted the study over the dry and wet seasons in year 2009 to 2012, and used 384-plex GBS for genotyping a population consisted of 363 elite breeding lines and was genotyped with 73,147 markers. Their study reported prediction accuracies based on the correlation between predicted GEBV and the phenotype in the testing population, were as high as 0.30 for grain yield using statistical method ridge regression best liner unbiased predictor (RR-BLUP), 0.63 for flowering time using multi-linear regression (MLR) and 0.34 for plant height using non-linear machine learning method that is Random Forest (RF). For rice recurrent selection programme consisting of F_1 to F_8 generations, Spindel et al. (2015) proposed using GEBV on two or more generations during fixation, such as F_3, F_6 and F_8 generations as resources allow to perform selection and the top performing fixed lines from F_8 generation can be proceeded to observational yield trials or cycled for crossing to continue to improve the population. Based on phenotype, the best lines from observational yield trials can be advanced to replicate yield trials, followed by multi-environment trials. Using GEBV, lines from multi-environment trials are selected as parents for the next generation of recurrent selection. Therefore, GEBV can be used to select traits that are costly, time and labour consuming, or traits that had to be evaluated after harvest, such as grain and nutritional qualities, whereas phenotypic selection can be conducted for other traits during the observational and replicated yield trial (Spindel et al. 2015).

Micronutrients in rice grain are low, and the micronutrient contents are further reduced after polishing (Mayer et al. 2008). Consumers that solely subsist on rice for vitamins and minerals are most likely to suffer from malnutrition. Lu et al. (2008) reported ten QTLs were identified for Fe, calcium (Ca), Zn, manganese (Mn) and copper (Cu) contents in rice grain, distributed on chromosome 1, 2, 4, 5, 7, 9 and 11. Garcia-Oliveira et al. (2009) mapped a total of 26 QTLs for Fe, Zn, Mn, Cu, Ca, Mg, phosphorus (P) and potassium (K), with 14 QTLs on chromosome 1, 9 and 12 accounting 45% of the phenotypic variation. Norton et al. (2010) identified QTLs for Zn and Fe, both have four QTLs located on different chromosomes. The QTLs for minerals are scattered across the genome and to introgress a few elements with many minor QTLs through MAS is not feasible. In addition, tests to determine the content of each element can only be done after harvesting the grain and also large amount of grain sample is needed, this process is deemed time and labour consuming. In this regard, genomic selection has the potential in improving rice grain nutrition, particularly the mineral elements that are controlled by many QTLs with small effects. By using the approach suggested by Spindel et al. (2015), selection using GEBV can be conducted at least twice at earlier breeding generation to develop and update the prediction model on the training population in order to use the GEBV for selection. GEBV could also be used to select fixed lines for later generation, yield trials or as parents for hybridization for next generation of breeding. In the population that GEBV is used, only a subset of approximately 300 plants are genotyped and phenotyped for the development and improvement of GS prediction model, while the rest of the plants can be genotyped. With GS approach and high GEBV, the proportion of superior individuals are increased in the breeding populations, thereby accelerates the gain from selection and reduces the time for developing improved rice variety (Heffner et al. 2009; Spindel et al. 2015).

10.4 Genetic Engineering

Genetic engineering is an alternative method to create variation in a breeding population. Genetic engineering is used particularly when the variation of a trait in a natural population is limited. Besides introducing proven genes along with suitable promoters into high-yielding rice variety, the utilization of genetic engineering methods must take into account of the stability of the traits by crossing the transgenic lines with different genotype background, and also ensuring the other traits are not negatively affected, especially yield-related traits.

Golden rice is the classical example of the application of genetic engineering for improving rice grain and nutritional qualities. The precursor of vitamin A, which is β-carotene, is absent in rice grain and the synthesis of β-carotene in rice grain requires plant enzymes, namely phytoene synthase (PSY), phytoene desaturase, ζ-carotene desaturase and lycopene β-cyclase to convert geranylgeranyl diphosphate to β-carotene (Ye et al. 2000; Bhullar and Gruissem 2013). The biosynthetic pathway of β-carotene is engineered in rice endosperm to express daffodil *psy* gene and bacterial carotene desaturase (*crtI*) from *Erwinia uredovora* by Ye et al. (2000) through *Agrobacterium*-mediated transformation and reported 1.6 μg/g of carotenoid in the rice grain. The study is further improved by Paine et al. (2005) and thereby developed Golden Rice 2 (GR2) with total carotenoid content in rice grain of 37 μg/g, in which 31 μg/g is β-carotene by introducing the combination of maize *psy* gene and *crtI* from *Erwinia uredovora*. They also stated that the presence of transgenes in GR2 plants did not evidently affect the plant phenotype, seed weight and seed germination (Paine et al. 2005). To date, GR2 has been bred into local rice varieties in India, the Philippines, Vietnam, and Bangladesh with field trials being conducted in the Philippines (www.goldenrice.org).

Rice micronutrient biofortification through genetic engineering depends on several factors, such as micronutrient uptake, transportation, accumulation and homeostasis in the plants (Sperotto et al. 2012). To increase the accumulation of iron (Fe) in the rice grain, Goto et al. (1999) used *Agrobacterium*-mediated transformation to introduce *ferritin* gene (*SoyferH1*) from soybean, which encodes an Fe storage protein into rice variety Kita-ake, with rice glutelin promoter, *GluB-1*. As a result, the transgenic rice seeds had three times more iron than the normal seeds (Goto et al. 1999). Paul et al. (2012) cloned and overexpressed rice endogenous *ferritin* gene, with the control of endosperm-specific *GlutelinA2* promoter and used modified biolistic method to transform a fragrant *indica* rice variety, Pusa-sugandhi II, resulting in 2.09-fold higher Fe content and 1.37-fold higher Zn content in the milled seed of T_3 plants than non-transgenic Pusa-sugandhi II. Another approach is to increase the Fe transportation in the rice plants by enhancing the expression of nicotianamine synthase (NAS) gene. NAS, followed by nicotianamine aminotransferase (NAAT) and deoxymugineic acid synthase (DMAS) are the enzymes involved in a pathway to synthesis 2′-deoxymugineic acid (DMA), the precursor of mugineic acid family phytosiderophores that involves in chelation and uptake of Fe and also Zn (Higuchi et al. 1999; Kobayashi and Nishizawa 2012). The overexpression of barley NAS gene, *HvNAS1* in transgenic *Japonica* rice, Tsukinohikari variety, using *Agrobacterium*-mediated transformation has showed an increment of endogenous nicotianamine and also an increment of Fe and Zn concentrations in polished and brown T_2 seeds (Masuda et al. 2009). Lee et al. (2009) overexpressed *OsNAS3* by introducing *35S* enhancer elements, used *Agrobacterium*-mediated transformation on *Japonica* rice, Dongjin variety and obtained 2.9- and 2.2-fold higher concentration of Fe and Zn, respectively in the seeds. Another approach is through the overexpression of Fe(II)-nicotianamine complex transporter gene, *OsYSL2*, along with sucrose transporter promoter, *OsSUT1*, which resulted in 4.4-fold higher Fe concentration in polished rice of transgenic Tsukinohikari rice variety, transformed by *Agrobacterium tumefaciens* (Ishimaru et al. 2010).

A combination of these approaches could also improve the content of micronutrients in rice. Masuda et al. (2012) introduced three genes, namely *SoyferH2*, *HvNAS1* and *OsYSL2* through *Agrobacterium*-mediated transformation into Tsukinohikari rice variety and had increased the Fe content up to 4.4-fold (4.0 μg/g) and the Zn content up to 1.6-fold in T_3 polished seeds compared to the non-transgenic rice, based on the result from paddy field cultivation. Similar approach was also conducted by Aung et al. (2013) to introduce the three genes into Myanmar high-quality rice variety, Paw San Yin, and reported that the Fe content in T_2 polished seeds was 3.4-fold (5.02 μg/g) higher than the non-transgenic line. Trijatmiko et al. (2016) introduced *OsNAS2* and *SferH-1* genes into IR64, an *indica* rice variety and the field evaluated T_2 had 15 μg/g Fe content in their polished grain.

Genetic engineering could also improve more than one nutritional quality traits in a rice variety. Singh et al. (2017) combined the three nutritional quality traits via single genetic locus into Nipponbare rice variety via *Agrobacterium*-mediated transformation. Their study has transformed Nipponbare rice variety with Arabidopsis *NAS* gene under the control of *CaMV 35S* promoter, bean *ferritin* gene under the control of *OsGLOBULIN* endosperm-specific promoter, and maize *psy* gene under the control of *OsGLUTELIN* endosperm-specific promoter. As a result, the transgenic lines showed improvement for the three nutritional quality traits, with endosperm Fe content ranging from 2.6 to 6.02 μg/g, β-carotene content in polished grains ranged from 1.9 and 3.4 μg/g and Zn content ranged from 24.9 to 29.7 μg/g.

Genetic engineering approach is also assisted by plant tissue culture techniques to select transformants, produce and regenerate transgenic plant (Sathyanarayana 2007). The callus induction and plant regeneration methods have to be optimized for successful transformation and development of transgenic lines. Generally, 2,4-dichlorophenoxyacetic acid (2,4-D) is added for callus induction and proliferation of rice ex-plants, cells and tissues (Jain and Jain 2000). For plant regeneration, hormone-free media are used or in combination with auxin (NAA) and cytokinin (BAP or kinetin) (Jain and Jain 2000). The hormone-free media or media containing NAA are used for shoots rooting (Jain and Jain 2000). Bishnoi et al. (2000) suggested using RZ medium containing 4% (w/v) maltose, 2,4-D, NAA and kinetin for callus induction and plant regeneration of *indica* and Basmati rice varieties, and the F_1 and F_2 plants from these two crosses.

Undeniably, there are numerous successful reports showing that genetic engineering has improved plant traits. However, the usage of genetic engineering is limited due to strict regulations and limited public acceptance. Nevertheless, genetic engineering has provided insights of the gene functions, which are useful information for further investigations and marker development.

10.5 Mutation Breeding and Genome Editing

Mutation breeding is an alternative approach to induce variation, apart from making crosses. Mutation breeding method creates variations of mutants using mutagen, and the mutants are detected by high-throughput screening technology. Chemical and physical mutagens are used for inducing mutagenesis in plants. Chemical mutagens that are commonly used are ethyl methanesulphonate (EMS), *N*-methyl-*N*-nitrosourea (MNU) and sodium azide (SA). Physical mutagens include gamma rays, X-rays, UV-rays, fast neutrons and ion beams. Many new mutant varieties of crops have developed through mutation breeding (https://mvd.iaea.org/).

Mutation breeding is widely used for improvement of rice grain and nutritional qualities. To date, more than 200 new mutant rice varieties of improved grain and nutritional qualities have been developed (https://mvd.iaea.org/). Bangladesh has recently developed a new mutant rice variety, Binadhan-19, through carbon ion beams at 40 Gy on the seed of NERICA-10 (https://mvd.iaea.org/#!Variety/4465). As a result, Binadhan-19 has long and slender grains with golden colour. Jeng et al. (2012) used SA to induce mutation in IR64 rice variety. In their study, the mutants M-IR-75 and M-IR-58 had Fe content of 28.10 and 27.26 mg/kg in their polished grain respectively, which was more than IR64 variety (3.90 mg/kg) and the Zn content in the polished grain of mutants M-IR-180, M-IR-49 and M-IR-175 (26.58, 28.95 and 26.16 mg/kg, respectively) was also more than IR64 variety (16.00 mg/kg). In another study, Tran and Ho (2017) have developed high iron fragrant rice by using physical mutagen and hybridization. Mutagenesis creates variation whereas hybridization with other varieties and selection would confirm the stable inheritance of the mutation in the gene of interest (Tadele 2016). Tran and Ho (2017) developed the mutant population by irradiating seed of Vietnamese local fragrant rice varieties with gamma rays from ^{60}Co facility at 120 and 150 Gy. The selected mutants with fragrance and high iron content were crossed with other varieties to improve agronomic traits, selfing and also phenotypic selection and genotypic screening for *BADH2* gene. Apart from fragrance and improved iron content, their new rice varieties also have slender grain, non-chalky endosperm, low amylose content, and also showing improved agronomic traits with yield more than 5 ton/ha.

The physical and chemical mutagens could induce a plethora of mutation in the plant genome at random position (Tadele 2016). The induced mutation can be detected through map-based system or MutMap (Abe et al. 2012). In MutMap, a mutant is crossed with its wild-type parent which is a variety with a reference genome sequence, then the resulting F_1 is self-pollinated, and the F_2 generation is grown in field for phenotype screening. The F_2 plants showing mutant phenotype are subjected to whole-genome sequencing and alignment to the reference sequence. MutMap allows rapid identification of mutation that affects quantitative traits in crop genomes. On the other hand, MutMap+, an extension of MutMap circumvents the need for crossing between the mutant and the parent variety (Fekih et al. 2013). MutMap+ uses M_3 generation, generated from selfing of an M_2 heterozygous individual and compares the whole genome sequences between the two bulks of mutant progeny and wild-type progeny to identify causal mutations. Nakata et al. (2017) used MutMap+ in their mutant Nipponbare population and identified two mutant alleles of *starch branching enzyme IIb* (*SBEIIb*) gene. SBEIIb plays an important role in forming short chains of amylopection cluster of degree of polymerization (DP) ≤13, and its activity level alters the fine structure of amylopectin (Tanaka et al. 2004). Nakata et al. (2017) discovered that the *altered gelatinizations 1 (ages1)* allele has a point mutation in the *SBEIIb* gene which caused SBEIIb deficiency whereas the *ages2* allele has a transposon insertion in the same gene that caused milder modification of starch properties. Their lines with *age1* and *age2* showed increased apparent amylose content, approximately 30% and 23%, respectively.

Targeted mutagenesis, also known as genome editing, that uses artificially engineered nucleases to replace, delete or insert genes has gained popularity among scientists in recent years (Tadele 2016). The nucleases that are used in genome editing include meganuclease, zinc finger nucleases (ZFNs), transcription activator-like effectors nucleases (TALENs) and the recently discovered nuclease that is clustered regularly interspaced short palindromic repeats (CRISPR)/CRISPR-associated (Cas) protein 9 (CRISPR/Cas9). These nucleases create specific double-stranded break at desired locus in the genome and modifies the gene by generating premature stop codons by non-homologous end joining (NHEJ) pathway, or to facilitate gene targeting through homologous recombination (HR) with a template (Miao et al. 2013). Shan et al. (2015) have demonstrated that TALENs can be used to develop fragrant rice. They engineered the TALENs to knockout the *OsBADH2* gene and successfully obtained homozygous mutants with 2-acetyl-1-pyrroline content, ranging from 0.5 to 0.75 mg/kg, which were higher than the positive control rice variety Daohuaxiang (0.5 mg/kg).

CRISPR/Cas systems are bacterial adaptive defence systems to fight against foreign nucleic acids (Barrangou et al. 2007). Cas9 protein acts as a nuclease and it is directed to a target site by engineered sequence-specific single guide RNA (gRNA) and cleaves each strand of DNA, creating double-stranded break at specific site with complementary to the 5′end of the gRNA (Jinek et al. 2012; Ma et al. 2015). Recently, Sun et al. (2017) have applied CRISPR/Cas9 system to develop high amylose rice. They constructed the CRISPR/Cas9 vectors with gRNAs targeting the first exon of *starch branching enzyme I* (*SBEI*) and the third exon of *SBEIIb*. The vectors are transformed into *japonica* rice variety, Kitaake through *Agrobacterium tumefaciens*. As a result, homozygous mutant T_0 lines with indels in *SBEI* and *SBEIIb* were obtained, at frequencies 26.7% and 40%, respectively. Furthermore, their T_1 plants generated from the self-pollination of homozygote T_0 plants are also homozygous for the same mutations. They also managed to recover homozygous plants carrying 1 bp deletion and the transgene-free plants in their T_1 generation. They also reported that the amylose content in *sbeIIb* mutant lines was 25%, higher than the *sbeI* mutant lines and wild types that both showed less than 15%. In addition, the grains of *sbeIIb* mutant lines also showed significantly higher resistant starch content (9.8%) compared to wild-type and *sbeI* mutant lines in which little or no resistant starch was found in the grains. Therefore, CRISPR/Cas9 technology is a useful method to improve the rice grain and nutritional qualities as CRISPR/Cas9

Fig. 10.3 General protocol for CRISPR/Cas9 for application in a breeding programme

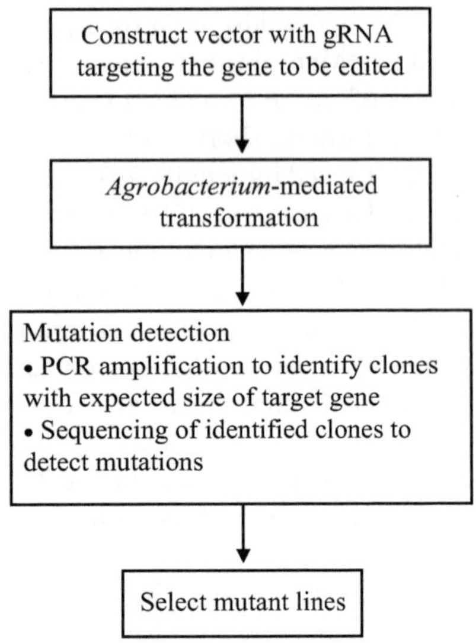

technology could modify the target gene accurately and able to generate transgene-free plants. A general protocol of CRISPR/Cas9 technology procedure is shown in Fig. 10.3.

10.6 Conclusion

The current breeding approaches have undeniably contributed to the successful outcome in improving rice grain and nutritional qualities (Table 10.1). Regardless of the technologies applied, prior to be released as a registered variety, it is important that the rice lines to be distinct, uniform and stable through evaluation at multiple locations and over the years. Therefore, phenotyping technology should advance in line with the genotyping technology. High-throughput imaging platforms would hasten the phenotyping process by reducing labour and time for collecting phenotypic data during field evaluation. In addition, high-throughput phenotyping technology could also be useful in breeding programmes to provide information on environmental effects or gene-by-environment interaction effects. Nevertheless, with the breeding approaches advance in expeditious rate, we envision that rice with grain and nutritional qualities tailored according to respective countries' consumers' preference and nutritional need can be developed and benefit the people.

Table 10.1 Examples of current breeding approaches to improve rice grain and nutritional qualities

Traits	Breeding approaches and genes/QTLs involved	Reference
Fragrance	MAS	
	BADH2 gene	Yi et al. (2009) Jin et al. (2010) Salgotra et al. (2012) Luo et al. (2016) Lau et al. (2017)
	Genetic engineering	
	BADH2 gene	Vanavichit et al. (2008)
	TALEN	
	BADH2 gene	Shan et al. (2015)
Amylose content	MAS	
	Wx gene	Yi et al. (2009) Jin et al. (2010) Wang et al. (2010b) Ni et al. (2011) Luo et al. (2014b)
	CRISPR/Cas9-mediated targeted mutagenesis	
	SBEI and *SBEIIb*	Sun et al. (2017)
Gelatinization temperature	MAS	
	SSIIa gene	Jin et al. (2010)
Gel consistency	QTL for amylose content and gel consistency	Yi et al. (2009)
Vitamin A	Genetic engineering	
	Daffodil *psy* gene and *crt1* from *Erwinia uredovora* Maize *psy* gene and *crt1* from *Erwinia uredovora*	Ye et al. (2000) Paine et al. (2005)
Iron and zinc	Genetic engineering	
	SoyferH1	Goto et al. (1999)
	Rice *ferritin* gene	Paul et al. (2012)
	HvNAS1	Masuda et al. (2009)
	OsNAS3	Lee et al. (2009)
	OsYSL2	Ishimaru et al. (2010)
	Rice *iron-regulated transporter 1* (*OsIRT1*) gene	Lee and An (2009)
Protein	Genetic engineering	
	Artificial synthesis of two new genes by fusing endogenous rice genes with lysine (K)/threonine (T) motif (TKTKK) coding sequences	Jiang et al. (2016)

References

Abe A, Kosugi S, Yoshida K et al (2012) Genome sequencing reveals agronomically important loci in rice using MutMap. Nat Biotechnol 30:174–178. https://doi.org/10.1038/nbt.2095

Ahn SN, Bollich CN, Tanksley SD (1992) RFLP tagging of a gene for aroma in rice. Theor Appl Genet 84:825–828. https://doi.org/10.1007/BF00227391

Amarawathi Y, Singh R, Singh AK et al (2008) Mapping of quantitative trait loci for basmati quality traits in rice (*Oryza sativa* L). Mol Breed 21:49–65. https://doi.org/10.1007/s11032-007-9108-8

Andersen JR, Lübberstedt T (2003) Functional markers in plants. Trends Plant Sci 8:554–560. https://doi.org/10.1016/j.tplants.2003.09.010

Arsenault JE, E a Y, Hossain MB et al (2010) The current high prevalence of dietary zinc inadequacy among children and women in rural Bangladesh could be substantially ameliorated by zinc biofortification of rice. J Nutr 140:1683–1690. https://doi.org/10.3945/jn.110.123059

Ashfaq M, Haider MS, Saleem I et al (2015) Basmati-rice a class apart (a review). J Rice Res. https://doi.org/10.4172/2375-4338.1000156

Aung MS, Masuda H, Kobayashi T et al (2013) Iron biofortification of Myanmar rice. Front Plant Sci. https://doi.org/10.3389/fpls.2013.00158

Barrangou R, Fremaux C, Deveau H et al (2007) CRISPR provides acquired resistance against viruses in prokaryotes. Science 315:1709–1712. https://doi.org/10.1126/science.1138140

Battenfield SD, Guzmán C, Gaynor RC et al (2016) Genomic selection for processing and end-use quality traits in the CIMMYT spring bread wheat breeding program. Plant Genome. https://doi.org/10.3835/plantgenome2016.01.0005

Bhat JA, Ali S, Salgotra RK et al (2016) Genomic selection in the era of next generation sequencing for complex traits in plant breeding. Front Genet. https://doi.org/10.3389/fgene.2016.00221

Bhullar NK, Gruissem W (2013) Nutritional enhancement of rice for human health: the contribution of biotechnology. Biotechnol Adv 31:50–57. https://doi.org/10.1016/j.biotechadv.2012.02.001

Bishnoi U, Jain RK, Rohilla JS et al (2000) Anther culture of recalcitrant indica × Basmati rice hybrids. Euphytica 114:93–101. https://doi.org/10.1023/A:1003915331143

Boopathi NM (2012) Genetic mapping and marker assisted selection: basics, practice and benefits. Springer, New Delhi

Bradbury LMT, Fitzgerald TL, Henry RJ et al (2005a) The gene for fragrance in rice. Plant Biotechnol J 3:363–370. https://doi.org/10.1111/j.1467-7652.2005.00131.x

Bradbury LMT, Henry RJ, Jin Q et al (2005b) A perfect marker for fragrance genotyping in rice. Mol Breed 16:279–283. https://doi.org/10.1007/s11032-005-0776-y

Bradbury LMT, Gillies SA, Brushett DJ et al (2008) Inactivation of an aminoaldehyde dehydrogenase is responsible for fragrance in rice. Plant Mol Biol 68:439–449. https://doi.org/10.1007/s11103-008-9381-x

Brar DS, Virk PS, Grewal D et al (2012) Breeding rice varieties with improved grain and nutritional quality. Qual Assur Saf Crop Foods 4:137–158. https://doi.org/10.1111/j.1757-837X.2012.00140.x

Braun J, Bos MS (2005) The changing economics and politics of rice: implications for food security, globalization, and environmental sustainability. In: Toriyama K, Heong K, Hardy B (eds) Proceedings of the world rice research conference. International Rice Research Institute (IRRI), Los Baños, pp 7–20

Buttery RG, Ling LC, Juliano B, Turnbaugh JG (1983) Cooked rice aroma and 2-acetyl-1-pyrroline. J Agric Food Chem 31:823–826

Calingacion M, Laborte A, Nelson A et al (2014) Diversity of global rice markets and the science required for consumer-targeted rice breeding. PLoS One 9:e85106. https://doi.org/10.1371/journal.pone.0085106

Champagne ET (2008) Rice aroma and flavor: a literature review. Cereal Chem 85:445–454. https://doi.org/10.1094/CCHEM-85-4-0445

Chen S, Wu J, Yang Y et al (2006) The *fgr* gene responsible for rice fragrance was restricted within 69kb. Plant Sci 171:505–514. https://doi.org/10.1016/j.plantsci.2006.05.013

Collard BCY, Vera Cruz CM, McNally KL et al (2008) Rice molecular breeding laboratories in the genomics era: current status and future considerations. Int J Plant Genomics 2008:1–25. https://doi.org/10.1155/2008/524847

Cordeiro GM, Christopher MJ, Henry RJ, Reinke RF (2002) Identification of microsatellite markers for fragrance in rice by analysis of the rice genome sequence. Mol Breed 9:245–250

Dreher K, Khairallah M, Ribaut J, Morris M (2003) Money matters (I): costs of field and laboratory procedures associated with conventional and marker-assisted maize breeding at CIMMYT. Mol Breed 11:221–234

Elshire RJ, Glaubitz JC, Sun Q et al (2011) A robust, simple genotyping-by-sequencing (GBS) approach for high diversity species. PLoS One 6:e19379. https://doi.org/10.1371/journal.pone.0019379

Farah A, Zainalabidin M, Ismail AL (2011) The influence of socio-demographic factors and product attributes on attitudes toward purchasing special rice among Malaysian consumers. Int Food Res J 18:1135–1142

Fekih R, Takagi H, Tamiru M et al (2013) MutMap+: genetic mapping and mutant identification without crossing in rice. PLoS One 8:e68529. https://doi.org/10.1371/journal.pone.0068529

Ferrero A, Nguyen N (2004) Constraints and opportunities for the sustainable development of rice-based production systems in Europe. FAO rice conference, pp 1–14.

Food and Agriculture Organization of the United Nations (FAO) (2017) FAOSTAT. http://www.fao.org/faostat/en/#data/CC

Garcia-Oliveira AL, Tan L, Fu Y, Sun C (2009) Genetic identification of quantitative trait loci for contents of mineral nutrients in rice grain. J Integr Plant Biol 51:84–92. https://doi.org/10.1111/j.1744-7909.2008.00730.x

Garland S, Lewin L, Blakeney A et al (2000) PCR-based molecular markers for the fragrance gene in rice (*Oryza sativa* L.). Theor Appl Genet 101:364–371. https://doi.org/10.1007/s001220051492

Garris AJ, Tai TH, Coburn J et al (2005) Genetic structure and diversity in *Oryza sativa* L. Genetics 169:1631–1638. https://doi.org/10.1534/genetics.104.035642

Global Rice Science Partnership (GRiSP) (2013) Rice almanac, 4th edn. International Rice Research Institute, Los Baños

Goto F, Yoshihara T, Shigemoto N et al (1999) Iron fortification of rice seed by the soybean ferritin gene. Nat Biotechnol 17:282–286. https://doi.org/10.1038/7029

Hayes BJ, Bowman PJ, Chamberlain AJ, Goddard ME (2009) Invited review: genomic selection in dairy cattle: progress and challenges. J Dairy Sci 92:433–443. https://doi.org/10.3168/jds.2008-1646

Heffner EL, Sorrells ME, Jannink J-L (2009) Genomic selection for crop improvement. Crop Sci 49(1). https://doi.org/10.2135/cropsci2008.08.0512

Higuchi K, Suzuki K, Nakanishi H et al (1999) Cloning of nicotianamine synthase genes, novel genes involved in the biosynthesis of phytosiderophores. Plant Physiol 119:471–480. https://doi.org/10.1104/PP.119.2.471

Hsu H, Chern W, Gale F (2001) How will rising income affect the structure of food demand. China Food Agric Issues 21st Cent 2001:10–13

Htet MK, Fahmida U, Thurnham DI et al (2016) Folate and vitamin B12 status and dietary intake of anaemic adolescent schoolgirls in the delta region of Myanmar. Br J Nutr 116:S36–S41. https://doi.org/10.1017/S0007114515001609

IRGSP (2005) The map-based sequence of the rice genome. Nature 436:793–800. https://doi.org/10.1038/nature03895

Ishimaru Y, Masuda H, Bashir K et al (2010) Rice metal-nicotianamine transporter, *OsYSL2*, is required for the long-distance transport of iron and manganese. Plant J 62:379–390. https://doi.org/10.1111/j.1365-313X.2010.04158.x

Jain RK, Jain S (2000) Transgenic strategies for genetic improvement of Basmati rice. Indian J Exp Biol 38:6–17

Jeng TL, Lin YW, Wang CS, Sung JM (2012) Comparisons and selection of rice mutants with high iron and zinc contents in their polished grains that were mutated from the *indica* type cultivar IR64. J Food Compos Anal 28:149–154. https://doi.org/10.1016/j.jfca.2012.08.008

Jiang SY, Ma A, Xie L, Ramachandran S (2016) Improving protein content and quality by over-expressing artificially synthetic fusion proteins with high lysine and threonine constituent in rice plants. Sci Rep. https://doi.org/10.1038/srep34427

Jin L, Lu Y, Shao Y et al (2010) Molecular marker assisted selection for improvement of the eating, cooking and sensory quality of rice (*Oryza sativa* L.). J Cereal Sci 51:159–164. https://doi.org/10.1016/j.jcs.2009.11.007

Jinek M, Chylinski K, Fonfara I et al (2012) A programmable dual-RNA-guided DNA endonuclease in adaptice bacterial immunity. Science 337:816–822. https://doi.org/10.1126/science.1225829

Juliano BO (2005) Trends in rice quality demand in Asia. Rice Ind Cult Environ 2005:43–53

Kelly M (2016) The nutrition transition in developing Asia: dietary change, drivers and health impacts. In: Jackson P, Spiess W, Sultana F (eds) Eating, drinking: surviving. SpringerBriefs in global understanding. Springer, Cham, pp 83–90

Khush GS, Juliano BO (1985) Rice grain quality and marketing: papers presented at the International Rice Research Conference, 1-5 June 1985. Int Rice Res Inst, pp 61–69

Kim S-R, Yang J, An G, Jena KK (2016) A simple DNA preparation method for high quality polymerase chain reaction in rice. Plant Breed Biotechnol 4:99–106. https://doi.org/10.9787/PBB.2016.4.1.099

Kobayashi T, Nishizawa NK (2012) Iron uptake, translocation, and regulation in higher plants. Annu Rev Plant Biol 63:131–152. https://doi.org/10.1146/annurev-arplant-042811-105522

Lau WCP, Rafii MY, Ismail MR et al (2015) Review of functional markers for improving cooking, eating, and the nutritional qualities of rice. Front Plant Sci. https://doi.org/10.3389/fpls.2015.00832

Lau WCP, Rafii MY, Ismail MR et al (2017) Development of advanced fragrant rice lines from MR269 × Basmati 370 through marker-assisted backcrossing. Euphytica 213:11. https://doi.org/10.1007/s10681-016-1794-z

Lee S, An G (2009) Over-expression of *OsIRT1* leads to increased iron and zinc accumulations in rice. Plant Cell Environ 32:408–416. https://doi.org/10.1111/j.1365-3040.2009.01935.x

Lee S, Jeon US, Lee SJ et al (2009) Iron fortification of rice seeds through activation of the *nicotianamine synthase* gene. Proc Natl Acad Sci 106:22014–22019. https://doi.org/10.1073/pnas.0910950106

Lipoeto NI, Lin KG, Angeles-agdeppa I (2012) Food consumption patterns and nutrition transition in South-East Asia. Public Health Nutr 16:1637–1643. https://doi.org/10.1017/S1368980012004569

Lorieux M, Petrov M, Huang N et al (1996) Aroma in rice: genetic analysis of a quantitative trait. Theor Appl Genet 93:1145–1151. https://doi.org/10.1007/BF00230138

Lu K, Li L, Zheng X et al (2008) Quantitative trait loci controlling Cu, Ca, Zn, Mn and Fe content in rice grains. J Genet 87:305–310. https://doi.org/10.1007/s12041-008-0049-8

Luo Y, Zakaria S, Basyah B et al (2014a) Marker-assisted breeding of Indonesia local rice variety Siputeh for semi-dwarf phonetype, good grain quality and disease resistance to bacterial blight. Rice. https://doi.org/10.1186/s12284-014-0033-2

Luo Y, Zakaria S, Basyah B et al (2014b) Marker-assisted breeding of Indonesia local rice variety Siputeh for semi-dwarf phonetype, good grain quality and disease resistance to bacterial blight. Rice 7:33. https://doi.org/10.1186/s12284-014-0033-2

Luo Y, Ma T, Zhang A et al (2016) Marker-assisted breeding of the rice restorer line Wanhui 6725 for disease resistance, submergence tolerance and aromatic fragrance. Rice 9:66. https://doi.org/10.1186/s12284-016-0139-9

Ma X, Zhang Q, Zhu Q et al (2015) A robust CRISPR/Cas9 system for convenient, high-efficiency multiplex genome editing in monocot and dicot plants. Mol Plant 8:1274–1284. https://doi.org/10.1016/j.molp.2015.04.007

Masuda H, Usuda K, Kobayashi T et al (2009) Overexpression of the barley nicotianamine synthase gene *HvNAS1* increases iron and zinc concentrations in rice grains. Rice 2:155–166. https://doi.org/10.1007/s12284-009-9031-1

Masuda H, Ishimaru Y, Aung MS et al (2012) Iron biofortification in rice by the introduction of multiple genes involved in iron nutrition. Sci Rep 2:543

Mayer JE, Pfeiffer WH, Beyer P (2008) Biofortified crops to alleviate micronutrient malnutrition. Curr Opin Plant Biol 11:166–170. https://doi.org/10.1016/j.pbi.2008.01.007

Meuwissen T (2007) Genomic selection: marker assisted selection on a genome wide scale. J Anim Breed Genet 124:321–322

Meuwissen THE, Hayes BJ, Goddard ME (2001) Prediction of total genetic value using genome-wide dense marker maps. Genetics 157:1819–1829

Miao J, Guo D, Zhang J et al (2013) Targeted mutagenesis in rice using CRISPR-Cas system. Cell Res 23:1233–1236. https://doi.org/10.1038/cr.2013.123

Minten B, Murshid KAS, Reardon T (2011) The quiet revolution in agrifood value chains in Asia: the case of increasing quality in rice markets in Bangladesh. IFPRI Discuss Pap 01141:1–25

Nakata M, Miyashita T, Kimura R et al (2017) MutMapPlus identified novel mutant alleles of a rice *starch branching enzyme IIb* gene for fine-tuning of cooked rice texture. Plant Biotechnol J. https://doi.org/10.1111/pbi.12753

Nguyen TTT, Bowman PJ, Haile-Mariam M et al (2016) Genomic selection for tolerance to heat stress in Australian dairy cattle. J Dairy Sci 99:2849–2862. https://doi.org/10.3168/jds.2015-9685

Ni D, Zhang S, Chen S et al (2011) Improving cooking and eating quality of Xieyou57, an elite indica hybrid rice, by marker-assisted selection of the *Wx* locus. Euphytica 179:355–362. https://doi.org/10.1007/s10681-011-0377-2

Norton GJ, Deacon CM, Xiong L et al (2010) Genetic mapping of the rice ionome in leaves and grain: Identification of QTLs for 17 elements including arsenic, cadmium, iron and selenium. Plant and Soil 329:139–153. https://doi.org/10.1007/s11104-009-0141-8

Paine JA, Shipton CA, Chaggar S et al (2005) Improving the nutritional value of golden cice through increased pro-vitamin A content. Nat Biotechnol 23:482–487. https://doi.org/10.1038/Nbt1082

Paul S, Ali N, Gayen D et al (2012) Molecular breeding of *Osfer2* gene to increase iron nutrition in rice grain. GM Crops Food 3:310–316

Peng S, Tang Q, Zou Y (2009) Current status of the research on high-yielding and high efficiency in resource use and improving grain quality in rice. Plant Prod Sci 12:3–8

Pingali P (2007) Westernization of Asian diets and the transformation of food systems: implications for research and policy. Food Policy 32:281–298. https://doi.org/10.1016/j.foodpol.2006.08.001

Pinkaew S, Winichagoon P, Hurrell RF, Wegmuller R (2013) Extruded rice grains fortified with zinc, iron, and vitamin A increase zinc status of Thai school children when incorporated into a school lunch program. J Nutr 143:362–368. https://doi.org/10.3945/jn.112.166058

Sakai H, Lee SS, Tanaka T et al (2013) Rice annotation project database (RAP-DB): an integrative and interactive database for rice genomics. Plant Cell Physiol e6(1-11):54. https://doi.org/10.1093/pcp/pcs183

Sakthivel K, Shobha Rani N, Pandey MK et al (2009) Development of a simple functional marker for fragrance in rice and its validation in Indian Basmati and non-Basmati fragrant rice varieties. Mol Breed 24:185–190

Salgotra RK, Millwood RJ, Agarwal S, Stewart CN (2011) High-throughput functional marker assay for detection of *Xa/xa* and *fgr* genes in rice (*Oryza sativa* L.). Electrophoresis 32:2216–2222. https://doi.org/10.1002/elps.201100196

Salgotra RK, Gupta BB, Millwood RJ et al (2012) Introgression of bacterial leaf blight resistance and aroma genes using functional marker-assisted selection in rice (*Oryza sativa* L.). Euphytica 187:313–323. https://doi.org/10.1007/s10681-011-0599-3

Sathyanarayana BN (2007) Plant tissue culture: practices and new experimental protocols. I. K. International Pvt Ltd, New Delhi

Shan Q, Zhang Y, Chen K et al (2015) Creation of fragrant rice by targeted knockout of the *OsBADH2* gene using TALEN technology. Plant Biotechnol J 13:791–800. https://doi.org/10.1111/pbi.12312

Shi W, Yang Y, Chen S, Xu M (2008) Discovery of a new fragrance allele and the development of functional markers for the breeding of fragrant rice varieties. Mol Breed 22:185–192. https://doi.org/10.1007/s11032-008-9165-7

Singh SP, Gruissem W, Bhullar NK (2017) Single genetic locus improvement of iron, zinc and β-carotene content in rice grains. Sci Rep. https://doi.org/10.1038/s41598-017-07198-5

Sperotto RA, Ricachenevsky FK, Waldow VDA, Fett JP (2012) Iron biofortification in rice: it's a long way to the top. Plant Sci 190:24–39. https://doi.org/10.1016/j.plantsci.2012.03.004

Spindel J, Begum H, Akdemir D et al (2015) Genomic selection and association mapping in rice (*Oryza sativa*): effect of trait genetic architecture, training population composition, marker number and statistical model on accuracy of rice genomic selection in elite, tropical rice breeding lines. PLoS Genet 11:e1004982. https://doi.org/10.1371/journal.pgen.1004982

Sun Y, Jiao G, Liu Z et al (2017) Generation of high-amylose rice through CRISPR/Cas9-mediated targeted mutagenesis of starch branching enzymes. Front Plant Sci. https://doi.org/10.3389/fpls.2017.00298

Suwannaporn P, Linnemann A (2008) Rice-eating quality among consumers in different rice grain preference countries. J Sens Stud 23:1–13

Tadele Z (2016) Mutagenesis and TILLING to dissect gene function in plants. Curr Genomics 17:499–508. https://doi.org/10.2174/1389202917666160520104158

Tanaka N, Fujita N, Nishi A et al (2004) The structure of starch can be manipulated by changing the expression levels of starch branching enzyme IIb in rice endosperm. Plant Biotechnol J 2:507–516. https://doi.org/10.1111/j.1467-7652.2004.00097.x

Timmer CP, Block S, Dawe D (2010) Long-run dynamics of rice consumption, 1960–2050. In: Pandey S, Byerlee D, Dawe D et al (eds) Rice in the global economy: strategic research and policy issues for food security. International Rice Research Institute, Los Baños, pp 139–174

Tomlins K, Manful J, Gayin J et al (2007) Study of sensory evaluation, consumer acceptability, affordability and market price of rice. J Sci Food Agric 87:1564–1575. https://doi.org/10.1002/jsfa.2889

Tran PT, Ho CQ (2017) Breeding new aromatic rice with high iron using gamma radiation and hybridization. In: Jankowicz-Cieslak J, Tai TH, Kumlehn J, Till BJ (eds) Biotechnologies for plant mutation breeding. Springer, New York, NY, pp 173–191

Trijatmiko KR, Duenãs C, Tsakirpaloglou N et al (2016) Biofortified indica rice attains iron and zinc nutrition dietary targets in the field. Sci Rep. https://doi.org/10.1038/srep19792

Vanavichit A, Tragoonrung S, Toojinda T et al (2008) Transgenic rice plants with reduced expression of *OS2AP* and elevated levels of 2-acetyl-1-pyrroline, vol 2, pp 1–17

Wang C, Zhang Y, Zhu Z, et al (2010a) Development of a new japonica rice variety Nan-jing 46 with good eating quality by marker assisted selection. Rice Genomics Genet. doi: https://doi.org/10.5376/rgg.2010.01.0003

Wang C, Zhang Y, Zhu Z et al (2010b) Development of a new *japonica* rice variety Nan-jing 46 with good eating quality by marker assisted selection. Rice Genomics Genet. https://doi.org/10.5376/rgg.2010.01.0003

Ye X, Al-Babili S, Klöti A et al (2000) Engineering the provitamin a (β-carotene) biosynthetic pathway into (carotenoid-free) rice endosperm. Science 287:303–305. https://doi.org/10.1126/science.287.5451.303

Yi M, Than K, Vanavichit A et al (2009) Marker assisted backcross breeding to improve cooking quality traits in Myanmar rice cultivar Manawthukha. F Crop Res 113:178–186. https://doi.org/10.1016/j.fcr.2009.05.006

Chapter 11
Quality Protein Maize for Nutritional Security

Firoz Hossain, Konsam Sarika, Vignesh Muthusamy, Rajkumar Uttamrao Zunjare, and Hari Shanker Gupta

11.1 Introduction

Though significant advances have been made in agricultural research and technological developments, malnutrition remains a widespread problem (Neeraja et al. 2017). The resource-poor mostly in the under-developed and developing world suffers from 'hidden hunger', a term now more often used to describe malnutrition that is primarily caused due to micronutrient deficiencies in staple diet (Bouis and Saltzman 2017). Two billion people suffer from micronutrient deficiency-related problems, while 815 million people are undernourished (Global Nutrition Report 2017). The extent is so widespread that 88% of the countries face one or the other forms of malnutrition. An estimated 45% deaths of children under the age 5 years is associated with malnutrition (Black et al. 2013). Stunting, wasting, and underweight are the core components of failure in the growth of a child due to malnutrition (Zimmerman et al. 2018). Malnutrition contributes to global burden of disease, and loss in annual gross domestic product (GDP) in Asia and Africa is to the extent of 11%, and poses severe socio-economic loss to the countries.

Higher ingestion of staple foods alone, poor quality food and low consumption of animal and fish products, nutritious fruits and vegetables among the poor people of the society are the major concerns in the developing world, as adequate essential nutrients are not met. Thus a supplement of balanced nutrition is required for healthy growth and development of humans, particularly essential amino acids, vitamins, and minerals (Bouis et al. 2011). Among several micronutrient deficiencies, PEM also known as protein energy undernutrition (PEU) accounted highest number of deaths worldwide during 2016 (Nyakurwa et al. 2017). PEM is defined as a deficit

F. Hossain (✉) · V. Muthusamy · R. U. Zunjare · H. S. Gupta
ICAR-Indian Agricultural Research Institute, New Delhi, India

K. Sarika
ICAR Research Complex for Northeast Hill Region, Umiam, Meghalaya, India

A. M. I. Qureshi et al. (eds.), *Quality Breeding in Field Crops*,
https://doi.org/10.1007/978-3-030-04609-5_11

of energy due to the insufficient intake of proteins (Jensen et al. 2013; Morley 2016). Protein is a critically important component for growth and development in humans. PEM leads to poor intellectual development and disorderly physical functions, and in acute cases, it may lead to mortality. Kwashiorkor and Marasmus are the two major manifestations of PEM. Kwashiorkor develops when fair to normal calories is consumed with insufficient protein, whereas inadequate consumption of both protein and calorie leads to Marasmus (Bain et al. 2013). Children under the age of 5 years, the elderly and pregnant women are the most vulnerable groups to PEM and thus warrants urgent attention (Muller and Krawinkel 2005; Mpofu et al. 2014).

In view of widespread effects of malnutrition, world leaders gathered at the United Nations and set eight 'Millennium Development Goals' (MDGs) at the beginning of this millennium, of which Goal 1 (eradication of extreme poverty and hunger), Goal 4 (reduction of child mortality) and Goal 5 (improvement of maternal health) pertain to providing healthy and nutritious food to the people worldwide. During 2015, global community further set new goals which are known as 'Sustainable Development Goals' (SDGs) to chart out a strategy toward meeting current human needs without compromising the future generations' needs. Of the 17 goals, 12 include indicators that are strongly pertinent to nutrition, reflecting the importance of nutrition in sustainable development. Well balanced nutrition provides a platform for progress in health, education, employment, female empowerment and poverty. It has been projected that alleviating malnutrition is a cost-effective step for every investment of $1 in proven nutrition programme that offers a benefit worth of $16 (IFPRI 2016). Thus, effort towards providing a balanced and nutritious diet assumes a great significance (Gupta et al. 2015a; Yadav et al. 2015).

Maize (*Zea mays* L.) is one of the important staple food crops for billions in South America, Africa and Asia (Yadav et al. 2015), with an estimated world production of 1060 million metric tonnes from 187 million hectare area distributed in as many as 168 countries (FAOSTAT 2016). Though a major portion of the produce worldwide goes for animal consumption, more than 900 million poor section of the society survive on it. One-third of the global cereal produce is maize and with rice and wheat, it provides nearly 30% of the food calories to more than 4.5 billion people (Shiferaw et al. 2011; Nuss and Tanumihardjo 2011). Maize is consumed in various ways which greatly differs from country to country, with maize flour and meal being two of the most popular products.

Maize protein possesses low nutritional significance to humans due to reduced content of essential amino acids like lysine and tryptophan. Lysine is the most important limiting amino acid in the maize endosperm protein, followed by tryptophan (Prasanna et al. 2001). Lysine content (2.7%) in maize protein is well below the recommendation by FAO (WHO/FAO/UN 1985) for human. Although its content (5.4%) is adequate in germ protein, the abundance of endosperm protein 'zeins' (average lysine content ~1.9%) which compose of 60–70% of endosperm protein reduces the overall level of lysine in the grain (Coleman and Larkins 1999). Similarly, the lack of tryptophan residues in zeins causes low level of tryptophan in endosperm protein. Therefore, approaches such as reduction of zeins and expression of tryptophan and lysine-rich proteins could modify and significantly improve the profile of grain protein. Studies reveal enhancement of tryptophan and lysine is significantly correlated (Vivek et al. 2008). Lysine therefore is generally the primary target for improving grain quality as simultaneous improvement of tryptophan content occurs.

11.2 Importance of Essential Amino Acids

Humans require 0.66 g protein/kg body weight a day to meet the requirement for proper growth and development (WHO/FAO/UN 2007). Twenty percent of our human body is made up of proteins which play crucial role in almost all biological processes. Twenty different amino acids are required and are usually incorporated into proteins (Lea and Azevedo 2003). Amino acids comprise the second largest component in our body in the form of proteins; water being the largest in human muscles, cells and other tissues (Latham and Michael 1997; Medici et al. 2004). Amino acids play critical role in processes such as biosynthesis and transport of neurotransmitter and are further classified into essential and non-essential amino acids. Humans and monogastric animals cannot synthesize 9 of the 20 amino acids that are found in proteins. These nine amino acids namely lysine, threonine, methionine, phenylalanine, tryptophan, isoleucine, leucine, valine, and histidine are essential amino acids and must be acquired through the diet (Galili et al. 2002). Amino acids are required in a specific ratio and lower levels of lysine or tryptophan affect the body's ability to use other amino acids. These amino acids are called limiting amino acids and their deficiency lead to reduced appetite, delayed growth, impaired skeletal development and aberrant behaviour (Moehn et al. 2004; Tome and Bos 2007). Red meat, chicken, turkey and dairy products are the rich sources of lysine and tryptophan. However, considering the poor affordability especially in the developing and underdeveloped world, enhancement of essential amino acids in staple diets holds great significance for proper metabolic functions (Table 11.1).

11.3 Storage Proteins in Maize

11.3.1 Zein (Prolamin) Fraction of Endospermic Protein

The maize grain constitutes endosperm (82%), germ (12%) and pericarp (6%). The main portion of the endosperm is starch as in other cereals, constituting an average 70% of the grain. Eight to ten percent of the grain is protein, of which 60% is composed of prolamins (Prasanna et al. 2001). Prolamins are soluble in 70% alcohol having high proline and glutamine content. It is known as zeins in maize and with different names in diverse crops viz., *gliadin* (*wheat*), *kafirin* (*sorghum*), *hordein* (*barley*), *secalin* (*rye*) and *avenin* (*oats*). The remaining fraction of protein,

Table 11.1 Requirements and functions of lysine and tryptophan in the human body

S. no.	Amino acid	Functions	Daily need	References
1.	Lysine	Key component of muscle and collagen, and also promotes the absorption and incorporation of calcium and into bone tissue	*Adult*: 30 mg/kg body weight *Children*: 35 mg/kg body weight per day	WHO/FAO/UN (2007); White and Broadley (2009)
2.	Tryptophan	Key component in synthesis of numerous metabolites such as kynurenine and serotonin. Important determinant of mood, cognition and behaviour	*Adult*: 4 mg/kg body weight a day *Children*: 4.8 mg/kg body weight	Harper and Yoshimura (1993); Riedel et al. (2003); WHO/FAO/UN (2007)

collectively known as non-zeins comprises glutelins (34%), globulins (3%) and albumins (3%) which are balanced in essential amino acids (Vasal 2000). However, zeins contain higher proportion of leucine (18.7%), phenylalanine (5.2%), isoleucine (3.8%), valine (3.6%) and tyrosine (3.5%). Deficiency of tryptophan and lysine results in relatively poor protein quality of maize grain. Prolamin being poor in quality, an inverse relationship is generally observed in its accumulation and protein quality in endosperm. Protein of normal maize has a biological nutritional value of 40% of that of milk (Bressani 1991) and hence requires complementary protein sources such as legumes or animal products.

Zeins, based on its aqueous solubility and ability to form disulphide bonds, are classified into four classes α-, γ-, β- and δ-zeins, consisting of complex closely related polypeptides of different molecular weight (Coleman and Larkins 1999) (Table 11.2). The most abundant α-zeins (19- and 22-kDa) are synthesized by four highly duplicated gene families distributing across six chromosomal locations consisting of more than 40 genes (Feng et al. 2009). On the contrary, the β-zein (15-kDa), γ-zein (16-, 27- and 50-kDa) and δ-zein (10- and 18-kDa) are encoded by single-copy genes (Xu and Messing 2009). However, the expression of these genes varies across different maize genetic backgrounds. The synthesis of zeins initiates after 12 days of pollination and most actively synthesized between 16 and 35 days, which can further continue up to 50 days after pollination (Wall and Bietz 1987). Due to their high levels of expression and complexity, zein synthesis serves as a model system to analyse coordinated genetic regulation of several genes expressed at very high levels at a specific developmental stage (Soave and Salamini 1984).

The synthesis of zein, specific to the maize grain is regulated by a basic leucine zipper (bZIP) transcription factor (TF) coded by *Opaque2* (*O2*) (Schmidt et al. 1990). The TF recognizes many motifs in the promoters of 22-kDa α-zeins and regulates 19-kDa α-zeins as well (Schmidt et al. 1987). Besides regulation of zein synthesis, *O2* also has non-zein transcriptional targets influencing DNA methylation and histone modification of zein genes showing a role in the *O2*-mediated activation (Locatelli et al. 2009). Studies showed other regulatory factors which are involved in the expression of zein such as a Dof-type *Prolamine-Box Binding Factor* (PBF) in 27-kDa γ-zein expression (Zhang et al. 2015); MADS-box protein (*ZmMADS47*) in α-zeins and the 50-kDa γ-zein genes expression (Qiao et al. 2016).

Table 11.2 Per cent proportion of each class of zein in total endospermic zein content

S. No.	Zein fraction	Protein fraction (kDa)	Total zein (%)
1.	α-Zein	22, 19	60–70
2.	γ-Zein	50, 27, 16	20–25
3.	β-Zein	15	5–10
4.	δ-Zein	18, 10	<5

11.3.2 Storage Protein Bodies (PB) of Maize Endosperm

Plant storage proteins are initially synthesized in polyribosomes of the rough endoplasmic reticulum (RER) and subsequently translocated into the lumen of the RER. In the case of maize, they are accumulated directly into PB accretions and remain within the ER lumen or, transported through the endomembrane system to protein storage vacuoles (PSVs) in other crop plant (Lending and Larkins 1992; Vitale and Denecke 1999; Choi et al. 2000). The efficient packaging of zeins fraction in PB defines the hard texture phenotype in the mature grain which is an essential characteristic of maize grains (Holding 2014). Maize grain has a glassy or vitreous periphery and a central opaque region in the mature grain. The peripheral vitreous region has a much more compactly packaged zeins protein in PBs than the soft, opaque centre where PBs are smaller and scarce (Tsai et al. 1978). A typical PB at 18–20 days after pollination is spherical in shape, discrete and a highly ordered arrangement of α- and δ-zeins in the centre, while γ- and β-zeins in the peripheral layer (Lending and Larkins 1989). Improper accumulation of zeins results in irregular PB shapes and opaque phenotype due to the weak packaging (Kim et al. 2006), as observed in opaque mutants where the central opaque region of endosperm extends to its periphery. RER membranes break down during the desiccation of seeds while exposing the zeins protein that has been mixed with the other content of the cell cytoplasm forming proteinaceous matrix. It acts as cementing glue and thereby providing an airtight interaction among the starch granules in normal vitreous endosperm of maize (Wu and Messing 2010; Wu et al. 2010).

11.4 High Tryptophan Mutants

11.4.1 Discovery of Opaque2 *(o2)*

Efforts and strategies were only limited to screening of elite maize germplasm and the breeding approach such as recurrent selection could not be easily employed in the absence of specific genes enhancing tryptophan and lysine until 1960s (Prasanna et al. 2001). The discovery of the *o2* mutant by Jones and Singleton in the 1920s provided a significant breakthrough. It caused nearly twofold enhanced accumulation of both tryptophan and lysine in the endosperm in comparison with the normal genotypes (Mertz et al. 1964). In addition, the content of other amino acids viz. histidine, arginine, aspartic acid and glycine increases, while a decline in glutamic acid, alanine and leucine was observed.

The recessive *o2* has been mapped on the chromosome 7L. The mutant *o2* gene encoding the modified bZIP transcriptional factor reduces α-zeins synthesis, besides down-regulating synthesis of lysine ketoglutarate reductase (LKR) which breaks down lysine into further component (Schmidt et al. 1990). It induces a general reduction in the accumulation of 22- and 19-kDa α-zeins to lesser extent with an overall 50–70% reduction of zeins. The reduced α-zeins results in abnormal morphological PBs with lesser size and number and thus, eventually a soft and starchy textured opaque grain. *o2* is generally accompanied with an increased non-zein proteins such as cytoskeleton-associated carbohydrate metabolizing enzymes and eEF1A, which are relatively rich in tryptophan and lysine (Azama et al. 2003; Lopez-Valenzuela et al. 2004; Jia et al. 2013).

11.4.2 Discovery of Other Mutants

Search for new mutants in maize that could enhance the endospermic amino acid profile led to the discovery of other genetic mutants such as *floury1 (fl1), floury2 (fl2), floury3 (fl3), opaque5 (o5), opaque6 (o6), opaque7 (o7), opaque15 (o15), Defective endosperm (Def-B30)*, and *Mucronate (Mc)* (Balconi et al. 2007). While all the other mutants confer a soft and opaque phenotype to the endosperm, *o2*, *fl2* and *Def-B30* alter zein content affecting different aspects of storage protein synthesis. These mutants have only ~35–55% storage proteins of total protein in wild genotype (Morton et al. 2016). On contrary, reduction of only ~10–20% could be observed in mutant *o9*, *o11* and *Mc* (Hunter et al. 2002). However, the precise mechanism on how these mutants affect a relatively small fraction of the zein polypeptides disrupting the formation of vitreous endosperm is still unclear (Soave and Salamani 1984). The *o1* mutation appears to have little effect on zein synthesis (Nelson et al. 1965).

11.4.3 Discovery of **Opaque16 (o16)**

However, recently, a recessive *opaque16* (*o16*) mutant located on chromosome 8 and isolated from Robertson's Mutator (Mu) stock was found to be associated with higher nutritional value in maize (Yang et al. 2005). Further, genotypes with *o16o16* possessed nearly twofold more tryptophan (0.072% in mutants; 0.035% in wild types) and lysine (0.247% in mutants; 0.125% in wild types) compared to normal maize (tryptophan 0.035% and lysine 0.125%) (Sarika et al. 2017). *o16* alone can be as good as *o2* for improving the nutritional quality of maize and provides a significant advantage to the breeders since *o16* possesses vitreous endosperm with equivalent grain hardness to wild line (Sarika et al. 2018). The availability of linked markers to *o16* offers additional advantage of adopting marker-assisted breeding along with *o2* in maize breeding programmes (Yang et al. 2005; Zhang et al. 2010, 2013).

11.5 Quality Protein Maize (QPM)

11.5.1 Endosperm Modifiers

The elation over the utility of *o2* in improving the nutritional quality was short-lived with the unearthing of pleiotropic effects such as opaque soft endosperm with increased susceptibility to damaged grains, pests and fungal diseases with sub-standard food processing (Shewry and Thatam 1990; Hossain et al. 2007, 2008a, b). Since starch granules are packed loosely with larger air spaces, grain weight is reduced due to low density per unit volume and correspondingly declining the yield to a tune of 10% or more (Singh and Venkatesh 2006). In an effort to overcome the problems, scientists at CIMMYT, Mexico, and University of Natal, South Africa, initiated careful examination of the nature of inherent problems involved with *o2* (Geevers and Lake 1992; Villegas et al. 1992). Its inferior quality could be overcome with the accumulation of endosperm modifiers without affecting the endospermic protein quality (Krivanek et al. 2006). This finding led to the development of hard endosperm-based high lysine and tryptophan maize popularly called as quality protein maize (QPM) (Vasal et al. 1984). The strategy adopted by CIMMYT researchers, the combined use of *o2* gene and endosperm modifiers in developing QPM, has proved to be successful and is followed in different countries' QPM research programme (Pandey et al. 2016). In appreciation of the interdisciplinary teamwork of CIMMYT researchers, the prestigious 'World Food Prize' was awarded to Surinder K. Vasal and Evangelina Villegas in the year 2000.

The genes regulating the development of vitreous endosperm in QPM have remained elusive, but the possible role of 27-kDa γ-zein in inducing the vitreous phenotype has been put forward since two- to three-fold higher accumulation of this particular zein could be observed in QPM (Wallace et al. 1990; Geetha et al. 1991; Lopes and Larkins 1995). The 27-kDa γ-zein appears to initiate and induce the formation of PBs, providing a cross-linking disulphide bonds during grain desiccation (Gibbon et al. 2003). Genetic mapping showed the linkage of *o2* endosperm modifiers loci in QPM with the locus encoding 27-kDa γ-zein located in chromosome 7. The finding was supported by Wu et al. (2010) demonstrating the supposition of γ-zeins over *o2* endosperm modification by RNAi silencing of 27- and 16-kDa γ-zein genes resulting in clumping of PBs and thus inducing opacity in QPM seeds. Besides, Holding et al. (2008, 2011), Gutierrez-Rojas et al. (2010) and Lebaka et al. (2013) identified several endosperm modifier loci in the maize genome. More recently, Liu et al. (2016) identified a quantitative trait locus (*qγ27*) which affects the expression of 27-kDa γ-zein. *qγ27* is mapped on chromosome 7 near the locus of 27-kDa γ-zein but to the locus as major *o2* modifier. *qγ27* is a result from a 15.26-kb duplication at 27-kDa γ-zein locus, which upsurges the level of its expression. Gradual increase of endosperm modification from the crown of the grain towards the tip could be observed due to selection and accumulation of endosperm modifiers in *o2* genetic background and have been used in breeding programmes worldwide. Various aspects of grain modification were reviewed by scientists at Purdue University and CIMMYT (Bauman 1975; Vasal et al. 1980, 1984; Glover and Mertz 1987; Bjarnason et al. 1988; Glover 1988; Prasanna et al. 2001).

Transgenic approach is an alternative approach to understand the relationship of zein synthesis and origin of the opaque phenotype by perturbing zein accumulation in endosperm.. Two research groups reported an increase of 15–20% lysine in transgenic lines developed by knocking down of 22- and 19-kDa α-zeins by RNA interference approach (RNAi) (Segal et al. 2003; Huang et al. 2004). However, this increase is well below the 100% increase often observed in *o2* genotypes but nevertheless, these experiments showed effective increase in grain lysine through transgenic approach. The second foremost finding of the studies was the reduction of α-zein protein synthesis sufficient enough to induce opacity in endosperm. RNAi lines with knockout 22-kDa α-zein showed a more pronounced opaque phenotype than the 19-kDa α-zein knockout RNAi lines. This confers a stronger interaction of 22-kDa α-zein with the γ- and β-zeins than that of the 19-kDa α-zein. Therefore, the insertion of 19-kDa α-zein into the centre of protein bodies was perturbed by the absence of the 22-kDa α-zein or it might enable direct contact of 19-kDa α-zein with the γ- and β-zeins. Such abnormalities in their arrangement/associations could have disrupted PB formation leading to an opaque endosperm phenotype. Different combinations of these mutants were used to enhance endospermic protein quality but were not successful because of severe large negative pleiotropic effects.

11.5.2 Nutritional Benefits of QPM

Overwhelming data demonstrating the nutritional superiority of QPM over normal maize are available. Various QPM feeding trials have been conducted where undernourished children given with QPM as the only protein source showed the same growth with those who were given modified cow milk formula in diet (Graham et al. 1990). Independent studies in different countries indicated a 12% weight gain in children consuming QPM over the conventional maize (Gunaratna et al. 2010). A study conducted in Guatemala demonstrated that *o2* maize nutritive value is 90% of milk protein as compared to the 40% in normal maize in young children (Prasanna et al. 2001; Bressani 1992). With twofold increase of tryptophan and lysine and doubling of biologically usable protein, QPM confers other nutritional benefits, i.e. better leucine/isoleucine ratio and higher niacin availability (Graham et al. 1990; Bressani 1992). The low leucine content in QPM helps in liberating more tryptophan for niacin biosynthesis even though QPM and normal maize have same levels of niacin. Thus, QPM reduces pellagra significantly (Vasal 2001). Meta-analysis of experiments conducted in different countries showed a strong implication about the nutritional benefits of QPM on gain of weight and height in infant and young children (Teklewold et al. 2015, Gunaratna et al. 2010). The palatability and cooking quality of traditional food prepared from QPM are more acceptable due to its softness, perceived sweetness and longer shelf life in eastern African countries (Akalu et al. 2010).

Several trials on animal feed were also conducted to assess the nutritional benefits and biological superiority of QPM. It was first demonstrated with rats where threefold increase in growth rate was observed when fed with a diet of 90% QPM (Mertz et al. 1964). Rats fed with QPM diet weighed more and were thicker, longer, denser and stronger than those fed with normal maize (Serna-Saldivar et al. 1992). QPM nutritional benefits were also extensively conducted in pigs. The weight gain was doubled in pigs raised on QPM as compared to the ones fed on only normal maize (Burgoon et al. 1992). Pigs fed with a diet of QPM alone with vitamins and minerals supplement grew faster twice the rate of those fed normal maize (Osei et al. 1994). In poultry too, QPM improves the growth performance of broilers resulting higher gain in weight (Onimisi et al. 2008). Diets with QPM can lead to significant reduction in the use of fishmeal and synthetic lysine additives which are also more economical (Qi et al. 2004).

11.6 Development of QPM Cultivars

11.6.1 Breeding Strategies Through Phenotypic Selection

In order to tackle the negative pleiotropic effects of *o2* without compromising the enhanced nutritional quality, genotypes with hard endosperm in *o2* genetic background were recombined to recover higher frequency of modifiers. Thus, several tropical populations with hard and vitreous texture with maintained protein quality of endosperm were developed viz. Composite K (H.E.*o2*), Ver.181-Ant.gp *o2* × Venzuela-1 *o2*, Thai *o2* Composite, PD (MS6) H.E.*o2*, Composite I, Yellow H.E.*o2* composite, White H.E.*o2* composite, etc. (Bjarnason and Vasal 1992). The repeated selection of 3–4 cycles in modified ears led to the development of a series of QPM donor stocks by the mid-1970s (Prasanna et al. 2001).

Several QPM populations, pools, inbreds, and hybrids were developed in CIMMYT through conventional conversion breeding approach which could adapt in subtropical and tropical environments and are widely used in the development of QPM cultivars in several countries in Africa, Asia and Latin America (Vasal 2001). Some of the very popular QPM cultivars that are worth mentioning are Obatanpa (Ghana), AMH760Q (Ethiopia), Longe-5 (Sudan), Yunrui-1 (China), Poshilo Makai-1 (Nepal), Chaskarpa (Bhutan), etc. In India during 1997, a nutritionally superior QPM composite with vitreous grain texture, 'Shakti-1', was released. Since 1998, intensive efforts in different breeding centres of country resulted in the release of QPM hybrids like Shaktiman-1 and HQPM-1. Bangladesh has also released QPM

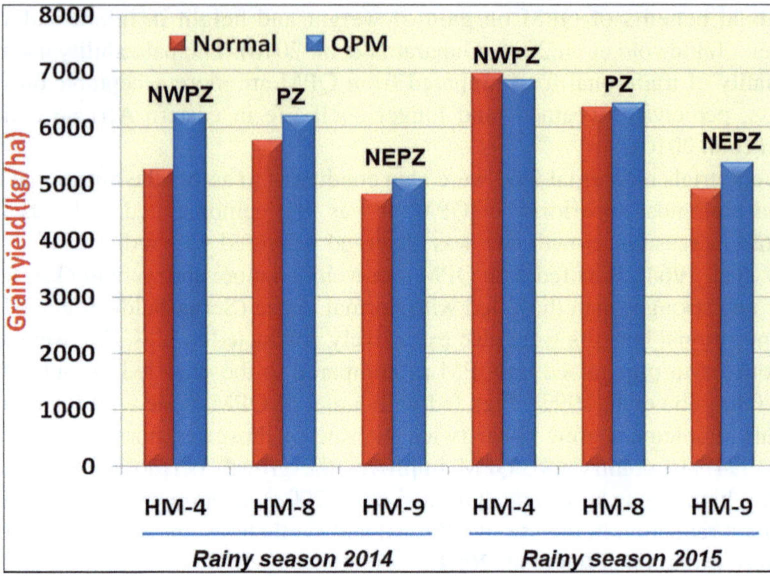

Fig. 11.1 Grain yield of normal hybrids (HM-4, HM-8 and HM-9) and their QPM version under AICRP-National Testing. *NWPZ* north western plain zone, *PZ* peninsular zone, *NEPZ* north eastern plain zone

hybrid, BARI Hybrid Maize-5. In Pakistan, QPM hybrids, QPHM200 and QPHM300 were released recently in 2017. QPM cultivation and grain yield potential are interchangeable with normal maize (Fig. 11.1, Hossain et al. 2018).

11.6.2 Breeding Strategies Through Marker-Assisted Selection

An accelerated breeding cycle is required to meet the increasing demand of maize grain and feed (Ribaut et al. 2003). The progress in plant biotechnology including molecular mapping of loci and the gene introgression through marker-assisted selection (MAS) offers options for enhancing selection efficiency in developing new cultivars with higher quality and yield potential (Varshney et al. 2012). Previously, only the conventional breeding procedures were used to convert lines into QPM version, even though the approach is tedious, laborious and time consuming. Conventional approach of introgression or pyramiding involving recessive genes such as *o2* and *o16* through backcrossing is not an easy procedure and the reasons being (1) recessive gene(s) needs to be identified in each selfed generation of cross/backcross generation, (2) a minimum six backcross generations is required to recover satisfactory levels of recurrent parent genome, and (3) involvement of rigorous biochemical tests for lysine and tryptophan in each breeding generation which are laborious (Collard et al. 2005).

Marker-assisted introgression using backcross breeding involves two steps: (1) foreground selection: targeting gene through marker, and (2) background selection: targeting uniformly distributed markers throughout the genome for recovering recurrent parental genome (RPG) (Hospital et al. 1992). This is an efficient way of transferring specific gene(s) to an otherwise superior parental lines or variety. Identification of gene of interest becomes precise through foreground selection, at the same time background selection expedites the rate of recovery of RPG with two backcrosses (Hospital et al. 1992; Visscher et al. 1996; Frisch et al. 1999). Easy available of reliable PCR-based gene-based or -linked markers has made MAS an effective option. Among different types of DNA sequence-based markers available, the microsatellite or Simple Sequence Repeat (SSR) markers are often the choice for its low cost, simplicity and effectiveness. SSR markers are codominant, robust, hypervariable, abundant and uniformly distributed across plant genomes (Powell et al. 1996). In maize, several thousand SSRs are mapped and available in the public domain (www.maizegdb.org).

The characterization of *O2* gene followed by detection of three gene-based SSR markers (*phi057*, *phi112* and *umc1066*) led to effective distinction of the dominant (*O2*) and recessive (*o2*) alleles (Schmidt et al. 1987; Motto et al. 1988; Kassahun and Prasanna 2003). These *o2*-specific SSR markers offer significant advantages in conversion of non-QPM lines into QPM (Babu et al. 2005; Pandey et al. 2018). Using MAS, several inbreds and eventually hybrids were converted into QPM version worldwide (Table 11.3). In India, Vivek Hybrid-9, an early maturing single cross normal hybrid was converted to QPM using MAS, and the improved hybrid named as 'Vivek QPM 9' was released (Gupta et al. 2013). Vivek QPM-9, similar yield potential with the original hybrid, possesses 41% and 30% more tryptophan and lysine over the original hybrid, respectively. It earned the distinction of being the first commercial MAS-based maize cultivar released in India. Recently, QPM

Table 11.3 Details of MAS undertaken for development of QPM genotypes worldwide

S. No.	Gene(s) introgressed	Inbred/hybrid	Country	Reference (s)
1.	*opaque2*	Inbred	India	Babu et al. (2005)
2.	*opaque2*	Inbred	Uganda	Manna et al. (2005)
3.	*opaque2*	Inbred	Kenya	Danson et al. (2006)
4.	*opaque2*	Inbred	Philippines	Magulama and Sales (2009)
5.	*opaque2*	Inbred and hybrid	Thailand	Jompuk et al. (2011)
6.	*opaque2*	Inbred	Serbia	Kostadinovic et al. (2016)
7.	*opaque2*	Inbred and hybrid	India	Gupta et al. (2013), Hossain et al. (2018)
8.	*opaque16*	Inbred	China	Yang et al. (2013), Zhang et al. (2010)
9.	*opaque16*	Inbred	India	Sarika et al. (2017)
10.	*opaque2, opaque16*	Inbred	China	Yang et al. (2005); Zhang et al. (2013)
11.	*opaque2, opaque16*	Inbred and hybrid	India	Sarika et al. (2018)

Fig. 11.2 Grain and ear type of QPM hybrid, 'Pusa HM-8 Improved' developed through MAS at IARI, New Delhi

version of three popular commercial hybrids viz., HM-4, HM-8 and HM-9 has been developed using MAS (Hossain et al. 2018) (Fig. 11.2). The reconstituted QPM hybrids have significant enhanced endospermic lysine (48–74%) and tryptophan (55–100%) with similar yield potential of the respective original hybrids. These three QPM hybrids viz. 'Pusa HM-4 Improved', 'Pusa HM-8 Improved' and'Pusa HM-9 Improved' have now been released for commercial cultivation in 2017 (Yadava et al. 2017). Similar efforts to introgress *o2* allele into normal inbreds through MAS have also been reported worldwide (Manna et al. 2006; Danson et al. 2006; Magulama and Sales 2009, Jompuk et al. 2011, Kostadinovic et al. 2016).

For further enhancing the nutritional quality attributes of grain, especially lysine and tryptophan in endosperm protein, the availability of suitable linked SSRs has offered a promising option of marker-assisted introgression of *o16*. In this context, linked SSRs, *umc1141* and *umc1149,* were successfully utilized for introgression of *o16* alone or pyramiding it in *o2* genetic background. The enhancement of protein quality (tryptophan and lysine) by *o16* over normal maize is comparable to *o2*-based QPM genotypes (Sarika et al. 2017). MAS has been employed for the improvement of parental lines and derived hybrids by pyramiding *o2* and *o16* in maize adapted to temperate regions at Guizhou Institute of Upland Food Crops, Guizhou Academy of Agricultural Sciences, China. A half-fold increase in lysine content among *o2* and *o16* pyramided progenies was reported (Yang et al. 2005; Zhang et al. 2010, 2013). At ICAR-Indian Agricultural Research Institute (IARI), New Delhi, researchers have successfully pyramided both the genes through marker-assisted backcrossed breeding by targeting the parental inbred lines of four *o2*-based QPM commercial hybrids viz., HQPM-1, HQPM-4, HQPM-5 and HQPM-7 (Sarika et al. 2018). The reconstituted hybrids possess an average of 0.13% tryptophan and 0.50% lysine compared to 0.08% and 0.37% in original hybrids, with an average enhancement was 60% and 49%, respectively. Moreover, *o16* alone does not possess much adverse effect on the endosperm modification, thereby providing the breeders significant advantage in the selection process (Sarika et al. 2018).

11.7 Enriching QPM with Micronutrients

In QPM genetic background, for further improving nutritional benefit, genes responsible for enhancing provitamin-A (proA) have been introgressed through MAS. Normal maize including QPM contains very low proA (1–2 ppm) compared to target level of 15 ppm (Zunjare et al. 2017). Muthusamy et al. (2014) improved Vivek QPM-9 by introgressing favourable allele of *crtRB1*, and the improved hybrid possessed significantly higher level of proA (8.15 ppm) at two months after harvest under normal storage conditions. The research effort led to world's first commercial release of proA enriched QPM hybrid, 'Pusa Vivek QPM-9 Improved' in India (Yadava et al. 2017). More recently, Zunjare et al. (2018) stacked *o2*, *crtRB1* and *lcyE* genes into the genetic background of popular QPM hybrids viz., HQPM-1, HQPM-4, HQPM-5 and HQPM-7, and reported the significant enhancement in proA (9.0–12.9 ppm). Further, proA enriched QPM parental lines of these hybrids have also been targeted for improvement of vitamin-E by marker-assisted introgression of *VTE4* favourable allele (Das et al. 2018). Further efforts on enrichment of iron and zinc in QPM genetic background have also been reported (Chakraborti et al. 2011a, b; Pandey et al. 2015; Gupta et al. 2015b, Mallikarjuna et al. 2015). These multinutrient rich high yielding QPM hybrids could be efficiently utilized in biofortification programmes of maize across the globe and hold great promise for nutritional security in a holistic manner.

11.8 Worldwide Acceptance of QPM

QPM offers a significant benefit and is of major interest to breeders and nutritionists. The micronutrient enriched or biofortified staple crops developed through breeding programme hold promise for sustainable and cost-effective solutions to fight micronutrient deficiencies (Pfeiffer and McClafferty 2007). It has been disseminated specially in the developing countries in Asia, Africa and Latin America. QPM hybrid development efforts have been progressed considerably in CIMMYT with simultaneous efforts in various countries' research breeding programmes converting local lines into QPM versions by effectively using donor stocks developed in CIMMYT. The successful deployment of QPM hybrids has an area of one million ha in sub-Saharan Africa, Ghana and Uganda together accounting for nearly 50% of the area, and the rest accounted by 18 other countries. QPM occupies 150,000 ha in 12 Latin American countries, especially Mexico, Venezuela, Bolivia, Guatemala, Honduras and Colombia, and in 250,000 ha in Asia, with 80–90% of this area in

China, and rest in other Asian countries, including India, Nepal, Philippines and Indonesia. Obatanpa is one of the most impactful QPM varieties. It is widely adapted across Africa and throughout the maize growing tropics and has been released in at least 16 African countries, ranging from South Africa to Ethiopia, and in non-African countries as diverse as Nicaragua and Philippines. Currently, it is grown in 41% of the maize area in Ghana (total maize area: 0.99 m ha) and in Uganda, under the name "Nalongo" occupying 30% maize area.

11.9 Challenges in Adoption of QPM

The successful and effective adoption of biofortified cultivars faces various challenges. The mistaken perception of low-yielding potential of nutritionally rich crops like QPM plays a crucial role in slowing down the pace of dissemination. Nutritional traits such as lysine, tryptophan, provitamin A, vitamin-E, Fe and Zn are phenotypically invisible and farmers face difficulty in convincing the trader regarding the extent of quality of his produce. Dilution of nutritional quality by contamination of foreign pollen grains from neighbouring maize fields is also a concern for the quality produce. Resistance in accepting nutritious foods with altered appearance (white vs. yellow maize) and lack of awareness on health benefits are also some of the challenges for dissemination of the QPM technology. The adoption of QPM has also been limited due to lack of profitable markets for commercial producers, unwillingness among maize food processors in marketing QPM as a premium product, and the absence of government incentive to encourage adoption by subsidizing the price of QPM seed. A qualitative value chain study conducted by Hellin and Erenstein (2009) highlighted some of associated challenges with the biofortified maize as poultry feed and also innate weaknesses in maize-poultry value chains in India, and they are (1) weak linkages between maize farmers and local poultry firms, (2) limited access to improved technology and channels of information and other business services for small scale maize and poultry producers, and (3) low prevalence of value chains with both growth and poverty reduction potential.

Development of diverse high yielding QPM hybrids in different maturity through broadening of germplasm base would provide wider opportunities for adaptation to different agro-ecologies (Hossain et al. 2016). Integration of modern *omic* approaches would further accelerate the breeding cycle to develop multinutrient- and stress-resilient QPM cultivars (Jha et al. 2015). Intensive efforts by public sector institutions and policy for intense promotional campaigns to ensure efficiency input–output linkage for QPM production can lead to significant increase in adoption and acceptance of QPM hybrid worldwide. Strengthening the seed chain to produce and supply good quality seeds is one of the important steps for the popularization of the biofortified maize. Providing subsidized seeds and other inputs would further contribute to the rapid dissemination of nutritionally improved cultivars among the farmers. Assurance of remunerative price through minimum support price and/or premium price for biofortified maize grains in the market will encourage the farmers to grow more biofortified maize (Gupta et al. 2015a). Inclusion of biofortified products in government-sponsored health benefit programmes especially for children, pregnant women and elderly people would help in their quick dissemination. Investment and extension activity would make the farmers, industry and consumers aware of the existence and benefits of QPM.

References

Akalu G, Tafesse S, Gunaratna NS, De Groote H et al (2010) The effectiveness of quality protein maize in improving the nutritional status of young children in the Ethiopian highlands. Food Nutr Bull 31:418–430

Azama K, Abe S, Sugimoto H, Davis E et al (2003) Lysine-containing proteins in maize endosperm: a major contribution from cytoskeleton-associated carbohydrate-metabolizing enzymes. Planta 217:628–638

Babu R, Nair SK, Kumar A, Venkatesh S, Sekhar JC, Singh NN, Srinivasan G, Gupta HS et al (2005) Two-generation marker-aided backcrossing for rapid conversion of normal maize lines to quality protein maize (QPM). Theor Appl Genet 111:888–897

Bain LE, Awah PK, Geraldine N, Kindong NP, Sigal Y, Bernard N et al (2013) Malnutrition in Sub-Saharan Africa: burden, causes and prospects. Pan Afr Med J 15:1–9. https://doi.org/10.11604/pamj.2013.15.120.253

Balconi C, Hartings H, Lauria M, Pirona R, Rossi V, Motto M et al (2007) Gene discovery to improve maize grain quality traits. Maydica 52:357–373

Bauman LF (1975) Germ and endosperm variability, mineral elements, oil content and modifier genes in *opaque-2* genotypes. In: High quality protein maize. Hutchinson Ross Publishing Co, Stroudsburg, PA, pp 143–147

Bjarnason M, Vasal SK (1992) Breeding of quality protein maize. Plant Breed Rev 9:181–216

Bjarnason M, Shoirt K, Vasal SK, Villegas E et al (1988) Genetic improvement of various quality protein maize (QPM) populations. Agronomy abstract. American Society of Agronomy, Madison, WI, p 74

Black RE, Victora CG, Walker SP et al (2013) Maternal and child under nutrition and overweight in low-income and middle-income countries. Lancet 382:427–451

Bouis HE, Saltzman A (2017) Improving nutrition through biofortification: a review of evidence from HarvestPlus, 2003 through 2016. Glob Food Sec 12:49–58. https://doi.org/10.1016/j.gfs.2017.01.009

Bouis HE, Christine H, Bonnie M et al (2011) Biofortification: a new tool to reduce micronutrient malnutrition. Improving diets nutrition food-based approaches, vol 32, pp 202–215

Bressani R (1991) Protein quality of high lysine maize for humans. Cereal Foods World 36:806–811

Bressani R (1992) Nutritional value of high-lysine maize in humans. In: Mertz ET (ed) Quality protein maize. American Association of Cereal Chemists, St. Paul, MN, pp 205–224

Burgoon KG, Hansen JA, Knabe DA, Bockholt AJ et al (1992) Nutritional value of quality protein maize for starter and finisher swine. J Anim Sci 70:811–817

Chakraborti M, Prasanna BM, Hossain F, Singh A et al (2011a) Evaluation of single cross quality protein maize (QPM) hybrids for kernel iron and zinc concentrations. Indian J Genet 71:312–319

Chakraborti M, Prasanna BM, Hossain F, Mazumdar S, Singh AM, Guleria SK, Gupta HS et al (2011b) Identification of kernel iron- and zinc-rich maize inbreds and analysis of genetic diversity using microsatellite markers. J Plant Biochem Biotechnol 20:224–233

Choi SB, Wang CL, Muench DG, Ozawa K, Franceschi VR, Wu YJ, Okita TW et al (2000) Messenger RNA targeting of rice seed storage proteins to specific ER subdomains. Nature 407:765–767

Coleman CE, Larkins BA (1999) The prolamins of maize. In: Shewry P, Casey R (eds) Seed proteins. Kluwer Academic, Dordrecht, pp 109–139. https://doi.org/10.1007/978-94-011-4431-5-6
Collard BCY, Jahufer MZZ, Brouwer JB, Pang ECK et al (2005) An introduction to markers, quantitative trait loci (QTL) mapping and marker-assisted selection for crop improvement. Euphytica 142:169–196
Danson J, Mbogori M, Kimani M et al (2006) Marker-assisted introgression of *opaque2* gene into herbicide tolerant elite maize inbred lines. Afr J Biotechnol 5:2417–2422
Das AK, Hossain F, Muthusamy V, Zunjare RU, Chauhan HS, Baveja A, Jaiswal SK, Bhata JS, Guleria SK, Saha S, Singh AK et al (2018) Genetic analyses of kernel tocopherols in maize possessing novel allele of *γ-tocopherol methyl transferase* (*ZmVTE4*). International symposium on biodiversity and biobanking – biodiverse 2018. 27–29 Jan 2018. IIT Guwahati, Guwahati, pp 112–113
FAO/WHO/UN (1985) Expert consultation. WHO techncal report series no. 724. WHO, Geneva
FAOSTAT (2016) http://www.fao.org/faostat/en/#data/QC/visualize
Feng LN, Zhu J, Wang G, Tang YP, Chen HJ, Jin WB et al (2009) Expressional profiling study revealed unique expressional patterns and dramatic expressional divergence of maize alpha-zein super gene family. Plant Mol Biol 69:649–659
Frisch M, Bohn M, Melchinger AE et al (1999) Comparison of selection strategies for marker-assisted backcrossing of a gene. Crop Sci 39:1295–1301
Galili G (2002) New insights into the regulation and functional significance of lysine metabolism in plants. Annu Rev Plant Biol 7:153–156
Geetha KB, Lending CR, Lopes MA, Wallace JC, Larkins BA et al (1991) Opaque-2 modifiers increase gamma-zein synthesis and alter its spatial distribution in maize endosperm. Plant Cell 3:1207–1219
Geevers HO, Lake JK (1992) Development of modified opaque2 maize in South Africa. In: Mertz ET (ed) Quality protein maize. American Association of Cereal Chemists, St. Paul, MN, pp 49–78
Gibbon BC, Wang X, Larkins BA et al (2003) Altered starch structure is associated with endosperm modification in quality protein maize. Proc Natl Acad Sci U S A 100:15329–15334
Global Nutrition Report (2017) http://www.globalnutritionreport.org
Glover DV (1988) Corn protein and starch – genetics, breeding, and value in foods and feeds. In: Proceedings of the 43rd annual corn and sorghum research conference. American Seed Trade Association, Chicago. IL, pp 106–130
Glover DV, Mertz ET (1987) Corn. In: Olson RA, Frey OJ (eds) Nutritional quality of cereal grains: genetic and agronomic improvement. ASA-CSSA-SSSA, Madison, WI, pp 183–236
Graham GG, Lembake J, Morales E et al (1990) Quality protein maize as the sole source of dietary protein and fat for rapidly growing young children. Pediatrics 85:85–91
Gunaratna NS, De Groote H, Nestel P, Pixley KV, McCabe GP et al (2010) A meta-analysis of community-level studies on quality protein maize. Food Policy 35:202–210
Gupta HS, Raman B, Agrawal PK, Mahajan V, Hossain F, Nepolean T et al (2013) Accelerated development of quality protein maize hybrid through marker-assisted introgression of *opaque2* allele. Plant Breed 132:77–82
Gupta HS, Hossain F, Muthusamy V et al (2015a) Biofortification of maize: an Indian perspective. Indian J Genet 75:1–22
Gupta HS, Hossain F, Nepolean T, Vignesh M, Mallikarjuna MG et al (2015b) Understanding genetic and molecular bases of Fe and Zn accumulation towards development of micronutrient-enriched maize. In: Rakshit A et al (eds) Nutrient use efficiency: from basics to advances. Springer, New Delhi. https://doi.org/10.1007/978-81-322-2169-2_17
Gutierrez-Rojas A, Betran J, Scott MP, Atta H, Menz M et al (2010) Quantitative trait loci for endosperm modification and amino acid contents in quality protein maize. Crop Sci 50:870–879
Harper AE, Yoshimura NN (1993) Protein quality, amino acid balance, utilization, and evaluation of diets containing amino acids as therapeutic agents. Nutrition 9:460–469

Hellin J, Erenstein O (2009) Maize poultry value chains in India: implications for research and development. J New Seeds 10:245–269

Holding DR (2014) Recent advances in the study of prolamin storage protein organization and function. Front Plant Sci 5:276. https://doi.org/10.3389/fpls.2014.00276

Holding DR, Hunter BG, Jung T, Gibbon BR, Clark CF, Bharti AR, Messing J, Hamaker BR, Larkins BA et al (2008) Genetic analysis of opaque-2 modifier loci in quality protein maize. Theor Appl Genet 117:157–170

Holding DR, Hunter BG, Klinger JP, Wu S, Gua X, Gibbon BC, Wu R, Schulze JM, Jung R, Larkins BA et al (2011) Characterization of opaque-2 modifier QTLs and candidate genes in recombinant inbred lines derived from K0326Y quality protein maize. Theor Appl Genet 22:783–794

Hospital F, Chevalet C, Mulsant P et al (1992) Using markers in gene introgression programs. Genetics 132:1199–1210

Hossain F, Prasanna BM, Sharma RK, Kumar P, Singh BB et al (2007) Evaluation of quality protein maize (QPM) genotypes for resistance to stored grain weevil, *Sitophilus oryzae*. Int J Trop Insect Sci 27:114–121

Hossain F, Prasanna BM, Kumar R, Singh BB et al (2008a) Genetic analysis of kernel modification in quality protein maize (QPM) genotypes. Indian J Genet 68:1–9

Hossain F, Prasanna BM, Kumar R, Singh BB et al (2008b) The genotype × pollination mode interaction affects kernel modification in quality protein maize (QPM) genotypes. Indian J Genet 68:132–138

Hossain F, Muthusamy V, Bhat JS, Jha SK, Zunjare R, Das A, Sarika K, Kumar R et al (2016) Maize: utilization of genetic resources in maize improvement. In: Singh M, Kumar S (eds) Broadening the genetic base of grain Cereals. Springer, New York, NY. https://doi.org/10.1007/978-81-322-3613-9_4

Hossain F, Muthusamy V, Pandey N, Vishwakarma AK, Baveja A, Zunjare RU et al (2018) Marker-assisted introgression of opaque2 allele for rapid conversion of elite hybrids into quality protein maize. J Genet. https://doi.org/10.1007/s12041-018-0914-z

Huang S, Adams WR, Zhou Q, Malloy KP, Voyles DA, Anthony J et al (2004) Improving nutritional quality of maize proteins by expressing sense and antisense zein genes. J Agric Food Chem 52:1958–1964

Hunter BG, Beatty MK, Singletary GW, Hamaker BR, Dilkes BP, Larkins BA et al (2002) Maize opaque endosperm mutations create extensive changes in patterns of gene expression. Plant Cell 14:2591–2612

IFPRI (2016) Global food policy report. International Food Policy Research Institute, Washington, DC

Jensen GL, Compher C, Sullivan DH et al (2013) Recognizing malnutrition in adults: definitions and characteristics, screening, assessment, and team approach. J Parenter Enter Nutr 37:802–807

Jha UC, Bhat JS, Patil BS, Hossain F, Barh D et al (2015) Functional genomics: applications in plant science. In: PlantOmics: the omics of plant science. Springer, New Delhi, pp 65–111. https://doi.org/10.1007/978-81-322-2172-2_4

Jia M, Wu H, Clay KL et al (2013) Identification and characterization of lysine-rich proteins and starch biosynthesis genes in the *opaque2* mutant by transcriptional and proteomic analysis. BMC Plant Biol 13:60

Jompuk C, Cheuchart P, Jompuk P et al (2011) Improved tryptophan content in maize with *opaque-2* gene using marker assisted selection (MAS) in backcross and selfing generations. Kasetsart J (Nat Sci) 45:666–674

Kassahun B, Prasanna BM (2003) Simple sequence repeat polymorphism in quality protein maize (QPM) lines. Euphytica 129:337–344

Kim CS, Gibbon BC, Gillikin JW, Larkins BA, Boston RS, Jung R et al (2006) The maize mucronate mutation is a deletion in the 16-kDa gamma zein gene that induces the unfolded protein response. Plant J 48:440–451

Kostadinovic M, Ignjatovic-Micic D, Vancetovic J et al (2016) Development of high tryptophan maize near isogenic lines adapted to temperate regions through marker assisted selection – impediments and benefits. PLoS One 11(12):e0167635

Krivanek A, Groote H, Gunaratna N, Diallo A, Freisen D et al (2006) Breeding and disseminating quality protein maize for Africa. Afr J Biotechnol 6:312–324

Latham MC (1997) Body composition, the functions of food, metabolism and energy. In: Human nutrition in the developing world. Food and nutrition series – no. 29. Food and Agriculture Organization of the United Nations, Rome

Lea PJ, Azevedo RA (2003) Primary products: plant amino acids. In: Thomas B, Murphy SJ, Murray BG (eds) Encyclopaedia of applied plant sciences, vol 3. Elsevier, Amsterdam, pp 871–883

Lebaka NG, Coors JG, Shaver RD, Bertics S, Gutierrez-Rojas A, Menz A, Betran J et al (2013) Quantitative trait loci for ruminal degradability in *opaque endosperm2* (*o2*) maize. Crop Sci 53:378–384

Lending CR, Larkins BA (1989) Changes in the zein composition of protein bodies during maize endosperm development. Plant Cell 1:1011–1023

Lending CR, Larkins BA (1992) Effect of the *floury2* locus on protein body formation during maize endosperm development. Protoplasma 171:123–133

Liu H, Shi J, Sun C, Gong H, Fan X, Qiu F, Huang X, Feng Q, Zheng X, Yuan X, Li C, Zhanga Z, Deng Y, Wang J, Pan G, Han B, Lai J, Wu Y et al (2016) Gene duplication confers enhanced expression of 27-kDa γ-zein for endosperm modification in quality protein maize. PNAS 113:4964–4969

Locatelli S, Piatti P, Motto M, Rossi V et al (2009) Chromatin and DNA modifications in the Opaque2-mediated regulation of gene transcription during maize endosperm development. Plant Cell 21:1410–1427

Lopes MA, Larkins BA (1995) Genetic analysis of opaque2 modifier gene activity in maize endosperm. Theor Appl Genet 91:274–281

Lopez-Valenzuela JA, Gibbon BC, Holding DR, Larkins BA et al (2004) Cytoskeletal proteins are coordinately increased in maize genotypes with high levels of eEF1A. Plant Physiol 135:1784–1797

Magulama EE, Sales EK (2009) Marker-assisted introgression of *opaque2* gene into elite maize inbred lines. USM R&D 17:131–135

Mallikarjuna MG, Thirunavukkarasu N, Hossain F, Bhat JS, Jha SK, Rathore A, Agrawal PK, Pattanayak A, Reddy SS, Gularia SK, Singh AM, Manjaiah KM, Gupta HS et al (2015) Stability performance of inductively coupled plasma mass spectrometry-phenotyped kernel minerals concentration and grain yield in maize in different agro-climatic zones. PLoS One. https://doi.org/10.1371/journal.pone.0139067

Manna R, Okello DK, Imanywoha J et al (2006) Enhancing Introgression of the *opaque-2* trait into maize lines using simple sequence repeats. Afr Crop Sci J 13:215–226

Medici LO, Azevedo RA, Smith RJ, Lea PJ et al (2004) The influence of nitrogen supply on antioxidant enzymes in plant roots. Funct Plant Biol 31:1–9

Mertz ET, Bates LS, Nelson OE et al (1964) Mutant gene that changes protein composition and increases lysine content of maize endosperm. Science 145:279–280

Moehn S, Bertolo RF, Pencharz PB, Ball RO et al (2004) Indicator amino acid oxidation responds rapidly to changes in lysine or protein intake in growing and adult pigs. J Nutr 134:836–841

Morley EJ (2016) Protein-energy undernutrition (PEU). https://www.merckmanuals.com/professional/nutritionaldisorders/undernutrition/proteinenergy-undernutrition-peu. Accessed 10 Oct 2017

Morton KJ, Jia S, Zhang C, Holding DR et al (2016) Proteomic profiling of maize opaque endosperm mutants reveals selective accumulation of lysine-enriched proteins. J Exp Bot 67:erv532. https://doi.org/10.1093/jxb/erv532

Motto M, Maddolini M, Panziani G, Brembilla M, Marrota R, Di Fonzo N et al (1988) Molecular cloning of the *o2-m5* allele *of Zea mays* using transposon tagging. Mol Gen Genet 121:488–494

Mpofu SJ, Msagati TAM, Krause RWM (2014) Cytotoxicity, phytochemical analysis and antioxidant activity of crude extracts from rhizomes of *Elephantorrhiza elephantina* and *Pentanisia prunelloides*. Afr J Tradit Complement Altern Med 11:34–52

Muller O, Krawinkel M (2005) Malnutrition and health in developing countries. CMAJ 173:279–286

Muthusamy V, Hossain F, Thirunavukkarasu N, Choudhary M, Saha S, Bhat JS et al (2014) Development of β-carotene rich maize hybrids through marker-assisted introgression of β-carotene hydroxylase allele. PLoS One 9:1–22

Neeraja CN, Ravindra BV, Ram S, Hossain F, Hariprasanna K, Rajpurohit BS, Prabhakar LT, Prasad KS, Sandhu JS, Datta SK et al (2017) Biofortification in cereals – progress and Prospects. Curr Sci 113:1050–1057

Nelson OE, Mertz ET, Bates LS et al (1965) Second mutant gene affecting the amino acid pattern of maize endosperm proteins. Science 150:1469–1470

Nuss ET, Tanumihardjo SA (2011) Quality protein maize for africa: closing the protein inadequacy gap in vulnerable populations. Adv Nutr 2:217–222

Nyakurwa CS, Gasura E, Mabasa S (2017) Potential for quality protein maize for reducing protein energy undernutrition in maize dependent sub-saharan African countries: a review. African Crop Sci J 25:521–537

Onimisi PA, Dafwang II, Omage JJ, Onyibe JE et al (2008) Apparent digestibility of feed nutrients, total tract and ileal amino acids of broiler chicken fed quality protein maize (obatampa) and normal maize. Int J Poult Sci 7:959–963

Osei SA, Okai DB, Ahenkora K, Dzah BD, Haag W, Twumasi-Afriyie S, Tua AK et al (1994) Quality protein maize as main source ofenergy and amino acids in the diets of starter pigs. Proc Ghanan Anim Sci Symp 22:31–36

Pandey N, Hossain F, Kumar K, Vishwakarma AK, Nepolean T, Vignesh M, Manjaiah KM, Agrawal PK, Guleria SK, Reddy SS, Gupta HS et al (2015) Microsatellite marker-based genetic diversity among quality protein maize (QPM) inbred lines differing for kernel iron and zinc. Mol Plant Breed 6:1–10

Pandey N, Hossain F, Kumar K, Vishwakarma AK, Muthusamy V, Saha S, Agrawal PK, Guleria SK, Reddy SS, Thirunavukkarasu N Gupta HS et al (2016) Molecular characterization of endosperm- and amino acids – modifications among quality protein maize inbreds. Plant Breed 135:47–54

Pandey N, Hossain F, Muthusamy V, Vishwakarma AK, Zunjare RU et al (2018) Haplotypes of recessive *opaque2* allele in exotic- and indigenous – quality protein maize inbreds. Indian J Agric Sci 88:253–259

Pfeiffer WH, McClafferty B (2007) HarvestPlus: breeding crops for better nutrition. Crop Sci 47:S88–S105

Powell WW, Machery GC, Provan J et al (1996) Polymorphism revealed by simple sequence repeats. Trends Genet 1:76–83

Prasanna BM, Vasal SK, Kassahun B, Singh NN et al (2001) Quality protein maize. Curr Sci 81:1308–1319

Qi G, Diao Q, Tu Y, Wu S, Zhang S et al (2004) Nutritional evaluation and utilization of quality protein maize (QPM) in animal feed. In: Protein sources for the animal feed industry. Expert consultation and workshop, 2002. FAO, Bangkok

Qiao Z, Qi W, Wang Q, Feng Y, Yang Q, Zhang N et al (2016) ZmMADS47 regulates zein gene transcription through interaction with *Opaque2*. PLoS Genet 12:e1005991. https://doi.org/10.1371/journal.pgen.1005991

Ribaut JM, Morris M, Dreher K, Khairallah M et al (2003) Money matters (II): costs of maize inbred line conversion schemes at CIMMYT using conventional and marker-assisted selection. Mol Breed 11:235–247

Riedel WJ, Sobczak S, Schmitt JA et al (2003) Tryptophan modulation and cognition. Adv Exp Med Biol 527:207–213

Sarika K, Hossain F, Muthusamy V, Baveja A, Zunjare R, Goswami R et al (2017) Exploration of novel *opaque16* mutation as a source for high -lysine and -tryptophan in maize 534 endosperm. Indian J Genet 77:59–64

Sarika K, Hossain F, Muthusamy V, Zunjare RU, Baveja A, Goswami R et al (2018) *Opaque16*, a high lysine and tryptophan mutant, does not influence the key physico-biochemical characteristics in maize kernel. PLoS One 13:e0190945. https://doi.org/10.1371/journal.pone.0190945

Schmidt RJ, Burr FA, Burr B et al (1987) Transposon tagging and molecular analysis of the maize regulatory locus opaque-2. Science 238. https://doi.org/10.1126/science.2823388

Schmidt RJ, Burr FA, Aukerman MJ, Burr B et al (1990) Maize regulatory gene opaque-2 encodes a protein with a leucine zipper motif that binds to zein DNA. Proc Natl Acad Sci U S A 87:46–50

Segal G, Song R, Messing J et al (2003) A new opaque variant of maize by a single dominant RNA-interference-inducing transgene. Genetics 165:387–397

Serna-Saldívar SO, Gomez MH, Islas-Rubio AR, Bockholt AJ, Rooney LW et al (1992) The alkaline processing properties of quality protein maize. In: Mertz E (ed) Quality protein maize. American Association of Cereal Chemists, Eagan, MN, pp 273–293

Shewry PR, Thatam AS (1990) The prolamin storage proteins of cereal seeds: structure and evolution. Biochem J 267:1–12

Shiferaw B, Prasanna BM, Hellin J, Banziger M et al (2011) Crops that feed the world 6. Past successes and future challenges to the role played by maize in global food security. Food Security 3:307–327

Singh NN, Venkatesh S (2006) Development of quality protein maize inbred lines. In: Kaloo G, Rai M, Singh M, Kumar S (eds) Heterosis in crop plants. Research Book Center, New Delhi, pp 102–113

Soave C, Salamini F (1984) Organization and regulation of zein genes in maize endosperm. Philos Trans R Soc Lond Ser B Biol Sci 304:341–343

Teklewold A, Wegary D, Tadesse A, Tadese B, Banta K, Friesen D et al (2015) Quality protein maize: a guide to the technology and its promotion in ethiopia. CIMMYT, El Batán

Tome D, Bos C (2007) Lysine requirement through the human life cycle. J Nutr 137:1642–1645

Tsai CY, Huber DM, Warren HL et al (1978) Relationship of kernel sink for N to maize productivity. Crop Sci 18:399–404

Varshney RK, Ribaut JM, Buckler ES, Tuberosa R, Rafalski JA, Langridge P et al (2012) Can genomics boost productivity of orphan crops? Nat Biotechnol 30:1172–1176

Vasal SK (2000) The quality protein maize story. Food Nutr Bull 21:445–450

Vasal SK (2001) High quality protein corn. In: Hallauer A (ed) Speciality corn, 2nd edn. CRC Press, Boca Raton, FL, pp 85–129

Vasal SK, Villegas E, Bajarnason M, Gelaw B, Geirtz P et al (1980) Genetic modifiers and breeding strategies in developing hard endosperm *opaque-2* materials. In: Pollmer WG, Philips RH (eds) Improvement of quality traits for silage use. Martinus Nijhoff Publ, The Hague, pp 37–71

Vasal SK, Villegas E, Tang CY et al (1984) Recent advances in the development of quality protein maize germplasm at CIYMMT. In: Cereal grain improvement. IAEA, Vienna, pp 167–189

Villegas E, Vasal SK, Bjarnason M et al (1992) Quality protein maize – what is it and how was it developed? In: Mertz ET (ed) Quality protein maize. American Association of Cereal Chemists, Eagan, MN, pp 27–48

Visscher PM, Haley CS, Thompson R et al (1996) Marker assisted introgression in backcross breeding programs. Genetics 144:1923–1932

Vitale AJ, Denecke J (1999) The endoplasmic reticulum gateway of the secretory pathway. Plant Cell 11:615–628

Vivek BS, Krivanek AF, Palacios-Rojas N, Twumasi-Afiriye S, Diallo AO et al (2008) Breeding quality protein maize (QPM) cultivars: protocols for developing QPM cultivars. CIMMYT, El Batán

Wall JS, Bietz JA (1987) Differences in corn endosperm protein in developing seeds of normal and *opaque-2* corn. Cereal Chem 64:275–280

Wallace JC, Lopes MA, Paiva E, Larkins BA et al (1990) New methods for extraction and quantitation of zeins reveal a high content of γ-Zein in modified *opaque2* maize. Plant Physiol 92:191–196

White PJ, Broadley MR (2009) Biofortification of crops with seven mineral elements often lacking in human diets – iron, zinc, copper, calcium, magnesium, selenium and iodine. New Phytol 182:49–84

WHO/FAO/UN (2007) Protein and amino acid requirements in human nutrition. 1764 report of a joint WHO/FAO/UN expert consultation, WHO technical report series, no 935. WHO, Geneva

Wu Y, Messing J (2010) RNA interference-mediated change in protein body morphology and seed opacity through loss of different zein proteins. Plant Physiol 153:337–347

Wu Y, Holding DR, Messing J et al (2010) γ-Zeins are essential for endosperm modification in quality protein maize. Proc Natl Acad Sci U S A 107:12810–12815

Xu JH, Messing J (2009) Amplification of prolamin storage protein genes in different subfamilies of the Poaceae. Theor Appl Genet 119:1397–1412

Yadav OP, Hossain F, Karjagi CG, Kumar B, Zaidi PH, Jat SL, Chawla JS, Kaul J, Hooda KS, Kumar O, Yadava P, Dhillon BS et al (2015) Genetic improvement of maize in India: retrospect and prospects. Agric Res 4:325–338

Yadava DK, Choudhury PK, Hossain F, Kumar D et al (2017) Biofortified varieties: sustainable way to alleviate malnutrition. Indian Council of Agricultural Research, New Delhi, pp 7–10

Yang W, Zheng Y, Zheng W, Feng R et al (2005) Molecular genetic mapping of a high-lysine mutant gene (opaque-16) and the double recessive effect with *opaque2* in maize. Mol Breed 15:257–269

Yang L, Wang W, Yang W et al (2013) Marker-assisted selection for pyramiding the *waxy* and *opaque16* genes in maize using cross and backcross schemes. Mol Breed 31:767–775

Zhang WL, Yang WP, Chen ZW, Wang MC, Yang LQ, Cai YL et al (2010) Molecular marker-assisted selection for o2 introgression lines with *o16* gene in corn. Acta Agron Sin 36:1302–1309

Zhang W, Yang W, Wang M, Wang W, Zeng G, Chen Z, Cai Y et al (2013) Increasing lysine content of waxy maize through introgression of *opaque2* and *opaque16* genes using molecular assisted and biochemical development. PLoS One 8:1–10

Zhang Z, Yang J, Wu Y et al (2015) Transcriptional regulation of zein gene expression in maize through the additive and synergistic action of *opaque2*, prolamine-box binding factor, and O2 heterodimerizing proteins. Plant Cell 27:1162–1172

Zimmerman AO, Millear AI, Stubbs RW et al (2018) Mapping child growth failure in Africa between 2000 and 2015. Nature 555:41–47

Zunjare RU, Hossain F, Muthusamy V, Baveja A, Chauhan HS, Thirunavukkarasu N, Saha S, Gupta HS et al (2017) Influence of rare alleles of β-carotene hydroxylase (*crtRB1*) and lycopene epsilon cyclase (*lcyE*) genes on accumulation of provitamin A carotenoids in maize kernels. Plant Breed. https://doi.org/10.1111/pbr.12548

Zunjare RU, Hossain F, Muthusamy V, Baveja A et al (2018) Development of biofortified maize hybrids through marker-assisted stacking of *β-carotene hydroxylase*, *lycopene-ε-cyclase* and *opaque2* genes. Front Plant Sci 9:1–12. https://doi.org/10.3389/fpls.2018.00178

Chapter 12
Genetic Improvement for End-Use Quality in Wheat

Hanif Khan

12.1 Introduction

The human population is growing exponentially, with current projections predicting a population of more than nine billion by 2050. A combination of improved agronomic practices and improved crop varieties will be imperative to ensure food and nutritional security in the coming decades. While overall production must increase, there is also growing demand to produce higher-quality, more nutritious food. Wheat is an important food grain consumed by humankind around the world. The more important modern wheat species are bread wheat (*Triticum aestivum* L.) and durum wheat (*T. turgidum* L. var. *durum*), which are different from one another in genomic make-up, in grain composition and in food end-use quality attributes. Wheat is an allopolyploid species that originated from a cross of the tetraploid species *Triticum turgidum* and the diploid species *Aegilops tauschii* (Coss) Schmalh. Wild tetraploid emmer wheat evolved from a hybridization of wild *Triticum urartu* tumanian ex Gandivan and an undiscovered species of the *Aegilops speltoides* Tausch lineage. Except for the very warm tropics, wheat adapts to all diverse climatic conditions prevailing in agricultural lands and, therefore, it is harvested in the world all year around. Wheat is used in a nearly countless variety of foods and is the leading cereal for human consumption. Bread wheat is a staple of many diets, with the milled flour used for a variety of products including leavened and unleavened breads, noodles, cookies, cakes, and pastries. Some major quality traits of wheat are:

- Kernel texture (hardness).
- Gluten strength.
- Polyphenol oxidase (PPO).
- Alpha-amylase (sprout/falling number).

H. Khan (✉)
ICAR-Indian Institute of Wheat and Barley Research, Karnal, Haryana, India

A. M. I. Qureshi et al. (eds.), *Quality Breeding in Field Crops*,
https://doi.org/10.1007/978-3-030-04609-5_12

Milling and end-use quality of wheat results from the composition of the kernel, and (especially) the endosperm. Kernel texture or hardness is a key determinant of wheat market class and end-use because hard wheats require more energy to mill, have higher levels of damaged starch, and larger particle size indices than soft wheats. A unique feature of wheat flour is that when it is mixed with water it forms dough, a material with complex rheological properties. The visco-elastic properties and gas holding capacity of wheat flour dough are the basis for the production of a wide range of products (bread, noodles, pasta, etc.), providing some of the most important classes of food around the world. Its wide adaptation to diverse environmental conditions, along with its unique characteristic of possessing a visco-elastic storage protein complex called gluten, are the main elements making wheat the most important food crop in the world. Wheat is considered an important source of energy. It provides between 1200 and 1450 kJ per 100 g of grain. In addition, it is an important supplier of different nutrients necessary for a healthy diet. The complex carbohydrates, mainly starch, are the major components of wheat (61–65%); but it is also an excellent source of dietary fiber (9–12%).

12.2 Wheat Types and Class

Global wheat production for 2016–2017 is projected to about 750 million tones (USDA 2018). Approximately 90–95% of the wheat produced in the world is common bread wheat (*T. aestivum*), which is better known as hard wheat or soft wheat, depending on grain hardness. Wheat is utilized mainly as flour (whole grain or refined) for the production of a large variety of flat and leavened breads, and for the manufacture of a large variety of other baking products. The rest is mostly durum wheat (*T. durum*), which is used to produce semolina (coarse flour), the main raw material of pasta making.

Wheat for the purpose of trading is classified into distinct classes of grain hardness (soft, medium-hard, and hard) and color (red, white, and amber). It may be further categorized into subclasses based on growing habit (winter or spring). Based on grain color and hardness bread wheat can be classified into:

- Soft White Winter.
- Soft White Spring.
- Club Wheat—winter and spring.
- Hard Red Winter.
- Hard Red Spring.
- Hard White—winter and spring.

Each wheat subclass may also be grouped into grades, which are usually done to adjust the basic price of a wheat stock by applying premiums or penalties. Wheat grades are indicators of the purity of a wheat class or subclass, the effects of external factors on grain soundness (rain, heat, frost, insect, and mold damage) and the cleanliness (chaff and foreign material) of the wheat lot. Grain protein content and alpha-amylase activity (enzymatic activity associated with the germination of the grain) are frequently considered as grading factors in wheat trading.

End-use quality is greatly dependent on flour protein content and composition. Variation in composition results from the interplay between the 'genotype' and the environment. These characteristics help to define the optimum end-product usage for specific wheat varieties and growing environments. The proteins in flour compete with other flour constituents for water making the overall dough more viscous. End-use quality in soft bread-wheat can be assessed by a wide array of measurements, generally categorized into grain, milling, and baking characteristics. Many different parameters within these three categories can be used to evaluate end-use quality (Wrigley and Morris 1996). Wheat grain contains albumins, globulins, prolamins and glutelins, the albumin and globulins are concentrated in the aleurone layer which is milled off in white flour. The prolamins are known as gliadins and the glutelins are known as glutenin in wheat. They make up the 70–90% of the total protein fraction of the endosperm of wheat (Koehler and Wieser 2012). Determining the significance of each parameter on the overall end-use quality as defined by the consumer product can prove very difficult given the genetic diversity of wheat and the extremely variable environments in which wheat is produced across the globe. It is critical to identify which quality measures provide the most accurate and stable assessment of end-use quality to optimize commercial utilization and consumer acceptance (Kiszonas et al. 2015). As a consequence, wheat processing industries now require various distinct wheat supplies possessing specific grain quality attributes. Thus, it is common to find that the value of a wheat crop in the market is generally determined by grain attributes associated with its processing quality.

Wheat grain is assessed for premilling characteristics, which impact marketing. These tests include kernel weight, weight per volume, color, hardness, vitreousness of the kernel, and total protein content. Many of these characteristics are strongly correlated with grain yield with varying levels of heritability (Battenfield et al. 2016). Soft grain that mills well has high flour extraction, high break flour percentage, and low flour ash.

12.3 Breeding Wheat for End-Use Quality

The definition of wheat quality has different meanings to different people, i.e., the 'quality' is in the eye of the consumer. There are quite large differences in grain composition and processing quality among wheat cultivars within a species. Hence, one cultivar may be suitable to prepare one food type but unsuitable to prepare a different one. Wheat cultivars must have suitable end-use quality for release and consumer acceptability. However, breeding for quality traits is often considered a secondary target relative to yield largely because of the quantity of seed needed and cost involved in breeding for quality traits. Quality differences among wheat cultivars have gained even more importance in grain trading due to important global economic and social trends. Recently, many countries have adopted, or are in the process of adopting, free-market economies, impacting positively on the income of the population, particularly of that concentrated in urban areas. However, without testing and selection, many undesirable materials are advanced, expending additional resources. Breeders must continue to develop and release new wheat cultivars, as older commercial cultivars tend to lose disease and pest resistance over three to five years. The major objectives of wheat breeding programs are to improve agronomic, disease and pest resistance, and milling and end-use quality traits.

12.3.1 Quality Indices and Tests

Wheat breeding programs with emphasis on end-use quality annually evaluates several thousand breeding samples for grain, milling, flour, dough, and the final product characteristics. The wheat quality laboratories generally conduct test on bread quality, cookies and cake quality for wheat breeding lines, pasta quality for Durum wheat lines. Diverse tests on various wheat class breeding programs in wheat lines for grain, milling, flour, dough, and bread characteristics are also conducted.

12.3.1.1 Grain Characteristics

Following are criteria and tests to determine grain characteristics: (a) Test weight, (b) Kernel hardness, (c) Kernel weight and size, (d) Wheat moisture, and (e) Wheat ash content.

12.3.1.2 Milling Characteristics

The criteria to determine milling characteristics are: (a) Flour yield and (b) Milling score.

12.3.1.3 Flour Characteristics

Flour quality is determined by following analyses: (a) Flour color, (b) Proximate analyses, (c) Sprout damage, (d) PPO activity, and (e) RVA pasting properties.

Acceptable end-use quality, rather than enhanced end-use quality is the goal in most of the wheat breeding programs because currently there are inadequate economic incentives to develop wheat cultivars only with enhanced end-use quality. Also, consistency of end-use quality is more desired by the baking industry than is altered quality. Analysis of chemical constituents of wheat flour which determine the functional properties of dough is required in the advanced lines. In order to characterize the role of chemical constituents and target processing quality to be modified, a precise description of the phenotype (processing quality) of a sample of grain has to be established. The technology to deal with this concern can be met with small-scale dough-testing equipment and various techniques/protocols are available to determine the milling, mixing, stretching, and end-product properties of small quantities of grain and/or flour.

12.3.2 Genetic Variability and Heritability for Quality Traits in Wheat

Grain end-use quality traits, such as milling yield, dough rheology, baking, and noodle traits are among the most important in wheat breeding. However, these traits are difficult to breed for as their assays require grain and flour quantities that can only be obtained in advanced generations in the breeding cycle, and are expensive. Consequently, these traits are an ideal target for genomic selection, where traits are predicted for potential wheat lines early in the breeding cycle using genome-wide marker effects for such traits (e.g. Guzman et al. 2016).

Traits with low heritability require much larger reference populations to achieve the same accuracy of prediction as traits with moderate or high heritability. Wide range of heritability in various traits of grain end-use quality traits in wheat have been reported. O'Brien and Ronalds (1987) reported broad-sense heritabilities in Australian wheat varieties from 0.15 for farinograph measured dough development time, to 0.88 for grain hardness, with an average heritability of 0.54 across 15 traits measured. Baker et al. (1971) assessed the heritability of 25 grain end-use quality traits in Canadian hard red spring wheat cultivars. Heritabilities ranged from 0.26 for diastatic activity to 0.89 for pigment content, with average heritabilities of milling, flour extensograph, farino-graph, and baking traits of 0.66, 0.71, 0.69, and 0.72, respectively. Pearson et al. (1981) reported that, in early segregating generations (F2 and F3) of a genetically diverse wheat population, pearling resistance, Pelshenke time and 1000 kernel weight had high heritabilities (0.72–0.80), flour yield had an intermediate heritability (0.57), while grain protein content had a low heritability (0.19).

All the above-mentioned estimates are broad-sense heritabilities, while the key parameter for influencing the accuracy of genomic selection for cumulative additive genetic gain is the narrow-sense (additive component) heritability. Narrow-sense heritability estimates were on average half that of the broad-sense heritabilities, and ranged from 0 for falling number to 0.71 for 1000 kernel mass (Barnard et al. 2002). The average narrow-sense heritability was 0.3. Narrow sense heritability can be estimated by diallel cross of parents or any other mating scheme of crossing using divergent parents for the traits and assessing grain end-use quality characteristics in their F2 progeny for a range of traits, and compare these to broad-sense heritability estimates. Heffner et al. (2011) compared the phenotypic and genomic prediction accuracy of genetic value for nine different grain quality traits within two bi-parental soft winter wheat (*Triticum aestivum* L.) populations, with only 96 lines in each population. Using a BLUP method of genomic prediction they found accuracies in cross-validation ranging from 0.36 for softness to 0.68 for pre-harvest sprouting, with an average of 0.52 across traits.

While the heritability of grain end-use quality traits is not likely to be the deterrent factor in genomic prediction of end-use quality, small reference population size may be a limiting factor. This size of reference set required depends on the genetic diversity, or effective population size of the selection sub-population.

12.3.3 Breeding Program and Selection for End-Use Quality

In general, the breeding program uses single and three-way crosses. At least one parent of the single cross should be a developed line from the zone and usually two parents of the three-way cross should be released cultivars and selected for good end-use quality. A single cross with an unadapted or poor quality line generally does not have progeny with sufficient adaptation or quality genes at a high enough frequency for successful selection.

Year 1: Make several crosses involving parents tested for end-use quality to produce F1 seed. In general, the breeding program uses single and three-way crosses.

Year 2: Grow the F1 seed in the greenhouse to avoid losses due to winterkilling if the seed was grown in the field. Harvest F2 seed from true F1s.

Year 3: Plant F2 seed in bulk populations in field. Select for monogenic and/or ologogenic traits like disease resistance, plant height, days to maturity, etc.

Year 4: Plant F3 seed in bulk populations in field. Infect plants with rust diseases. F3 can be grown at off-season nursery location or hot spot for important disease. Select disease-resistant single heads of about 500–2000 from each cross.

Year 5: Plant F4 head rows at main breeding location in winter season. On the basis of plant type and disease resistance, select and harvest 1000–1500 heads from best rows. Evaluate harvested seed for quality traits and select 500–1000 lines for advancement.

Year 6: Plant F5 plots in the field and simultaneously screen the lines in the greenhouse for rust disease. On the basis of plant type, yield, and disease resistance, harvest 200–300 plots. Evaluate harvested seed using microquality analyses (flour protein and Mixograph) and select 100–150 lines for advancement that have acceptable end-use quality.

Year 7: Plant 100–150 F6 lines and five checks in unreplicated trials at several locations. Send seed to Rust Laboratory for rust resistance testing. On the basis of plant type, yield, disease resistance, and end-use quality select about 40 lines for advancement. Evaluate harvested seed using a full milling and baking.

Year 8: Plant 40 F7 lines and two checks in replicated and observation trials at several locations. Send seed to Rust Laboratory for rust testing. On the basis of plant type, yield, end-use quality, and disease resistance, select about 20 lines for advancement. Evaluate harvested seed using a full milling and baking procedure in the quality laboratory.

Year 9: Plant 50 F8–F12 lines in replicated and observation trials at multiple locations. The 50 lines include 10 check lines, 20 lines retained from the previous year's trials and the 20 newly advanced lines. On the basis of plant type, yield, end-use quality, and disease resistance, select 30–35 lines (including checks) for retention. Evaluate harvested seed using a full milling and baking procedure in wheat quality laboratory. Increase seed of approximately ten lines for advancement. Submit 5–10 lines to regional quality nurseries and collect multilocationl data. Submit 2–4 lines to Zonal/state cultivar multilocation testing.

Year 10–12: Repeat the procedure of year 9. Continue seed increase of advanced lines performing better than checks. If performance warrants release, release one line as a new cultivar.

A breeding program is a continuum; thus lines are constantly added and dropped from list of elite lines. Of the 20 lines advanced in year 8, only 8–12 be retained in year 9, 5–8 can be retained in year 10, 5 should be retained in year 11, and 1 or 2 in year 12. On average, over 5,000 lines should be looked at to find a cultivar. Over 1,000 yield plots should be harvested each year. It takes a minimum of 12 years to create a new wheat cultivar. A cultivar should be tested in over 100 location-years before we get enough information to release it. End-use quality assays begin in year 6. Due to a limited amount of seed (average plot yields ranges between 500 and 900 g/plot), the initial tests are microquality analyses.

As for other traits, the key parameters determining the accuracy of genomic predictions for grain end-use quality traits will be the size of the reference or experimental population where the marker effects are predicted, the extent of linkage disequilibrium between markers and the loci/QTLs influencing the trait, and the heritability of the trait.

Assembling large reference sets for grain end-use quality traits is not easy, as their assays require large amounts of flour and being expensive. Historically, such data have been collected only on a limited number of lines. Consequently, predictors of grain end-use quality, such as nuclear magnetic resonance (NMR) and near infrared (NIR) may provide a solution. These techniques require much lesser quantities of flour and can have considerably lower cost. One prospective approach to building a reference set large enough for accurate genomic predictions of end-use quality would be to develop predictions for these traits based on NMR and/or NIR, evaluate a large number of lines with these assays, and combine with existing end-use quality data in a multi-trait analysis.

12.3.4 Breeding in Early Generation for Quality Traits

During the first stage of breeding, the complete progenies of a cross may be assessed for its processing value. The procedures to assess quality traits are briefly described below:

External quality: this property is relatively easy to determine visually and does not require any special equipment.

Protein content: exceptionally high or low values generally refer to abnormal growth or ripening conditions. A number of techniques may be used to determine protein content. The NIR method is very handy because it combines analyses of protein content and grain hardness.

Grain hardness: as a rule, bread-making wheat should have hard, vitreous grains; biscuit-making wheat, on the contrary, has grains with soft, opaque kernels. Several simple methods exist for the determination of grain hardness.

Protein strength: the breeder may choose from three tests to determine gluten quality: Zeleny sedimentation, SDS sedimentation, or a mixograph test. All three are quick and simple methods and suitable for large numbers of samples.

Additional assessments: For some particular reasons certain analyses may be done on a limited number of samples.

Alpha-amylase activity. If the α-amylase activity is too low, the miller can readily increase it through the addition of malt flour or fungal amylases, but if it is too high there is no way of reducing it. Thus, α-amylase level in wheat has become a key quality factor. The analysis of the falling number is only useful if one or both of the parents of a cross has high natural α-amylase activity.

SDS-polyacrylamide gel-electrophoresis. With this technique it is possible to determine the HMW glutenin subunits. A guide for wheat breeders who wish to develop varieties with improved bread-making quality is therefore to cross genotypes that have complementary good quality subunits and to select progeny with an increased quality score.

12.3.5 Breeding for Quality in Advanced Generations

In the final stage of quality breeding, the milling and baking characteristics of a limited number of promising lines have to be evaluated on an industrial scale, at the time multilocation coordinated trials and before releasing them for commercial cultivation. It is common practice for breeding companies in developed economies to cooperate with the milling and baking industry, which can give an expert opinion on the practical value of a new cultivar.

Each of these products demands wheat with a specific best-fit quality profile specifically considering the protein concentration, grain hardness, and gluten strength (Pena 2002). Processing and end-use quality for wheat is a combination of many defined parameters. Multiple phenotypic traits of the grain, flour, dough, and final products must be assessed to determine an overall quality and best end-use product. Typically, hard grain with high protein and strong and extensible gluten is acceptable for making leavened breads and industrial pan bread, hard or medium hard grain with intermediate levels of protein and good dough extensibility make good flat-breads/chapattis, and soft grain with low protein and weak and extensible gluten is used for cookies, cakes, and pastries.

Due to the high cost and seed quantity demands associated with wheat end-use quality, indirect selection is a desirable option for many breeding programs. In the case of soft white wheat end-use quality, initial selections in F2 and F3 generations using genomic selection, followed by indirect selection with small scale tests such as the flour solvent retention capacity (SRC) tests in F4 and F5 generations, should allow many lines with poor end-use quality attributes to be discarded from the breeding program (Jernigan et al. 2017). Hence, breeding lines advanced to yield trials should have the favorable alleles needed for better end-use quality, which can be confirmed through full milling and baking tests. Selection on a relatively small number of loci can have great impacts on the improvement of soft wheat end-use quality.

These days, work to improve wheat happens all over wheat growing parts of the world, at public universities, government research centers, and within private companies. In these programs, wheat breeders work with molecular geneticists, quality specialists, plant pathologists entomologists, and extension workers, who share scientists' findings with farmers.

The wide variety of food products prepared from wheat flour has resulted in ongoing demand from the wheat processing industry for wheat with particular quality attributes. In addition, dietary deficiencies of essential micronutrients such as zinc (Zn) and iron (Fe) are a major health worry in developing countries especially for pregnant women and children under age 5. An estimated 17.3% of the world's population is at risk for scarce zinc intake, a factor highly correlated with stunted growth in children (Wessells and Brown 2012). Genetic biofortification harnessing natural genetic variation present in wild relatives, synthetics, and landraces for micronutrient uptake from the soil and translocation into wheat grain is a sustainable resolution that can supplement micronutrient-deficient rural people with limited availability from formal markets or healthcare programs (Velu et al. 2014). Evaluation of landraces and secondary gene pools (i.e., tetraploid and diploid progenitors of hexaploid wheat) for higher micronutrient concentration identified *T. dicoccoides*, *A. tauschii*, *T. monococcum*, and *T. boeticum* Boiss as the most promising sources for improving Zn and Fe content in grain (Cakmak et al. 2000).

Large-scale screening of wheat genetic resources at CIMMYT and various national-wheat improvement programs have identified einkorn wheat, wild emmer wheat, and landraces with high amounts of Fe and Zn in grain (Ortiz-Monasterio et al. 2007). The available genetic variation in wild emmer (*T. dicoccoides*), *T. spelta*, *T. dicoccum* species is being used to develop nutrient-enriched wheat cultivar. In India, recently WB-2 cultivar of wheat having significantly higher Zn and Fe content has been released for cultivation. The stocks (*T. turgidum* ssp. *diccocum*/*A. tauschii*) are also being used for genetic biofortification of Zn and Fe wheat breeding programs. At CIMMYT, evaluation of a representative subset of Mexican and Iranian landraces under Zn-enriched soil conditions in Cd. Obregon, Mexico, showed more than a twofold variation for Zn (40–96 mg/kg) and Fe (27–56 mg/kg). A major locus affecting Fe and Zn concentration, *Gpc-B1* (250 kb-locus), was mapped by Distelfeld et al. (2007) and found to encode a NAC transcription factor (NAM-B1) that accelerates senescence and increases nutrient remobilization from leaves to grain (Uauy et al. 2006).

Various research programs are currently emphasizing to screen for genetic variability of the bioactive compounds. High heritability for some of these compounds such as tocols, sterols and arabionoxylan fiber and available genetic diversity increases the possibility of utilizing the variation for improving nutritional quality in wheat. Grain bran is particularly rich in dietary fiber, vitamins (folic acid), and phytochemicals, which have been associated with a protective role for many chronic ailments including type 2 diabetes and cardiovascular diseases (de Munter et al. 2007).

Grain proteins are one of the important components that determine end-use quality. Studies have found higher grain protein content in landraces than in modern wheat (Dotlacil et al. 2010) which indicate landraces and wild relatives could be a potential source to improve protein content. Understanding the molecular basis for the visco-elasticity of wheat gluten proteins is an important prerequisite for manipulating their properties in order to improve the quality for traditional uses and to develop new properties for novel uses. While grain protein content is important, gluten quality is equally important. Gluten, an essential component of dough, is a complex protein network formed mainly by two kinds of proteins, monomeric gliadins and polymeric glutenins, which in turn are divided into high molecular weight glutenins (HMWGs) and low molecular weight glutenins (LMWGs). Payne and Lawrence (1983), using SDS-PAGE, described the HMWGs using a numbering system, with "subunit 1" having the slowest mobility and the highest molecular weight. Many processing qualities of wheat are determined, or modified, by seed storage proteins. Decades of efforts to relate gluten–protein composition to dough quality were eventually rewarded by relating the composition of the HMWGs. Subsequent studies have extended these predictive relationships also to include the LMW subunits. Additionally, it has become clear that the balance of gliadin and glutenin characteristics is fundamental to meaningful predictive models. This knowledge has since formed a basis for screening progeny in breeding trials, thus to discard those lines that would not suit dough-quality targets. In addition, our knowledge about the direct effects of the gluten polypeptides has benefited from their direct inclusion in the glutenin polymer, and the opportunities to manipulate dough properties by conventional breeding as well as by novel approaches to genetic engineering.

Seed storage protein sequence analysis has shown that a large amount of sequence variation exists among and between family members (Shewry et al. 1994). Variation in glutenin and gliadin functionality is controlled by several well characterized genes. The high molecular weight glutenins (*Glu-A1*, *Glu-B1*, and *Glu-D1*) are located on the long arm of group 1 homoeologous chromosomes (Payne 1987). The genes for ω- and γ-gliadins (*Gli-A1*, *Gli-B1*, and *Gli-D1*) and the low molecular weight glutenins are on the short arm of group 1 chromosomes (Payne et al. 1984; Payne 1987). The genes for a- and b-gliadins (*Gli-A2*, *Gli-B2* and *Gli-D2*) are located on the short arms of group 6 chromosomes (Payne et al. 1984). A listing of HMWGs and LMWGs, giving a reference variety for each allele, has been compiled by McIntosh et al. (2013). The Wheat Gene Catalog currently describes 56 alleles for *Glu-A1*, 83 for *Glu-B1*, 75 for *Glu-D1*, 54 for *Glu-A3*, 28 for *Glu-B3*, and 13 for *Glu-D3*. The total flour protein content and glutenin to gliadin ratios independently influence dough strength and extensibility (Uthayakumaran et al. 1999). The major alleles that differentiate 'strong' from 'weak' gluten in wheat have been selected for by breeders of both hard and soft wheat but variation in protein functionality exists within these allelic classes. Additional research is needed to determine the genetic architecture of protein functionality, especially as proteins interact other endosperm components. Recently, a novel allele HMW glutenin allele was identified from *A. longissimma* Schweir and Muschl through the use of a Chinese Spring substitution line CS-1S (1B) that could potentially improve dough- and bread-making quality (Wang et al. 2013). Large-scale sequence analysis of the cDNA libraries generated from wheat mid-endosperm development (Clarke et al. 2000) was consistent with these observations in showing extensive variation within and between the gliadin and LMW glutenin families. The knowledge coming from the elucidation of gliadin and glutenin chemistry and genetics has provided a range of valuable practical uses. The first of these benefits, some decades ago, was for the identification of wheat varieties on the basis of their gliadin composition. It was found that the gliadin proteins could be readily extracted and fractionated by gel electrophoresis, providing patterns that were unique for specific varieties, yet unaffected by differences in growth conditions.

Perhaps the utmost value of the *Glu-1* score to wheat breeders has been in the selection of dedicated crosses (Cornish et al. 2006). Often the only way to improve yield and disease resistance in bread-quality breeding programs is to make crosses with elite braid- or biscuit-quality varieties or with ill-adapted varieties from other countries/regions. The problem then is to restore all the components of bread quality in the selection process. SDS-PAGE can be used at early generations to great advantage by selecting for high *Glu-1* scores and homozygous glutenin-subunit alleles among the progeny, leaving the breeder more freedom to select for other, targeted traits at early and succeeding generations. Analysis of the progeny of dedicated crosses, in which the allelic composition of the parents is known, is a more targeted way to determine the effects of specific alleles on quality. However, by having a narrow range of genotypes, the information is limited to the effects of the specific alleles in a particular genetic background. It is necessary to confirm the allele quality rankings in other backgrounds before generalized conclusions can be reached.

Plant breeders generally use a successful wheat variety as the basis for future crosses due to the inbuilt disease resistance, yield potential, and quality of this variety. In selecting future lines based on this recurrent parent for quality, the breeder is unintentionally selecting for the same "good quality" glutenin alleles found in the parent. Hence, when wheat varieties within a country are categorized and grouped according to their glutenin allelic composition, lines with similar pedigree are grouped together. This concept of "protein families" has proved useful to group varieties of similar glutenin alleles and pedigrees. Within a protein family, one generally finds wheat varieties with very similar rheological properties and hence common end uses. This approach has been used to understand the major wheat-quality groups in Australia (Cornish et al. 2006). Ram et al. (2015) evaluated 240 diverse set of wheat cultivars released in India during the last several decades for HMW and LMW glutenin alleles, for assessing their diversity and effect on sedimentation volume and mixograph parameters. Both SDS-PAGE and PCR-based markers were employed in identifying alleles encoded at *Glu-1* and *Glu-3* loci. Extensive allelic variation was observed at both the *Glu-1* and *Glu-3* loci. There was prevalence of *Glu-A1b*, *Glu-B1i*, *Glu-D1a*, *Glu-A3c*, *Glu-B3b*, *Glu-B3g*, and *Glu-D3b*. The alleles *Glu-A1b*, *Glu-B1i*, *Glu-D1d*, *Glu-A3b*, *Glu-B3g/h*, and *Glu-D3b* exhibited high SDS sedimentation volume. *Glu-B1i* and *Glu-D1d* showed highly significant positive effect ($p < 0.001$) on sedimentation volume and also had additive effects. However, overall there was decline in the frequency of *Glu-B1i* allele during last two decades in Indian wheat breeding and not a single 1B/1R translocation Indian cultivar possessed this allele. This information can be useful in designing breeding program for the improvement of Indian bread wheat quality.

For soft wheat, gluten must not contribute to poor texture. Kernel hardness is measured as break flour percentage or particle size. Most of the genetic variation for kernel hardness (between soft and hard wheats) is determined by allelic variation at the puroindoline genes, *Pina* and *Pinb*, located at the hardness locus (*Ha*) on the short arm of chromosome 5D (Morris 2002). Selection for soft or hard alleles at the *Ha* locus is relatively easy but considerable variation for kernel texture due to uncharacterized additional loci still exists within these classifications (Souza et al. 2002).

12.3.5.1 Puroindoline Gene Expression Cascade

→ puroindoline genes expressed → puroindoline proteins in endosperm → kernel texture: hard or soft → milling performance and flour yield → flour starch damage/flour granularity → dough water absorption → processing and baking performance → end-product quality.

While the genetic control of wheat processing quality such as dough rheology is well understood, limited information is available relating to the genetic control of baking parameters, particularly sponge and dough (S&D) baking. A study by Mann et al. (2009) highlighted the inconsistent genetic control of protein content across the test sites, with only two loci (3A and 7A) showing QTL at three of the five sites. Dough rheology QTLs were highly consistent across the five sites, with major effects associated with the *Glu-B1* and *Glu-D1* loci. The *Glu-D1* 5 + 10 allele had consistent effects on S&D properties across sites; however, there was no evidence for a positive effect of the high dough strength *Glu-B1-al* allele at *Glu-B1*. In the absence of robust predictive tests, high heritability values for S&D demonstrated that direct selection is the best option for achieving genetic gain in this product category.

12.4 Conclusion

Small-scale tests for evaluation of grain quality in the early generations have been used by wheat breeders for many years. Breeders use grain or flour protein (FP%), sedimentation (SDS), and high molecular weight (HMW) glutenin subunit data to truncate their breeding populations early in the breeding process. This allows breeders to economically utilize resources by testing genotypes in the later generations that have release potential. However, there should be optimal balance between selection for grain quality using these easy-to-measure characters and the need to maintain germplasm with high yield potential. Breeding populations should not be trimmed on the basis of FP% alone if grain yield is the main objective in the breeding program. The ratio SDS/FP% provided the best predictive estimates of both grain quality and grain yield. The HMW glutenin subunit composition should not be the sole basis upon which populations are curtailed for grain quality and yield as considerable numbers of lines with suboptimal HMW glutenin subunits are high yielding with superior grain quality. An integrated approach to studying the genetic basis of grain, rheological, and baking properties not only provides a powerful mechanism to identify QTL governing quality traits, but also provides a mechanism for re-evaluating assumptions concerning the relationships between predictive tests and end product quality.

References

Baker RJ, Tipples KH, Campbell AB (1971) Heritabilities of and correlations among quality traits in wheat. Can J Plant Sci 51:441–448

Barnard A, Labuschagne M, van Niekerk H (2002) Heritability estimates of bread wheat quality traits in the Western Cape province of South Africa. Euphytica 127:115

Battenfield SD, Guzmán C, Gaynor RC, Singh RP, Peña RJ, Dreisigacker S et al (2016) Genomic selection for processing and end-use quality traits in the CIMMYT spring bread wheat breeding program. Plant Genome 9. https://doi.org/10.3835/plantgenome2016.01.0005

Cakmak I, Ozkan H, Braun HJ, Welch RM, Romheld V (2000) Zinc and iron concentrations in seeds of wild, primitive and modern wheats. Food Nutr Bull 21:401–403. https://doi.org/10.1177/156482650002100411

Clarke BC, Hobbs M, Skylas D, Appels R (2000) Genes active in developing wheat endosperm. Funct Integr Genomics 1(1):44–55

Cornish GB, Békés F, Eagles HA, Payne PI (2006) Prediction of dough properties for bread wheats. In: Wrigley CW, Békés F, Bushuk W (eds) Gliadin and glutenin. The unique balance of wheat quality. AACCI Press, St Paul, MN, pp 243–280

de Munter JSL, Hu FB, Spiegelman D, Franz M, van Dam RM (2007) Whole-grain, bran, and germ intake and risk of type 2diabetes: a prospective cohort study and systematic review. PLoS Med 4:e261. https://doi.org/10.1371/journal.pmed.0040261

Distelfeld A, Cakmak I, Peleg Z, Ozturk L, Yazici AM, Budak H (2007) Multiple QTL-effects of wheat Gpc-B1 locus on grain protein and micronutrient concentrations. Physiol Plant 129:635–643. https://doi.org/10.1111/j.1399-3054.2006.00841

Dotlacil L, Hermuth J, Stehno Z, Dvoracek V, Bradova J, Leisova L (2010) How can wheat landraces contribute to present breeding? Czech J Genet Plant Breed 46:70–74

Guzman C, Peña RJ, Singh R, Autrique E, Dreisigacker S, Crossa J, Rut-koski J, Poland J, Battenfield S (2016) Wheat quality improvement at CIMMYT and the use of genomic selection on it. Appl Transl Genom 11:3–8

Heffner EL, Jannink J, Sorrells ME (2011) Genomic selection accuracy using multifamily prediction models in a wheat breeding program. Plant Genome 4:65–75

Jernigan KL, Morris CF, Zemetra R, Chen J, Garland Campbell KA, Carter AH (2017) Genetic analysis of soft white wheat end-use quality traits in a club by common wheat cross. J Cereal Sci 76:148–156

Kiszonas AM, Fuerst EP, CF M (2015) Modeling end-quality in United States soft wheat germplasm. Cereal Chem 92:57–64

Koehler P, Wieser H (2012) Chemistry of cereal grains. In: Handbook on sourdough biotechnology. Springer, New York, NY, pp 11–45. https://doi.org/10.1007/978-1-4614-5425-0_2

Mann G, Diffey S, Cullis B, Azanza F, Martin D, Kelly A et al (2009) Genetic control of wheat quality: interactions between chromosomal regions determining protein content and composition, dough rheology, and sponge and dough baking properties. Theor Appl Genet 118:1519–1537

McIntosh RA, Yamazaki Y, Dubcovsky J, Rogers J, Morris C, Somers J, Appels R, Devos KM (2013) Catalogue of gene symbols for wheat. In: KOMUGI-integrated wheat science database at http://www.shigen.nig.ac.jp/wheat/komugi/genes/download.jsp

Morris CF (2002) Puroidolines: the molecular genetic basis of wheat grain hardness. Plant Mol Biol 48:633–647

O'Brien L, Ronalds JA (1987) Heritabilities of small-scale and standard measures of wheat quality for early generation selection. Aust J Agric Res 38:801–808

Ortiz-Monasterio I, Palacios-Rojas N, Meng E, Pixley K, Trethowan R, Pena RJ (2007) Enhancing the mineral and vitamin content of wheat and maize through plant breeding. J Cereal Sci 46:293–307. https://doi.org/10.1016/j.jcs.2007.06.005

Payne PI (1987) Genetics of wheat storage proteins and the effect of allelic variation on bread-making quality. Annu Rev Plant Physiol 38:141–153

Payne PI, Holt LM, Jackson EA, Law CN (1984) Wheat storage proteins: their genetics and their potential for manipulation by plant breeding. Philos Trans R Soc Lond B 304:359–371

Payne PI, Lawrence GJ (1983) Catalogue of alleles for the complex loci Glu-A1, Glu-B1 and Glu-D1 which code for high-molecular-weight subunits of glutenin in hexaploid wheat. Cereal Res Commun 11:29–35

Pearson DC, Rosielle AA, Boyd WJR (1981) Heritabilities of five wheat quality traits for early generation selection. Aust J Exp Agric 21:512–515

Pena R (2002) Wheat for bread and other foods. Bread wheat improvement and production. Food and Agriculture Organization of the United Nations, Rome, pp 483–542

Ram S, Sharma S, Sharma I (2015) Allelic diversity of HMW and LMW glutenins in Indian wheats and their relationship with sedimentation volume and mixograph parameters. Cereal Res Commun 43:492–503

Shewry PR, Miles MJ, Tatham AS (1994) The prolamin storage proteins of wheat and related cereals. Prog Biophys Mol Biol 61:37-59

Souza EJ, Graybosch RA, Guttieri MJ (2002) Breeding wheat for improved milling and baking quality. J Crop Prod 5:39–74

Uauy C, Distelfeld A, Fahima T, Blechl A, Dubcovsky J (2006) A NAC gene regulating senescence improves grain protein, zinc, and iron content in wheat. Science 314:1298–1301. https://doi.org/10.1126/science.1133649

USDA (2018) World agricultural supply and demand estimates. Report of February 8, 2018. USDA, Washington, DC, p 4 https://www.usda.gov/oce/commodity/wasde/Secretary_Briefing.pdf

Uthayakumaran S, Gras PW, Stoddard FL, Bekes F (1999) Effect of varying protein content and glutenin-to-gliadin ratio on the functional properties of wheat dough. Cereal Chem 76:389–394

Velu G, Ortiz-Monasterio I, Cakmak I, Hao Y, Singh RP (2014) Biofortification strategies to increase grain zinc and iron concentrations in wheat. J Cereal Sci 59:365–372. https://doi.org/10.1016/j.jcs.2013.09.001

Wang S, Yu Z, Cao M, Shen X, Li N, Li X et al (2013) Molecular mechanisms of HMW glutenin subunit from 1S genome of *Aegilops longissima* positively affecting wheat bread making quality. PLoS One 8:58947. https://doi.org/10.1371/journal.pone.0058947

Wessells KR, Brown KH (2012) Estimating the global prevalence of zinc deficiency: results based on zinc availability in national food supplies and the prevalence of stunting. PLoS One 7:e50568. https://doi.org/10.1371/journal.pone.0050568

Wrigley CW, Morris CF (1996) Breeding cereals for quality improvement. In: Henry RJ, Kettlewell PS (eds) Cereal grain quality. Chapman and Hall, London, pp 321–369

Chapter 13
Use of Modern Molecular Biology and Biotechnology Tools to Improve the Quality Value of Oilseed Brassicas

S. K. Rai, Vanya Bawa, Zahoor Ahmad Dar, N. R. Sofi, S. S. Mahdi, and Asif M. Iqbal Qureshi

13.1 Introduction

The brassicas, commonly known as rapeseed mustard, are an important source of edible oils, including vegetable crops, in almost all countries of the world. Oils extracted from plants have been used by mankind since ancient times and have been manipulated for welfare in many ways. The use and exploitation of plants that produce oil are predominantly aimed at production of edible oils. Rapeseed mustard is the third most important source of vegetable oil in the world, is grown in more than 50 countries across the globe and is known for its quality and good fatty acid profile. Dietary fat, a significant source of energy, supplies calories and carries fat-soluble vitamins. Many by-products of oilseed crops are used as cakes to feed cattle, and other by-products include manure. The use of oilseed crops is not limited to consumption; it also covers other industrial and domestic uses in surfactants, soaps, detergents, lubricants, solvents, paints, inks, chemical feedstocks and cosmetics.

There is a wide range of oil crops grown worldwide. Among them, soybean, peanut, rapeseed mustard, sunflower, safflower, *Sesamum*, linseed, castor and cotton seed are

S. K. Rai (✉) · V. Bawa
Division of Plant Breeding and Genetics, Sher-e-Kashmir University of Agricultural Sciences and Technology, Jammu, Jammu and Kashmir, India

Z. A. Dar
Genetics, Plant Breeding & Biotechnology, DARS, Budgam, Sher-e-Kashmir University of Agricultural Sciences & Technology of Kashmir, Srinagar, Jammu and Kashmir, India

N. R. Sofi · S. S. Mahdi
MRCFC, Khudwani, Sher-e-Kashmir University of Agricultural Sciences & Technology, Kashmir, Jammu and Kashmir, India

A. M. I. Qureshi
Genetics, Plant Breeding & Biotechnology, MRCFC, Khudwani, Sher-e-Kashmir University of Agricultural Sciences & Technology of Kashmir, Srinagar, Jammu and Kashmir, India

A. M. I. Qureshi et al. (eds.), *Quality Breeding in Field Crops*,
https://doi.org/10.1007/978-3-030-04609-5_13

the predominant sources of edible oil. The world's total oilseed production from major oil crops was 423.55 million tonnes (Mt) from 205.08 million hectares (MHa) area during 2009–2010 (http://www.fas.usda.gov/psdonline). The leading countries in oilseed production are the USA, Brazil, Argentina, China and India. The greater the demand and consumption, the more manipulations are needed for greater and better-quality production. It is to be noted that in the past few years, the adoption and manipulation of oilseed crops has grown significantly because of consumption patterns and industrial demand pertaining to the composition of seed oils, particularly their quality parameters. Seed oils are made up of a wide range of fatty acids, with six predominant types: 16- or 18-carbon palmitic, stearic, oleic, linoleic and linolenic acids, and 12-carbon lauric acid, as well as other unusual fatty acids produced by wild plant species. However, there are various associated constraints. A significant problem with mustard oil is that it contains large amounts of erucic acid (50%) and glucosinolates (80–160 μmol/g); on the other hand, it is stable in terms of containing less polyunsaturated fatty acids (PUFAs) (Wendlinger et al. 2014. However, the need for improvement in quality is still a matter of concern.

Use of molecular breeding strategies in association with efficient biotechnological tools can serve to overcome quality barriers for improvement in a short span of time. In fact, biotechnology includes several agricultural and food-manufacturing tools and techniques. The Convention on Biological Diversity (CBD) defines biotechnology as "any technological application that uses biological systems, living organisms, or derivatives thereof, to make or modify products or processes for specific uses". The potential of modern biotechnology is widely known, as it makes use of the tools and techniques offered by recombinant DNA technology to generate modified organisms, making plants more useful and suitable for several potential applications in terms of low erucic acid content, low glucosinolate content, high oil content, production of hormones, xenotransplantation and bioremediation. Recently used techniques include genetic editing, post-transcriptional modifications and protein profiling. Various observed morpho-physiological variations in different wild alleles can serve as vital sources of novel alleles and can be utilized in crop improvement programmes (Pratap and Gupta 2009). Genetic improvement of oilseed crops using modern biotechnology has been rapidly replacing plant breeding to incorporate desirable characteristics that are impossible to achieve by only using conventional breeding approaches. This technology enables us to overcome shortcomings of the species being promoted, especially where exogenous genes are needed because there are characters that are difficult to improve and incorporate by traditional breeding and those characters are tissue specific, or where temporal expression or suppression of endogenous genes would be valuable for quality enhancement. Through the intervention of biotechnological tools, yield quantitative trait loci (QTLs) can be identified and can be introgressed into improved backgrounds, using marker-assisted selection (MAS) (Yadava et al. 2012). For breaking the barrier, a population improvement programme involving diverse parents of high potential can be incorporated into non-conventional breeding programmes to achieve more fruitful results in a short duration of time.

For oilseed crops, the desire is to use modern biotechnology for the production of plants with specific fatty acid content and low erucic acid and glucosinolate profiles in the seed oil content. In this chapter we discuss the roles of various techniques in improving quality parameters and their roles in breeding for long-term global food security.

13.1.1 Novel Breeding Technologies

In the modern era, to find and introgress quality traits, it is necessary to identify and utilize novel technologies such as transgenic technology, which gives breeders additional tools to manipulate genomes by using recombinant DNA methods that are continually being improved and refined. It is not worth using transgenic technology individually. More specifically, transgenesis is used comprehensively in various ways, including combinations with other breeding technologies such as hybrid creation, mutation breeding, backcrossing, tissue culture and MAS. Novel oil-related traits are now available in transgenic crops and have been produced via non-transgenic breeding approaches such as mutagenesis or wide crosses. Although transgenic methods have been successful in producing high oleic oil content in several major crops, it has proved much more difficult to engineer commercially relevant levels of this kind of novel improvement in different areas. Transgenic technology can be improved and made more efficient to widen the scope of dissection and improvement of quality traits. Indeed there are many cases where breeders actually want to knock out a particular gene to create a favourable phenotype in terms of quality. In several cases where the identity of the target gene is unknown, the use of mutagenesis/TILLING (Targeting Induced Local Lesions in Genomes) is an effective method in which large populations can be treated and screened for those knockouts (Gilchrist et al. 2013). In plant molecular biology, there is an increasing list of traits for which the identity of the target gene is known to a high degree of probability. In such cases, transgene-induced gene mutation is useful, and two transgenic technologies can be used here: zinc-finger mutagenesis and RNA interferase (RNAi). In comparison with previous antisense approaches, RNAi is usually much more effective in reducing levels of target gene transcripts. Other advantages include the fact that RNAi-encoding transgenes are inherited in a much more stable way, and the degree of down-regulation of the target gene can be modulated by varying the strength of the transgene promoter. Some reports have described the use of a seed-specific napin A promoter to drive the knockdown of BnFAE1 in transgenic CY2. Using Southern blotting, the results confirmed the presence of the transgene. Reverse transcription polymerase chain reaction (RT-PCR) analysis confirmed that the levels of BnFAE1 (the gene for erucic acid) were greatly decreased in BnFAE1-Ri lines compared with the CY2 cultivar. Knockdown of BnFAE1 sharply decreased the levels of erucic acid (to less than 3%), increased the content of oleic acid (by more than 60%) and slightly increased the PUFA content. In comparison, the F1 obtained from parents of high erucic acid showed dramatical decrease in erucic acid content i.e with decreased expression of BnFAE1 (Shi et al. 2015). Also, unlike other technologies, RNAi can knock out all members of a multigene family. A fusion fragment was used to assemble unique intron-spliced hairpin interfering constructs. In the transgenic plant FFRP4-4, expression of fatty acid (Delta12)–desaturase 2 (BnaFAD2) and fatty acid elongase 1 (BnaFAE1) genes was completely inhibited, and it was observed that by a single transformation, a fusion product

Table 13.1 Studies involving quality trait analysis in brassicas

S. No.	Research undertaken	Results	Reference
1.	Analysis of a mutant *Brassica napus* (canola) population for identification of new genetic diversity via TILLING and next-generation sequencing	432 unique mutations screened in 26 different genes	Gilchrist et al. (2013)
2.	A rich TILLING resource for studying gene function in *Brassica rapa*	3072 M_2 plants provided an average of 68 mutations	Stephenson et al. (2010)
3.	Production of a high-efficiency TILLING population through polyploidization	In *Arabidopsis thaliana*, 528 individuals were screened for induced mutations in 15 genes	Tsai et al. (2013)
4.	Assessment of FAE1 polymorphisms in three *Brassica* species, using EcoTILLING, and their association with differences in seed erucic acid content	*FAE1* homologues indicated that 18 SNPs differed between the A and C genomes in nine accessions of *Brassica rapa*	Wang et al. (2010)
5.	Soybean *GmbZIP123* gene enhancement of lipid content in the seeds of transgenic *Arabidopsis* plants	87 transcription factor genes with higher abundance at the stage of lipid accumulation	Song et al. (2013)

SNP single-nucleotide polymorphism, *TILLING* targeting induced local lesions in genomes
[a]EcoTILLING is a method that uses TILLING techniques to look for natural mutations in individuals, usually for population genetics analysis

was formed by linking two fragments, which were amplified from BnaFAD2 and BnaFAE1 genes of *Brassica napus* (Peng et al. 2010). Finally, DNA constructs are relatively easy to employ and can be used efficiently in the latest generation of transgene vectors. In practical use, RNAi technology has already been demonstrated to be useful in generating variations in many important lipid-related traits in oilseed crops (Cheng et al. 2013; Gayen et al. 2014). However, it is worth pointing out that RNAi is not a radically novel technology in terms of creating genetic variation but can be used as an alternative technology for generation of knockout mutations, which can also be generated in many ways. Another method, zinc-finger mutagenesis, is an even more recently refined method for targeting knockout of plant genes (Shukla et al. 2009). There are many more techniques to be understood, as the growing availability of whole-genome sequences is spurring functional gene studies in plants. However, functional genetic technologies are often challenging and expensive. Different types of studies have been performed for various quality traits in oilseed brassicas (see Table 13.1).

13.1.2 New Genetic Variant Discovery and Quantitative Trait Locus Analysis

Variation and selection are the two keys to an efficient breeding programme, as they govern the principle for the selection of variants in high-resolution QTL mapping strategies in which, the number of identified recombination events that signifies the population in linkage disequilibrium are utilized for analysing the marker density and the trait complexity further. Sufficient recombination events in QTL intervals can be identified in species where a large progeny can be generated easily (Darvasi 1998). This approach is useful when a large population of diverse nature is available in which genetic recombinations can be harnessed using various mapping approaches. Alternative fine-mapping strategies, e.g. association mapping, have been devised for such populations using "historical recombination events" (Xiong and Guo 1997), which are reflected by haplotype frequencies. For any mapping approach it is important to give a lot of attention to the phenotyping of traits. It increases the accuracy, level of precision and throughput at all levels of biological organization while reducing costs and minimizing labour via automation, remote sensing, data integration and experimental design (Cobb et al. 2013). QTL analysis has now been applied to many other crop species and their wild relatives (Agarwal et al. 2008; Bernardo 2008; Guimarães et al. 2007), and in oil crops, identification of complex lipid-related traits such as fatty acid composition and erucic acid contents has also been performed. With advancements in the area of QTL analysis in the recent past, tremendous achievements have occurred in the area of identification of QTLs and genes (Montoya et al. 2013). It is bliss for researchers that prediction of the function and location of a gene with maximum certainty is now possible. However, the precision of prediction of genes and their functions through genomics needs to be enhanced by making use of high-throughput genomic tools and technologies. This will help us to harness the full benefits/potential of genomics for crop improvement programmes. Meaningful QTL/gene discovery programmes using either QTL mapping or association mapping require accurate and precise phenotyping data on complex traits. It is relevant to mention here that valid and applicable results reported with non-conventional approaches so far have not yielded the expected results in terms of finding major QTLs and genes, despite the huge volumes of molecular genotypic data generated during the last few years (Edmeades et al. 2004; Araus et al. 2008; Collins et al. 2008; Xu and Crouch 2008; Passioura 2010). As far as oleiferous crops are concerned, remarkable work has been done in the recent past to decipher the genes underlying quality traits, as more than 40 QTLs. have been observed. The identification of two QTLs involved in erucic acid content in rapeseed was localized by interval and multiple QTL mapping (MQM) analyses using MapQTL, showing only an additive effect, whereas they separately explained 43% and 31%, respectively, of the variation in erucic acid content. Further, these findings were in agreement with those of previous studies showing that erucic acid content is controlled by two loci that have additive effects (Harvey and Downey 1964; Barret et al. 1998; Fourmann et al. 1998), whereas in association studies, trait associations were observed in *Brassica napus* and more than 40 QTLs were identified using a single-nucleotide polymorphism (SNP) array (Sun et al. 2016) (Fig. 13.1). Similar studies are shown in Tables 13.2 and 13.3.

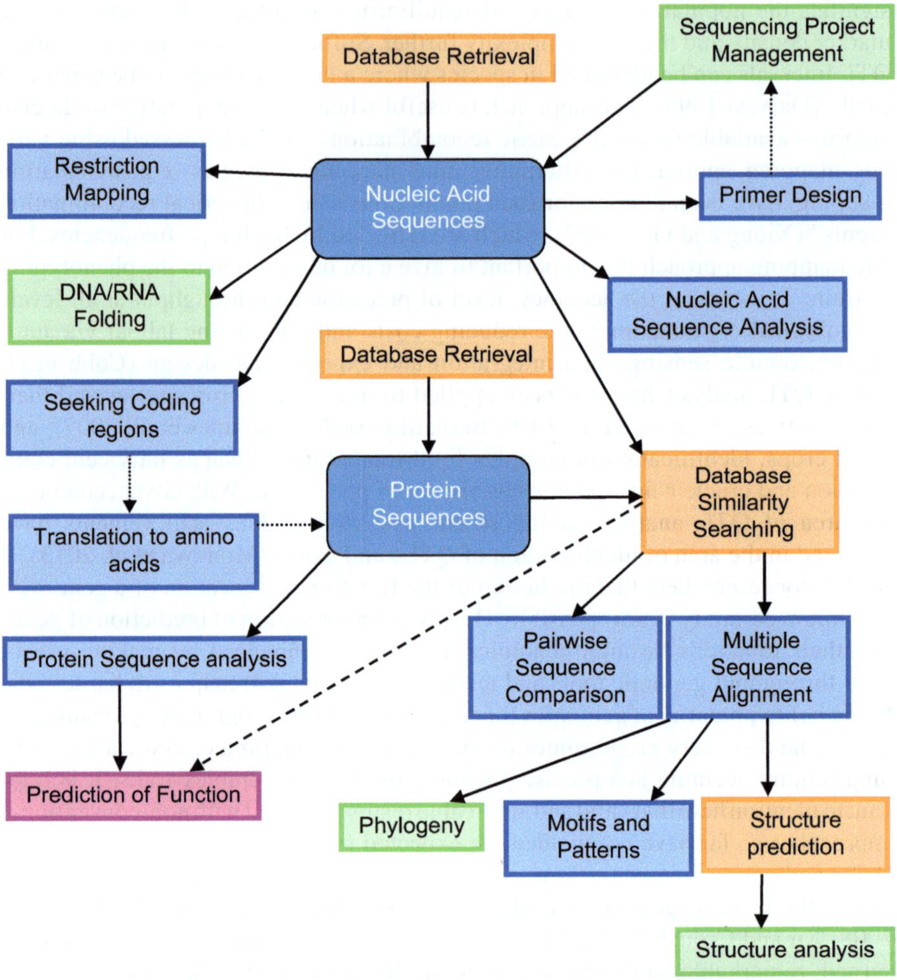

Fig. 13.1 Overview of the sequence analysis pipeline

13.1.3 Use of Bioinformatics as a Tool in Oilseed Brassicas

The field of bioinformatics allows us to tackle and utilize the vast volumes of data being generated by genome-sequencing projects. Efficient tools are needed to organize the data and to make it available in the public domain in the most user-friendly format. With studies on genome regulation and structure, bioinformatics covers many areas, including databases on regulatory sequences, regulation of gene expression, analysis and recognition of genomic sequences, gene structure prediction, modelling of transcriptional and translational control, and large-scale genomes. For better utilization it is necessary for databases to facilitate free association across

Table 13.2 Quantitative trait locus (*QTL*) mapping studies in oilseed brassicas

S. No.	Research undertaken	Population used	QTLs identified	Reference
1.	Analysis of QTLs for erucic acid and oil content in seeds on the A8 chromosome and the linkage drag between the alleles for the two traits in *Brassica napus*	Doubled haploids	7 QTLs for oil content, 2 QTLs for erucic acid content	Cao et al. (2010)
2.	Identification of stable QTLs for seed oil content by combined linkage and association mapping in *Brassica napus*	F_2	40 QTLs	Sun et al. (2016)
3.	Molecular tagging of the erucic acid trait in oilseed mustard (*Brassica juncea*) by QTL mapping and SNPs in the FAE1 gene	Doubled haploids	2 QTLs	Gupta et al. (2004)
4.	Genome-wide association study revealing novel elite allelic variations in the seed oil content of *Brassica napus*	Inbred lines	60 K SNP array; 50 loci identified	Liu et al. (2016)
5.	Prediction of seed quality traits with high accuracy in *Brassica napus* by use of genomic data	Doubled haploids and inbreds	9 QTLs	Zou et al. (2016)
6.	A comparative linkage map of oilseed rape and its use for QTL analysis of seed oil and erucic content	Doubled haploids		Qiu et al. (2006)
7.	QTL identification in two genetic systems for rapeseed glucosinolate and erucic acid content over two seasons	Doubled haploids, BC1F1 1	9 loci for glucosinolate content, 3 loci for erucic acid content	Xu et al. (2015)

SNP single-nucleotide polymorphism

genomes with respect to gene sequences, putative functions or genetic map positioning, and they should be interlinked for better updates in terms of new submissions. In new breeding strategies, breeders routinely use computer models to formulate predictive hypotheses to create phenotypes of interest from complex allele combinations, and then construct those combinations by scoring large populations for very large numbers of genetic markers (Schranz et al. 2006; Hu et al. 2011). *Arabidopsis thaliana* has become universally recognized as a model plant for study. It is a small flowering plant and belongs to the Brassicaceae family, which includes broccoli, cauliflower, cabbage and radish. This kind of cross-genome referencing will lead to a convergence of economically relevant breeding information with basic molecular genetic information. Specific phenotypes of commercial interest are expected to be dramatically improved by these advances, and each and every component is linked; hence, a pipeline of tools and software are associated sequentially from identification of sequences of nucleotides, as shown in Fig. 13.2.

Table 13.3 Association-mapping studies in oilseed *Brassica*

S. No.	Research undertaken	Markers used	Reference
1.	Molecular markers for seed oil content in Indian mustard	219 RAPDs	Sharma et al. (1999)
2.	Association of gene-linked SSR markers with seed glucosinolate content in oilseed rape (*Brassica napus* ssp. *napus*)	104 SSRs	Hasan et al. (2008)
3.	Genome-wide association mapping and identification of candidate genes for fatty acid composition in *Brassica napus* L., using SNP markers	SNP array	Qu et al. (2017)
4.	Association of gene-linked SSR markers with seed glucosinolate content in oilseed rape (*Brassica napus* ssp. *napus*)	1 QTL for *Alternaria* blight resistance, 2 QTLs for low glucosinolate content	Hasan et al (2008)
5.	Identification and mapping of candidate genes and QTLs involved in the fatty acid desaturation pathway in *Brassica napus*	34 QTLs for fatty acid content of seed oil: 14 in the A genome and 20 in the C genome	Smooker et al. (2011)
6.	QTL analysis of an intervarietal set of substitution lines in *Brassica napus*: seed oil content and fatty acid composition	13 QTLs affected the fatty acid composition of the seed and were distributed among linkage groups 1, 3, 6, 7, 8, 11, 13, 14, 18 and 19	Burns et al. (2003)
7.	Analysis of QTLs for seed oil content in *Brassica napus* by association mapping and QTL mapping	19 QTLs for seed oil content	Fu et al. (2017)

QTL quantitative trait locus, *RAPD* random amplified polymorphic DNA, *SNP* single-nucleotide polymorphism, *SSR* simple sequence repeat

Many brassica databases have been built for better understanding and use of genomic data sets for brassica species. These databases include Brassica.info (see http://www.brassica.info/), BrassEnsembl (see http://plants.ensembl.org/index.html), BrassicaDB (see http://brassica.nbi.ac.uk/BrassicaDB/), CropStoreDB (see http://www.cropstoredb.org/), Bolbase (see http://ocri-genomics.org/bolbase/) and the Brassica Database (BRAD; see http://brassicadb.org/brad/). These databases all have different bases of emphasis.

Brassica.info integrates information about genomic resources and releases news of projects and activities in brassica genome studies, where as the anoted and deciphered genomic information can be retrieved from the provided downloading services.

BrassEnsembl visualizes different sets of brassica genomic data under a single frame.

CropStoreDB provides a practical approach for managing crop genetic data.

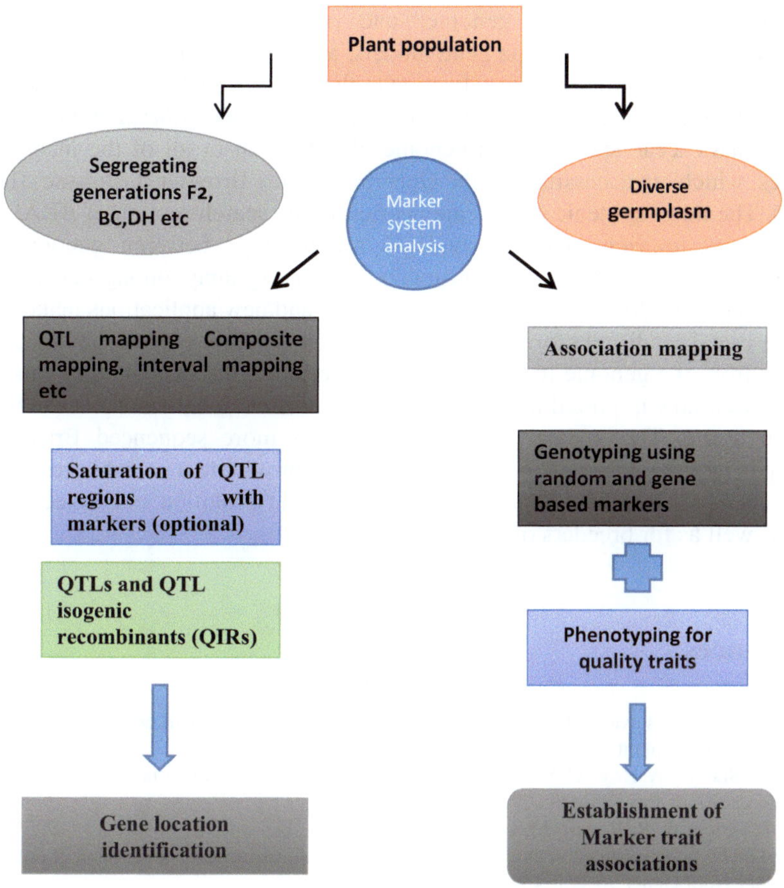

Fig. 13.2 Different steps involved in quantitative trait locus (*QTL*) mapping and association mapping

Bolbase is the first resource platform for the *Brassica oleracea* genome and for genomic structure comparisons with its relatives (Yu 2013). It enables researchers to better study the function and evolution of brassica genomes and enhances molecular breeding research in a more comprehensive manner. This database is updated regularly with new features, improvements in genome annotation and new genomic sequences as they become available.

The Brassica Database (BRAD) assists researchers and plant breeders to use recently released Brassicaceae genome sequences efficiently in scientific investigations and breeding applications (Wang 2015). Unlike the other databases mentioned above, BRAD uses information from genomic studies and gene function studies in the model species *Arabidopsis thaliana* to annotate newly sequenced genomes of brassica species. It has recently been updated to version 2.0 (V2.0), which is a substantial improvement on BRAD V1.0, and its latest release (BRAD V2.9) incorporates more Brassicaceae genomes and provides comprehensive functional annotations of all Brassicaceae gene models, and genome and gene-level syntenic data sets, in association with visualization tools. Its updated data sets and their functions include all syntenic gene pairs between *Arabidopsis thaliana* and most of the other Brassicaceae species, with genome visualization of all of the incorporated species, which are accessible via the Generic Genome Browser (GBrowse) (Donlin 2007). The new "syntenic figure" application in the search section of BRAD V2.0 allow users to view pairwise syntenic relationships between genomes. The GBrowse_syn module is used to visualize multiple synteny among genomes. The inclusion of bulk Brassicaceae genome data sets and new applications make BRAD V2.0 a user-friendly platform from which to conveniently retrieve genomic information from the genome to gene levels. Furthermore, the genome and annotation information now imported into GBrowse allow all functional elements can be seen in one frame. The update of BRAD integrates more sequenced Brassicaceae genomes into the database in the public domain, providing a valuable resource for researchers working on comparative genomics, plant evolution and molecular biology, as well as for breeders of Brassicaceae crops.

References

Agarwal M, Shrivastava N, Padh H (2008) Advances in molecular marker techniques and their applications in plant sciences. Plant Cell Rep 27:617–631

Araus JL, Slafer GA, Royo C, Serret MD (2008) Breeding for yield potential and stress adaptation in cereals. Crit Rev Plant Sci 27(6):377–412

Barret P, Delourme R, Renard M, Domergue F, Lessire L, Delseny M, Roscoe TJ (1998) A rapeseed FAE1 gene is linked to the E1 locus associated with variation in the content of erucic acid. Theory Appl Genet 96:177–186

Bernardo R (2008) Molecular markers and selection for complex traits in plants: learning from the past 20 years. Crop Sci 48:1649–1664

Burns MJ, Barnes SR, Bowman JG, Clarke MHE, Werner CP, Kearsey MJ (2003) QTL analysis of an intervarietal set of substitution lines in *Brassica napus*: (i) seed oil content and fatty acid composition. Heredity 90(1):39

Cao Z, Tian F, Wang N, Jiang C, Lin B, Xia W, Shi J, Long Y, Zhang C, Meng J (2010) Analysis of QTLs for erucic acid and oil content in seeds on A8 chromosome and the linkage drag between the alleles for the two traits in *Brassica napus*. J Genet Genomics 37(4):231–240

Cheng J, Zhu LH, Salentijn EM (2013) Functional analysis of the omega-6 fatty acid desaturase (CaFAD2) gene family of the oil seed crop *Crambe abyssinica*. BMC Plant Biol 13:146

Cobb JN, DeClerck G, Greenberg A, Clark R, McCouch S (2013) Next-generation phenotyping: requirements and strategies for enhancing our understanding of genotype–phenotype relationships and its relevance to crop improvement. Theor Appl Genet 126(4):867–887

Collins NC, Tardieu F, Tuberosa R (2008) Quantitative trait loci and crop performance under abiotic stress: where do we stand? Plant Physiol 147(2):469–486

Darvasi A (1998) Experimental strategies for the genetic dissection of complex traits in animal models. Nat Genet 18(1):19

Donlin MJ (2007) Using the generic genome browser (GBrowse). Curr Protoc Bioinformatics 17(1):9

Edmeades GO, McMaster GS, White JW, Campos H (2004) Genomics and the physiologist: bridging the gap between genes and crop response. Field Crop Res 90(1):5–18

Fourmann M, Barret P, Renard M, Pelletier G, Delourme R, Brunel D (1998) The two genes homologous to *Arabidopsis* FAE1 co-segregate with the two loci governing erucic acid content in *Brassica napus*. Theor Appl Genet 96:852–858

Fu Y, Zhang D, Gleeson M, Zhang Y, Lin B, Hua S, Ding H, Frauen M, Li J, Qian W, Yu H (2017) Analysis of QTL for seed oil content in *Brassica napus* by association mapping and QTL mapping. Euphytica 213(1):17

Gayen D, Ali N, Ganguly M, Paul S, Datta K, Datta KS (2014) RNAi mediated silencing of lipoxygenase gene to maintain rice grain quality and viability during storage. Plant Cell Tissue Organ Cult 118:229–243

Gilchrist EJ, Sidebottom CH, Koh CS, MacInnes T, Sharpe AG, Haughn GW (2013) A mutant Brassica napus (Canola) population for the identification of new genetic diversity via TILLING and next generation sequencing. PLoS One 8(12):e84303

Guimarães EP, Ruane J, Scherf B, Sonnino A, Dargie J (eds) (2007) Marker-assisted selection: current status and future perspectives in crops, livestock, forestry and fish. Food & Agriculture Organization of the United Nations, Rome

Gupta V, Mukhopadhyay A, Arumugam N, Sodhi YS, Pental D, Pradhan AK (2004) Molecular tagging of erucic acid trait in oilseed mustard (Brassica juncea) by QTL mapping and single nucleotide polymorphisms in FAE1 gene. Theor Appl Genet 108:743–749

Harvey BL, Downey RK (1964) The inheritance of erucic acid content in rapeseed (*Brassica napus*). Can J Plant Sci 44(1):104–111

Hasan M, Friedt W, Kühnemann Pons J, Freitag NM, Link K, Snowdon RJ (2008) Association of gene-linked SSR markers to seed glucosinolate content in oilseed rape (*Brassica napus* ssp. napus). Theor Appl Genet 116(8):1035–1049

Hu TT, Pattyn P, Bakker EG, Cao J, Cheng JF, Clark RM, Fahlgren N, Fawcett JA, Grimwood J, Gundlach H, Haberer G (2011) The *Arabidopsis lyrata* genome sequence and the basis of rapid genome size change. Nat Genet 43(5):476

Liu S, Fan CC, Li JN, Cai GQ, Yang QY, Wu J, et al (2016) A genome-wide association study reveals novel elite allelic variations in seed oil content of Brassica napus. Theor Appl Genet 129:1203–1215. https://doi.org/10.1007/s00122-016-2697

Montoya C, Lopes R, Flori A, Cros D, Cuellar T, Summo M, Espeout S, Rivallan R, Risterucci AM, Bittencourt D, Zambrano JR (2013) Quantitative trait loci (QTLs) analysis of palm oil fatty acid composition in an interspecific pseudo-backcross from *Elaeis oleifera* (HBK) Cortés and oil palm (*Elaeis guineensis* Jacq.). Tree Genet Genomes 9(5):1207–1225

Passioura JB (2010) Scaling up: the essence of effective agricultural research. Funct Plant Biol 37(7):585–591

Peng Q, Hu Y, Wei R, Zhang Y, Guan C, Ruan Y, Liu C (2010) Simultaneous silencing of FAD2 and FAE1 genes affects both oleic acid and erucic acid contents in *Brassica napus* seeds. Plant Cell Rep 29(4):317–325

Pratap A, Gupta SK (2009) Biology and ecology of crucifers. In: Gupta SK (ed) Biology and breeding of crucifers. CRC, Boca Raton, FL

Qiu D, Morgan C, Shi J, Long Y, Liu J, Li R, Weihmann T (2006) A comparative linkage map of oilseed rape and its use for QTL analysis of seed oil and erucic acid content. Theor Appl Genet 114(1):67–80

Qu C, Jia L, Fu F, Zhao H, Lu K, Wei L, Xu X, Liang Y, Li S, Wang R, Li J (2017) Genome-wide association mapping and identification of candidate genes for fatty acid composition in *Brassica napus* L. using SNP markers. BMC Genomics 18(1):232

Schranz ME, Lysak MA, Mitchell-Olds T (2006) The ABC's of comparative genomics in the Brassicaceae: building blocks of crucifer genomes. Trends Plant Sci 11(11):535–542

Sharma R, Mohapatra T, Mukherjee AK, Sharma RP (1999) Molecular markers for seed oil content in Indian mustard. J Plant Biochem Biotechnol 8(2):99–102

Shi J, Lang C, Wu X, Liu R, Zheng T, Zhang D, Chen J, Wu G (2015) RNAi knockdown of fatty acid elongase1 alters fatty acid composition in *Brassica napus*. Biochem Biophys Res Commun 466(3):518–522

Shukla VK, Doyon Y, Miller JC (2009) Precise genome modification in the crop species *Zea mays* using zinc-finger nucleases. Nature 459:437–441

Smooker AM, Wells R, Morgan C, Beaudoin F, Cho K, Fraser F, Bancroft I (2011) The identification and mapping of candidate genes and QTL involved in the fatty acid desaturation pathway in *Brassica napus*. Theor Appl Genet 122(6):1075–1090

Song QX, Li QT, Liu YF, Zhang FX, Ma B, Zhang WK, Man WQ, Du WG, Wang GD, Chen SY, Zhang JS (2013) Soybean GmbZIP123 gene enhances lipid content in the seeds of transgenic *Arabidopsis* plants. J Exp Bot 64(14):4329–4341

Stephenson P, Baker D, Girin T, Perez A, Amoah S, King GJ, Østergaard L (2010) A rich TILLING resource for studying gene function in *Brassica rapa*. BMC Plant Biol 10(1):62

Sun F, Liu J, Hua W, Sun X, Wang X, Wang H (2016) Identification of stable QTLs for seed oil content by combined linkage and association mapping in *Brassica napus*. Plant Sci 252:388–399

Tsai H, Missirian V, Ngo K, Tran RK, Chan S, Sundaresan V, Comai L (2013) Production of a high efficiency TILLING population through polyploidization. Plant Physiol 161(4):1604–1614

Wang N, Shi L, Tian F, Ning H, Wu X, Long Y, Meng J (2010) Assessment of FAE1 polymorphisms in three Brassica species using EcoTilling and their association with differences in seed erucic acid contents. BMC Plant Biol 10:137–148

Wang X (2015, pii) Brassica database (BRAD) version 2.0: integrating and mining Brassicaceae species genomic resources. Database (Oxford):bav093. https://doi.org/10.1093/database/bav093

Wendlinger C, Hammann S, Vetter W (2014) Various concentrations of erucic acid in mustard oil and mustard. J Food Chem 153:393–397

Xiong M, Guo SW (1997) Fine-scale genetic mapping based on linkage disequilibrium: theory and applications. Am J Hum Genet 60(6):1513–1531

Xu Y, Crouch JH (2008) Marker-assisted selection in plant breeding: from publications to practice. Crop Sci 48:391–407

Xu JF, Long Y, Wu JG, Xu HM, Zhao ZG, Wen J, Meng JL, Shi CH (2015) QTL identification on two genetic systems for rapeseed glucosinolate and erucic acid contents over two seasons. Euphytica 205(3):647–657

Yadava DK, Vasudev S, Naveen S, Mohapatra T, Prabhu KV (2012) Breeding major oil crops: present status and future research needs. In: Gupta SK (ed) Technological innovations in major world oil crops, volume 1: breeding, vol 1. Springer, New York, NY, pp 17–51

Yu J (2013) Bolbase: a comprehensive genomics database for *Brassica oleracea*. BMC Genomics 14:664. https://doi.org/10.1186/1471-2164-14-664

Zou J, Zhao Y, Liu P, Shi L, Wang X, Wang M, Meng J, Reif JC (2016) Seed quality traits can be predicted with high accuracy in *Brassica napus* using genomic data. PLoS One 11(11):e0166624

Index

A. M. I. Qureshi et al. (eds.), *Quality Breeding in Field Crops*,
https://doi.org/10.1007/978-3-030-04609-5

MIX
Papier aus verantwortungsvollen Quellen
Paper from responsible sources
FSC® C105338

If you have any concerns about our products,
you can contact us on
ProductSafety@springernature.com

In case Publisher is established outside the EU,
the EU authorized representative is:
Springer Nature Customer Service Center GmbH
Europaplatz 3, 69115 Heidelberg, Germany

Printed by Libri Plureos GmbH
in Hamburg, Germany